MATHEMATICAL FOUNDATIONS OF QUANTUM MECHANICS

PRINCETON LANDMARKS IN MATHEMATICS AND PHYSICS

Non-standard Analysis, *by Abraham Robinson*

General Theory of Relativity, *by P.A.M. Dirac*

Angular Momentum in Quantum Mechanics, *by A. R. Edmonds*

Mathematical Foundations of Quantum Mechanics, *by John von Neumann*

Introduction to Mathematical Logic, *by Alonzo Church*

Convex Analysis, *by R. Tyrrell Rockafellar*

Riemannian Geometry, *by Luther Pfahler Eisenhart*

The Classical Groups: Their Invariants and Representations, *by Hermann Weyl*

Topology from the Differentiable Viewpoint, *by John W. Milnor*

Algebraic Theory of Numbers, *by Hermann Weyl*

Continuous Geometry, *by John von Neumann*

Linear Programming and Extensions, *by George B. Dantzig*

Operator Techniques in Atomic Spectroscopy, *by Brian R. Judd*

The Topology of Fibre Bundles, *by Norman Steenrod*

Mathematical Methods of Statistics, *by Harald Cramér*

MATHEMATICAL FOUNDATIONS OF QUANTUM MECHANICS

By JOHN VON NEUMANN

translated from the German edition by

ROBERT T. BEYER

PRINCETON UNIVERSITY PRESS

PRINCETON, NEW JERSEY

Published by Princeton University Press, 41 William Street,
Princeton, New Jersey 08540
In the United Kingdom: Princeton University Press,
Chichester, West Sussex

Copyright © 1955 by Princeton University Press; copyright renewed
© 1983 by Princeton University Press

All Rights Reserved

Library of Congress Card No. 53-10143
ISBN 0-691-08003-8
ISBN 0-691-02893-1 (pbk.)

Princeton University Press books are printed on acid-free paper and meet the guidelines for permanence and durability of the Committee on Production Guidelines for Book Longevity of the Council on Library Resources

Sixth printing, 1971
Twelfth printing, and first paperback printing for the Princeton Landmarks in Mathematics and Physics Series, 1996

http://pup.princeton.edu

Printed in the United States of America

10 9 8 7 6 5 4

TRANSLATOR'S PREFACE

This translation follows closely the text of the original German edition. The translated manuscript has been carefully revised by the author so that the ideas expressed in this volume are his rather than those of the translator, and any deviations from the original text are also due to the author.

The translator wishes to express his deep gratitude to Professor von Neumann for his very considerable efforts in the rendering of the ideas of the original volume into a translation which would convey the same meanings.

Robert T. Beyer

Providence, R. I.
December, 1949

PREFACE

The object of this book is to present the new quantum mechanics in a unified representation which, so far as it is possible and useful, is mathematically rigorous. This new quantum mechanics has in recent years achieved in its essential parts what is presumably a definitive form: the so-called "transformation theory." Therefore the principal emphasis shall be placed on the general and fundamental questions which have arisen in connection with this theory. In particular, the difficult problems of interpretation, many of which are even now not fully resolved, will be investigated in detail. In this context the relation of quantum mechanics to statistics and to the classical statistical mechanics is of special importance. However, we shall as a rule omit any discussion of the application of quantum mechanical methods to particular problems, as well as any discussion of special theories derived from the general theory -- at least so far as this is possible without endangering the understanding of the general relationships. This seems the more advisable since several excellent treatments of these problems are either in print or in process of publication.[1]

---

[1]There are, among others, the following comprehensive treatments: Sommerfeld, Supplement to the 4th edition of <u>Atombau und Spektrallinien</u>, Braunschweig, 1928; Weyl, <u>The</u>

On the other hand, a presentation of the mathematical tools necessary for the purposes of this theory will be given, i.e., a theory of Hilbert space and the so-called Hermitean operators. For this end, an accurate introduction to unbounded operators is also necessary, i.e., an extension of the theory beyond its classical limits (developed by Hilbert and E. Hellinger, F. Riesz, E. Schmidt, O. Toeplitz). The following may be said regarding the method employed in this mode of treatment: as a rule, calculations should be performed with the operators themselves (which represent physical quantities) and not with the matrices, which, after the introduction of a (special and arbitrary) coordinate system in Hilbert space, result from them. This "coordinate free," i.e., invariant, method, with its strongly geometric language, possesses noticeable formal advantages.

Dirac, in several papers, as well as in his recently published book,[2] has given a representation of quantum mechanics which is scarcely to be surpassed in brevity and elegance, and which is at the same time of invariant character. It is therefore perhaps fitting to advance a few arguments on behalf of our method, which deviates considerably from that of Dirac.

The method of Dirac, mentioned above, (and this is overlooked today in a great part of quantum mechanical

---

Theory of Groups and Quantum Mechanics (tr. by H. P. Robertson), London, 1931; Frenkel, Wave Mechanics, Oxford, 1932; Born and Jordan, Elementare Quantenmechanik, Berlin, 1930; Dirac, The Principles of Quantum Mechanics, 2nd ed., Oxford, 1936.

[2]Cf. Proc. Roy. Soc. London, 109 (1925) and the following issues, especially 113 (1926). Independently of Dirac, P. Jordan, Z. Physik 40 (1926) and F. London, Z. Physik 40 (1926) gave similar foundations for the theory.

literature, because of the clarity and elegance of the theory) in no way satisfies the requirements of mathematical rigor -- not even if these are reduced in a natural and proper fashion to the extent common elsewhere in theoretical physics. For example, the method adheres to the fiction that each self-adjoint operator can be put in diagonal form. In the case of those operators for which this is not actually the case, this requires the introduction of "improper" functions with self-contradictory properties. The insertion of such a mathematical "fiction" is frequently necessary in Dirac's approach, even though the problem at hand is merely one of calculating numerically the result of a clearly defined experiment. There would be no objection here if these concepts, which cannot be incorporated into the present day framework of analysis, were intrinsically necessary for the physical theory. Thus, as Newtonian mechanics first brought about the development of the infinitesimal calculus, which, in its original form, was undoubtedly not self-consistent, so quantum mechanics might suggest a new structure for our "analysis of infinitely many variables" -- i.e., the mathematical technique would have to be changed, and not the physical theory. But this is by no means the case. It should rather be pointed out that the quantum mechanical "Transformation theory" can be established in a manner which is just as clear and unified, but which is also without mathematical objections. It should be emphasized that the correct structure need not consist in a mathematical refinement and explanation of the Dirac method, but rather that it requires a procedure differing from the very beginning, namely, the reliance on the Hilbert theory of operators.

In the analysis of the fundamental questions, it will be shown how the statistical formulas of quantum mechanics can be derived from a few qualitative, basic assumptions. Furthermore, there will be a detailed discussion of the problem as to whether it is possible to trace the statistical character of quantum mechanics to an ambiguity (i.e., incompleteness) in our description of nature. Indeed, such an interpretation would be a natural concomitant of the general principle that each probability statement arises from the incompleteness of our knowledge. This explanation "by hidden parameters," as well as another, related to it, which ascribes the "hidden parameter" to the observer and not to the observed system, has been proposed more than once. However, it will appear that this can scarcely succeed in a satisfactory way, or more precisely, such an explanation is incompatible with certain qualitative fundamental postulates of quantum mechanics.[3]

The relation of these statistics to thermodynamics is also considered. A closer investigation shows that the well known difficulties of classical mechanics, which are related to the "disorder" assumptions necessary for the foundation of thermodynamics, can be eliminated here.[4]

---

[3]Cf. Ch. IV. and Ch. VI.3.

[4]Cf. Ch. V.

C O N T E N T S

CHAPTER V

General Considerations

CHAPTER VI

The Measuring Process

MATHEMATICAL FOUNDATIONS OF QUANTUM MECHANICS

CHAPTER I

INTRODUCTORY CONSIDERATIONS

1. THE ORIGIN OF THE TRANSFORMATION THEORY

This is not the place to point out the great success which the quantum theory attained in the period from 1900 to 1925, a development which is dominated by the names of Planck, Einstein and Bohr.[5] At the end of this period of development, it was clear beyond doubt that all elementary processes, i.e., all occurrences of an atomic or molecular order of magnitude, obey the "discontinuous" laws of quanta. There were also available quantitative, quantum theoretical methods in almost all directions, which for the most part yielded results in good or at least fair agreement with experiment. And, what was fundamentally of greater significance, was that the general

---

[5]Its chief stages were: The discovery of the quantum laws by Planck for the case of black body radiation [cf. Planck's presentation in his book, Theory of Heat Radiation (tr. by M. Masius), Philadelphia, 1914]; the hypothesis of the corpuscular nature of light (theory of light quanta) by Einstein [Ann Phys. [4] 17 (1905)], wherein the first example was given of the dual form: Wave-corpuscle, which, we know today, dominates all of microcosmic physics; the application of these two sets of rules to the atomic model by Bohr [Phil. Mag. 26 (1913); Z. Physik 6 (1920)].

opinion in theoretical physics had accepted the idea that the principle of continuity ("natura non facit saltus"), prevailing in the perceived macrocosmic world, is merely simulated by an averaging process in a world which in truth is discontinuous by its very nature. This simulation is such that man generally perceives the sum of many billions of elementary processes simultaneously, so that the leveling law of large numbers completely obscures the real nature of the individual processes.

Nevertheless, up to the time mentioned there existed no mathematical-physical system of quantum theory which would have embodied everything known up to that time in a unified structure, let alone one which could have exhibited the monumental solidity of the system of mechanics, electrodynamics and relativity theory (which system was disrupted by the quantum phenomena). In spite of the claim of quantum theory to universality, which had evidently been vindicated, there was lacking the necessary formal and conceptual instrument; there was a conglomeration of essentially different, independent, heterogeneous and partially contradictory fragments. The most striking points in this respect were: the correspondence principle, belonging half to classical mechanics and electrodynamics (but which played a decisive role in the final clarification of the problem); the self-contradictory dual nature of light (wave and corpuscular, cf. Note 5 and Note 148); and finally, the existence of unquantized (aperiodic) and quantized (periodic or multiple periodic) motions.[6]

---

[6]The quantum laws (added on to the laws of mechanics) for multiple periodic motions were first developed by Epstein-Sommerfeld [cf. e.g. Sommerfeld, Atombau and Spektrallinien, Braunschweig (1924)]. On the other hand, it was ascertained that a freely moving mass point or a planet on an hyperbolic orbit (in contrast to those on elliptical orbits) is "unquantized." The reader will find a complete

The year 1925 brought the resolution. A procedure initiated by Heisenberg was developed by Born, Heisenberg, Jordan, and a little later by Dirac, into a new system of quantum theory, the first complete system of quantum theory which physics has possessed. A little later Schrödinger developed the "wave mechanics" from an entirely different starting point. This accomplished the same ends, and soon proved to be equivalent to the Heisenberg, Born, Jordan and Dirac system (at least in a mathematical sense, cf. I. 3, 4).[7] On the basis of the Born statistical interpretation of the quantum theoretical description[8] of nature, it was possible for Dirac and Jordan[9] to join the two theories into one, the "transformation theory," in which the two are united in complementary fashion, and in which they make possible a grasp of physical problems which is especially simple mathematically.

It should also be mentioned (although it does not belong to our particular subject) that after Goudsmit and Uhlenbeck had discovered the magnetic moment and the spin of the electron, almost all the difficulties of the earlier quantum theory disappeared, so that today we are in possession of a mechanical system which is almost entirely satisfactory. To be sure, the great unity with

---

treatment of this phase of quantum theory in the books by Reiche, The Quantum Theory (tr. by H. S. Hatfield and H. L. Brose), New York, 1922; and Landé, Fortschritte der Quantentheorie, Dresden, 1922.

[7]This was proved by Schrödinger, Ann. Physik [4] 79 (1926).

[8]Z. Physik 37 (1926).

[9]Cf. the articles mentioned in Note 2. Schrödinger's papers have been published in book form, Collected Papers on Wave Mechanics (tr. by J. F. Shearer and W. M. Deans), London, 1928.

electrodynamics and relativity theory mentioned earlier has not yet been recovered, but at least there is a mechanics which is universally valid, where the quantum laws fit in a natural and necessary manner, and which explains satisfactorily the majority of our experiments.[10]

2. THE ORIGINAL FORMULATIONS OF QUANTUM MECHANICS

In order to obtain a preliminary view of the problem, let us set forth briefly the basic structure of the Heisenberg-Born-Jordan "matrix mechanics" and the Schrödinger "wave mechanics."

In both theories, a classical mechanical problem is initially proposed, which is characterized by a Hamiltonian function $H(q_1, \dots, q_k, p_1, \dots, p_k)$. (This means the following, as may be found in greater detail in textbooks of mechanics: Let the system have k degrees of freedom, i.e., its existing state is determined by giving the numerical values of k coordinates $q_1, \dots, q_k$. The energy is a given function of the coordinates and their time derivatives:

$$E = L(q_1, \dots, q_k, \dot{q}_1, \dots, \dot{q}_k) ,$$

[10] The present state of affairs may be described in this way, that the theory, so far as it deals with individual electrons or with electronic shells of atoms or molecules, is entirely successful, as it is also whenever it deals with electrostatic forces and with electromagnetic processes connected with the production, transmission and transformation of light. On the other hand, in problems of the atomic nucleus, and in all attempts to develop a general and relativistic theory of electromagnetism, in spite of noteworthy partial successes, the theory seems to lead to great difficulties, which apparently cannot be overcome without the introduction of wholly new ideas.

and, as a rule, is a quadratic function of the derivatives $\dot{q}_1, \ldots, \dot{q}_k$. The "conjugate momenta" $p_1, \ldots, p_k$ of the coordinates $q_1, \ldots, q_k$ are introduced by the relations

$$p_1 = \frac{\partial L}{\partial \dot{q}_1}, \ldots, p_k = \frac{\partial L}{\partial \dot{q}_k}$$

In the case of the above assumption on L, these depend linearly on the $q_1, \ldots, q_k$. If need be, we can eliminate the $\dot{q}_1, \ldots, \dot{q}_k$ from L by the use of the $p_1, \ldots, p_k$, so that

$$E = L(q_1, \ldots, q_k, \dot{q}_1, \ldots, \dot{q}_k) = H(q_1, \ldots, q_k, p_1, \ldots, p_k)$$

This H is the Hamiltonian function.) In both theories, we must now learn as much as possible from this Hamiltonian function about the true, i.e. quantum mechanical, behavior of the system. Primarily, therefore, we must determine[11]

---

[11]Motion, according to classical mechanics, is determined (as is well known) by the Hamiltonian function, since it gives rise to the equations of motion

$$\dot{q}_l = \frac{\partial H}{\partial p_l}, \quad \dot{p}_l = -\frac{\partial H}{\partial q_l} \qquad (l = 1, \ldots, k)$$

Before the discovery of quantum mechanics, the attempt was made to explain the quantum phenomena, while retaining these equations of motion, by the formulation of supplementary quantum conditions (cf. Note 6). For each set of values of $q_1, \ldots, q_k, p_1, \ldots, p_k$, given at the time $t = 0$, the equations of motion determined the further time variation, or "orbit" of the system in the $2k$ dimensional "phase space" of the $q_1, \ldots, q_k, p_1, \ldots, p_k$. Any additional condition, therefore, results in a limitation of all possible initial values or orbits to a certain discrete set. (Then, corresponding to the few admissible orbits, there is only a smaller number of possible energy levels.) Even

the possible energy levels, then find out the corresponding "stationary states," and calculate the "transition probabilities," etc.[12]

The directions which the matrix theory gives for the solution of this problem run as follows: We seek a system of k matrices $Q_1, \dots, Q_k$, $P_1, \dots, P_k$,[13] which in the first place satisfy the relations

---

though quantum mechanics has broken completely with this method, it is nevertheless clear from the outset that the Hamiltonian function must still play a great role in it. Indeed, common experience proves the validity of the Bohr correspondence principle, which asserts that the quantum theory must give results in agreement with those of classical mechanics in the so-called limiting case of large quantum numbers.

[12]The three latter concepts are taken from the old quantum theory developed principally by N. Bohr. Later we shall analyze these ideas in detail from the point of view of quantum mechanics. Cf. the Dirac theory of radiation given in III.6. Their historical development can be followed in Bohr's papers on the structure of the atom, published from 1913 to 1916.

[13]As a more detailed mathematical analysis would show, this is necessarily a problem of infinite matrices. We shall not go any further here into the properties of such matrices, since we shall consider them thoroughly later on. For the present it suffices that the formal algebraic calculation with these matrices is to be understood in the sense of the known rules of matrix addition and multiplication. By 0, 1, we mean the null matrix and the unit matrix respectively (with all elements identically zero, and with elements equal to one on the main diagonal and zero everywhere else, respectively).

$$\left.\begin{array}{l} Q_mQ_n - Q_nQ_m = 0,\ P_mP_n - P_nP_m = 0 \\ P_mQ_n - Q_nP_m \begin{cases} = 0 & \text{for } m \neq n \\ = \dfrac{h}{2\pi i}\,1 & \text{for } m = n \end{cases} \end{array}\right\} (m, n = 1,\ldots,k)$$

and for which, in the second place, the matrix

$$W = H(Q_1,\ldots,Q_k,\ P_1,\ldots,P_k)$$

becomes a diagonal matrix. (We shall not go into the details here of the origin of these equations, especially the first group, the so-called "commutation rules" which govern the whole non-commutative matrix calculus of this theory. The reader will find exhaustive treatments of this subject in the works cited in Note 1. The quantity h is Planck's constant.) The diagonal elements of W, say $w_1, w_2, \ldots$, are then the different allowed energy levels of the system. The elements of the matrices $Q_1,\ldots,Q_k$ -- $q_{mn}^{(1)},\ldots,q_{mn}^{(k)}$ -- determine in a certain way the transition probabilities of the system (from the mth state with the energy w_m into the nth state with the energy w_n, $w_m > w_n$) and the radiation thereby emitted.

In addition, it should be noted that the matrix

$$W = H(Q_1,\ldots,Q_k,\ P_1,\ldots,P_k)$$

is not completely determined by $Q_1,\ldots,Q_k$, $P_1,\ldots,P_k$ and the classical mechanical Hamiltonian function

$$H(q_1,\ldots,q_k,\ p_1,\ldots,p_k)$$

inasmuch as the Q_l and P_l do not all commute with one another (in multiplication), while it would be meaningless to distinguish between say p_1q_1 and q_1p_1 for $H(q_1,\ldots,q_k,\ p_1,\ldots,p_k)$ in the classical mechanical sense.

We must therefore determine the order of the variables q_l and p_l in the terms of H, beyond the classical meaning of this expression. This process has not been carried out with complete generality, but the appropriate arrangements are known for the most important special cases. (In the simplest case, whenever the system under investigation consists of particles, and therefore has $k = 3\nu$ coordinates $q_1, \ldots, q_{3\nu}$ -- such that e.g. $q_{3\mu-2}, q_{3\mu-1}, q_{3\mu}$ are the three cartesian coordinates of the μth particle, $\mu = 1, \ldots, \nu$ -- in which the interaction of these particles is given by a potential energy $V(q_1, \ldots, q_{3\nu})$, there is no such doubt. The classical Hamiltonian function is then

$$H(q_1, \ldots, q_{3\nu}, p_1, \ldots, p_{3\nu}) = \sum_{\mu=1}^{\nu} \frac{1}{2m_\mu} (p_{3\mu-2}^2 + p_{3\mu-1}^2 + p_{3\mu}^2) + V(q_1, \ldots, q_{3\nu})$$

where m_μ is the mass of the μth particle, and $p_{3\mu-2}, p_{3\mu-1}, p_{3\mu}$ are the components of its momentum. What this means after the substitution of the matrices

$$Q_1, \ldots, Q_{3\nu}, P_1, \ldots, P_{3\nu}$$

is perfectly clear; in particular, the potential energy introduces no difficulties, since all the $Q_1, \ldots, Q_{3\nu}$ commute with each other.) It is important that only Hermitean matrices are permitted, i.e. such matrices $A = \{a_{mn}\}$, for which $a_{mn} = \overline{a_{nm}}$ holds identically (the elements a_{mn} may be complex!). Therefore,

$$H(Q_1, \ldots, Q_k, P_1, \ldots, P_k)$$

must be Hermitean, whenever all the $Q_1, \ldots, Q_k, P_1, \ldots, P_k$ are such. This involves a certain restriction in the problem of the order of the factors which was mentioned above. However, the restriction is not sufficient to determine the $H(Q_1, \ldots, Q_k, P_1, \ldots, P_k)$ uniquely from the classical $H(q_1, \ldots, q_k, p_1, \ldots, p_k)$.[14]

On the other hand, the directions of wave mechanics are the following: First we form the Hamiltonian function $H(q_1, \ldots, q_k, p_1, \ldots, p_k)$ and then the differential equation

$$H(q_1, \ldots, q_k, \frac{h}{2\pi i}\frac{\partial}{\partial q_1}, \ldots, \frac{h}{2\pi i}\frac{\partial}{\partial q_k})\psi(q_1, \ldots, q_k) = \lambda\psi(q_1, \ldots q_k)$$

for an arbitrary function $\psi(q_1, \ldots, q_k)$ in the configuration space of the system (and not in the phase space, i.e. the $p_1, \ldots, p_k$ do not enter into ψ). In this way,

$$H(q_1, \ldots, q_k, \frac{h}{2\pi i}\frac{\partial}{\partial q_1}, \ldots, \frac{h}{2\pi i}\frac{\partial}{\partial q_k})$$

is interpreted simply as a functional operator. For example, this operator in the case mentioned above,

$$H(q_1, \ldots, q_{3\nu}, p_1, \ldots, p_{3\nu}) = \sum_{\mu=1}^{\nu} \frac{1}{2m_\mu}(p_{3\mu-2}^2 + p_{3\mu-1}^2 + p_{3\mu}^2) + V(q_1, \ldots, q_{3\nu})$$

---

[14] If Q_1, P_1 are Hermitean, neither Q_1P_1 nor P_1Q_1 is necessarily Hermitean, but it is true that $\frac{1}{2}(Q_1P_1 + P_1Q_1)$ is always Hermitean. In the case of $Q_1{}^2P_1$, we should also consider $\frac{1}{2}(Q_1{}^2P_1 + P_1Q_1{}^2)$ as well as $Q_1P_1Q_1$ (however, these two expressions happen to be equal for $P_1Q_1 - Q_1P_1 = \frac{h}{2\pi i}1$). In the case of $Q_1{}^2P_1{}^2$, we should also consider $\frac{1}{2}(Q_1{}^2P_1{}^2 + P_1{}^2Q_1{}^2)$, $Q_1P_1{}^2Q_1$, $P_1Q_1{}^2P_1$, etc. (these expressions do not all coincide in the special case mentioned above). We shall not discuss this further at present, since the operator calculus to be developed later will permit these relations to be seen much more clearly.

transforms the function $\psi(q_1, \ldots, q_3)$ into

$$\sum_{\mu=1}^{\nu} \frac{1}{2m_\mu}\left(\frac{h}{2\pi i}\right)^2\left(\frac{\partial^2}{\partial q_{3\mu-2}^2}\psi + \frac{\partial^2}{\partial q_{3\mu-1}^2}\psi + \frac{\partial^2}{\partial q_{3\mu}^2}\psi\right) + V\psi$$

(we have omitted the variables $q_1, \ldots, q_{3\nu}$). Since the operation

$$q_1 \frac{h}{2\pi i}\frac{\partial}{\partial q_1}$$

is different from the operation[15]

$$\frac{h}{2\pi i}\frac{\partial}{\partial q_1} q_1$$

there is here, too, an uncertainty because of the ambiguity of the order of the terms q_m and p_m in

$$H(q_1, \ldots, q_k, p_1, \ldots, p_k)$$

However, Schrödinger has pointed out how this uncertainty can be eliminated, by reduction to a definite variation principle, in such a way that the resulting differential

---

[15]We have

$$\frac{h}{2\pi i}\frac{\partial}{\partial q_1}(q_1\psi) = q_1 \frac{h}{2\pi i}\frac{\partial}{\partial q_1}\psi + \frac{h}{2\pi i}\psi$$

Consequently,

$$\frac{h}{2\pi i}\frac{\partial}{\partial q_1} \cdot q_1 - q_1 \cdot \frac{h}{2\pi i}\frac{\partial}{\partial q_1} = \frac{h}{2\pi i} 1$$

where 1 is the identity operator (transforming ψ into itself), i.e. $\frac{h}{2\pi i}\frac{\partial}{\partial q_1}$ and q_1 satisfy the same commutation rules as the matrices P_1 and Q_1.

equation becomes self-adjoint.[16]

Now this differential equation (the "wave equation") has the character of an eigenvalue problem in which λ is to be interpreted as an eigenvalue parameter, and in which the vanishing of the eigenfunction $\psi = \psi(q_1, \ldots, q_k)$ at the boundaries of the configuration space (the space of the $q_1, \ldots, q_k$) -- together with the conditions of regularity and single-valuedness -- is required. In the sense of the wave theory, the eigenvalues of λ (both discrete and continuous spectra)[17] are the allowed energy levels. And even the corresponding (complex) eigenfunctions are related to the corresponding (stationary, in the Bohr sense) states of the system. For a ν-electron system ($k = 3\nu$, cf. supra, e is the charge of the electron) the charge density of the μth electron, measured at the point x, y, z, is given by the expression

$$e \underbrace{\int \cdots \int}_{(3\nu-3)\text{fold}} |\psi(q_1 \cdots q_{3\mu-3}\, xyz\, q_{3\mu+1} \cdots q_{3\nu})|^2 dq_1 \cdots dq_{3\mu-3}\, dq_{3\mu+1} \cdots dq_{3\nu}$$

i.e. according to Schrödinger, this electron is to be thought of as "smeared" over the entire x, y, z ($= q_{3\mu-2}, q_{3\mu-1}, q_{3\mu}$) space. (In order that the total charge be e, ψ must be normalized by the condition

$$\underbrace{\int \cdots \int}_{(3\nu)\text{fold}} |\psi(q_1 \cdots q_{3\nu})|^2 dq_1 \cdots dq_{3\nu} = 1 .$$

The integration is over all 3ν variables. The same

---

[16]Cf. his first two articles, in the book mentioned in Note 9. [also Ann. Phys. [4] 79 (1926)].

[17]Cf. the first of the works of Schrödinger mentioned in Note 16. A precise definition of the spectrum and its parts will be given later, in II.6. to II.9.

equation obtains for each $\mu = 1,\ldots,\nu$.)

In addition, the wave mechanics can also make observations on systems which are not in Bohr stationary states,[18] in the following way: If the state is not stationary, i.e. if it changes with time, then the wave function $\psi = \psi(q_1,\ldots,q_k;t)$ contains the time t , and it varies according to the differential equation

$$- H(q_1,\ldots,q_k, \frac{h}{2\pi i}\frac{\partial}{\partial q_1},\ldots,\frac{h}{2\pi i}\frac{\partial}{\partial q_k})\psi(q_1,\ldots,q_k;t) = \frac{h}{2\pi i}\frac{\partial}{\partial t}\psi(q_1,\ldots,q_k;t) \quad ^{19}$$

That is, ψ can be given arbitrarily for $t = t_0$, and it is then determined uniquely for all t . Even the stationary ψ are really time dependent, as a comparison of the two Schrödinger differential equations will show, but the dependence on t is given by

$$\psi(q_1,\ldots,q_k;t) = e^{-\frac{2\pi i}{h}\lambda t}\psi(q_1,\ldots,q_k;t)$$

That is, t appears only in a factor of absolute value 1 , independent of $q_1,\ldots,q_k$ (i.e., constant in configuration space), so that, for example, the charge density distribution defined above does not change. (We shall

---

[18] In the original framework of matrix mechanics (cf. our presentation above), such a general state concept, of which the stationary states are special cases, was not given. Only the stationary states, arranged according to the eigenvalues of the energy, were the object of that theory.

[19] $H = H(q_1,\ldots,q_k, p_1,\ldots,p_k)$ may also contain the time t explicity. Naturally, there will then be in general no stationary states at all.

suppose generally -- and we shall find this confirmed later by more detailed considerations -- that a factor of absolute value one and constant in configuration space is, in the case of ψ, essentially unobservable).

Since the eigenfunctions of the first differential equation form a complete orthogonal set,[20] we can develop each $\psi = \psi(q_1, \ldots, q_k)$ in terms of this set of functions. If $\psi_1, \psi_2, \ldots$ are the eigenfunctions (all independent of time), and $\lambda_1, \lambda_2, \ldots$ their respective eigenvalues, the development becomes

$$\psi(q_1, \ldots, q_k) = \sum_{n=1}^{\infty} a_n \psi_n(q_1, \ldots, q_k) \quad {}^{21}$$

If ψ is still time dependent, then t is introduced in the coefficients a_n (the eigenfunctions $\psi_1, \psi_2, \ldots$ on the other hand are to be understood both here and in everything which follows as independent of time). Therefore, if the $\psi = \psi(q_1, \ldots, q_k)$ at hand is actually $\psi = \psi(q_1, \ldots, q_k; t_0)$, then it follows with regard to

$$\psi = \psi(q_1, \ldots, q_k; t) = \sum_{n=1}^{\infty} a_n(t) \psi_n$$

that

$$H\psi = \sum_{n=1}^{\infty} a_n(t) H\psi_n = \sum_{n=1}^{\infty} \lambda_n a_n(t) \psi_n ,$$

$$\frac{h}{2\pi i} \frac{\partial}{\partial t} \psi = \sum_{n=1}^{\infty} \frac{h}{2\pi i} \dot{a}_n(t) \psi_n$$

---

[20] Provided that only a discrete spectrum exists. Cf. II.6.

[21] These, as also all the following series expansions, converge "in the mean." We shall consider this again in II.2.

and by equating the coefficients of the second differential equation:

$$\frac{h}{2\pi i}\dot{a}_n(t) = -\lambda_n a_n(t), \quad a_n(t) = c_n e^{-\frac{2\pi i}{h}\lambda_n t}$$

i.e.,

$$a_n(t) = e^{-\frac{2\pi i}{h}\lambda_n(t-t_0)}\, a_n(t_0) = e^{-\frac{2\pi i}{h}\lambda_n(t-t_0)}\, a_n,$$

$$\psi = \psi(q_1,\ldots,q_k;t) = \sum_{n=1}^{\infty} e^{-\frac{2\pi i}{h}\lambda_n(t-t_0)}\, a_n\psi_n(q_1,\ldots,q_k)$$

Therefore, if ψ is not stationary, i.e., if all a_n except one do not vanish, then ψ (for variable t) no longer changes only by a space constant factor of absolute value unity. Therefore, in general, the charge densities also change, i.e., real electrical oscillations occur in space.[22]

We see that the initial concepts and the practical methods of the two theories differ considerably. Nevertheless, from the beginning they have always yielded the same results, even where both gave details differing from the older concept of quantum theory.[23] This extraordinary situation was soon clarified,[24] as mentioned in I.1., with the proof of Schrödinger of their mathematical equivalence. We shall now turn our attention to this equivalence proof, and at the same time explain the

---

[22]That such oscillations fail to occur for the stationary states, and only for these, was one of the most important postulates of Bohr in 1913. Classical electrodynamics is in direct contradiction to this.

[23]Cf. the second work of Schrödinger mentioned in Note 16.

[24]Cf. Note 7.

Dirac-Jordan general transformation theory (which combines the two theories).

3. THE EQUIVALENCE OF THE TWO THEORIES: THE TRANSFORMATION THEORY

The fundamental problem of the matrix theory was to find the matrices $Q_1, \ldots, P_1, \ldots, P_k$ such that first, the commutation rules of I.2 (page 9) are satisfied, and second, that a certain function of these matrices, $H(Q_1, \ldots, Q_k, P_1, \ldots, P_k)$ becomes a diagonal matrix. This problem had already been divided into two parts by Born and Jordan in their first publication:

First, matrices $\overline{Q}_1, \ldots, \overline{Q}_k, \overline{P}_1, \ldots, \overline{P}_k$ were sought which have only to satisfy the commutation rules. This could easily be accomplished;[25] then, in general,

$$\overline{H} = H(\overline{Q}_1, \ldots, \overline{Q}_k, \overline{P}_1, \ldots, \overline{P}_k)$$

would not be a diagonal matrix. Then the correct solutions were expressed in the form

$$Q_1 = S^{-1}\overline{Q}_1S, \ldots, Q_k = S^{-1}\overline{Q}_kS, P_1 = S^{-1}\overline{P}_1S, \ldots, P_k = S^{-1}\overline{P}_kS$$

where S could be an arbitrary matrix (except that it must be one which possesses an inverse S^{-1} with the properties $S^{-1}S = SS^{-1} = 1$). Since the validity of the commutation rules for $Q_1, \ldots, Q_k, P_1, \ldots, P_k$ follows (identically in S!) from the validity of the rules for

$$\overline{Q}_1, \ldots, \overline{Q}_k, \overline{P}_1, \ldots, \overline{P}_k$$

---

[25] Cf., for example, §§ 20, 23 of the book of Born and Jordan mentioned in Note 1.

and since

$$\overline{H} = H(\overline{Q}_1, \ldots, \overline{Q}_k, \overline{P}_1, \ldots, \overline{P}_k) \text{ goes over into}$$

$$H = H(Q_1, \ldots, Q_k, P_1, \ldots, P_k) \text{ with } S^{-1}\overline{H}S = H$$ [26]

the only requirement put on S is that $S^{-1}\overline{H}S$ be a diagonal matrix where $\overline{H}$ is given. (Of course, one would also have to see to it that $S^{-1}\overline{Q}_1 S$, etc. be Hermitian, just as the $\overline{Q}_1$, etc. were. However, it can be shown upon closer examination that this additional condition on S can always be satisfied later, and it shall therefore not be considered in these initial observations).

Consequently there is need of transforming a given $\overline{H}$ to the diagonal form by means of the scheme $S^{-1}\overline{H}S$. Let us formulate therefore precisely what this means.

Let the matrix $\overline{H}$ have the elements $h_{\mu\nu}$, the desired matrix S the elements $s_{\mu\nu}$, and the (also unknown) diagonal matrix H the diagonal elements w_μ, i.e., the general element $w_\mu \delta_{\mu\nu}$.[27] Now $H = S^{-1}\overline{H}S$ is

---

[26] Because

$$S^{-1}.1.S = 1, \qquad S^{-1}.aA.S = a.S^{-1}AS,$$

$$S^{-1}.(A + B).S = S^{-1}AS + S^{-1}BS, \qquad S^{-1}.AB.S = S^{-1}AS.S^{-1}BS,$$

therefore for each matrix polynomial $P(A, B, \ldots)$,

$$S^{-1}P(A, B, \ldots)S = P(S^{-1}AS, S^{-1}BS, \ldots)$$

If we choose for P the left sides of the commutations relations, therefore their invariance follows from this; if we choose H for P, then we get $S^{-1}\overline{H}S = H$.

[27] $\delta_{\mu\nu} = 1$ for $\mu = \nu$ and $= 0$ for $\mu \neq \nu$ is the well known Kronecker delta.

the same as $SH = \bar{H}S$, and this means (if we equate the corresponding elements to one another on both sides of the equation, according to the well known rules of matrix multiplication):

$$\sum_\nu s_{\mu\nu} \cdot w_\nu \delta_{\nu\rho} = \sum_\nu h_{\mu\nu} \cdot s_{\nu\rho}$$

i.e.,

$$\sum_\nu h_{\mu\nu} \cdot s_{\nu\rho} = w_\rho \cdot s_{\mu\rho}$$

The individual columns $s_{1\rho}, s_{2\rho}, \ldots$ of the matrix $S (\rho = 1, 2, \ldots)$ and the corresponding diagonal elements w_ρ of the matrix H are therefore solutions of the so-called eigenvalue problem, which runs as follows:

$$\sum_\nu h_{\mu\nu} x_\nu = \lambda \cdot x_\mu \qquad (\mu = 1, 2, \ldots)$$

(The trivial solution $x_1 = x_2 = \ldots = 0$ is naturally excluded). In fact, $x_\nu = s_{\nu\rho}$, $\lambda = w_\rho$ is a solution. ($x_\nu \equiv 0$ i.e., $s_{\nu\rho} \equiv 0$ [for all ν] is inadmissible, because then the ρth column of S would vanish identically, although S possesses an inverse S^{-1}!) . It is worth mentioning that these are essentially the only solutions.

Indeed, the above equation means the following: the transform of the vector $x = \{x_1, x_2, \ldots\}$ by the matrix $\bar{H}$ is equal to its multiple by the number λ . We transform $x = \{x_1, x_2, \ldots\}$ by S^{-1} , and there results a vector $y = \{y_1, y_2, \ldots\}$. If we transform y by H , then this is the same as the transform of x by

$$HS^{-1} = S^{-1}\bar{H}SS^{-1} = S^{-1}\bar{H}$$

Hence it is a transform of λx by S^{-1} , and therefore the result is λy . Now Hy has the components

$$\sum_{\nu} w_{\mu} \delta_{\mu\nu} y_{\nu} = w_{\mu} y_{\mu},$$

λy the components λy_{μ} . Therefore it is required that $w_{\mu} y_{\mu} = \lambda y_{\mu}$ for all $\mu = 1, 2, \ldots$, i.e., $y_{\mu} = 0$ whenever $w_{\mu} \neq \lambda$. If we call η^{ρ} that vector whose ρth component is 1 , but all the others of which are 0 , then this means: y is a linear combination of those η^{ρ} for which $w_{\rho} = \lambda$ -- in particular it is zero if there are none such. The value x results from the application of S to y , therefore it is a linear combination of the η^{ρ} from before, transformed with S . The μth component of $S\eta^{\rho}$ is (since the νth component of η^{ρ} was $\delta_{\nu\rho}$)

$$\sum_{\nu} s_{\mu\nu} \delta_{\nu\rho} = s_{\mu\rho}$$

If we then interpret the ρth column of $S, s_{1\rho}, s_{2\rho}, \ldots$ as a vector, then x is a linear combination of all columns for which $w_{\rho} = \lambda$ -- in particular is zero if this does not occur. Consequently our original assertion has been proved: the $w_1, w_2, \ldots$ are the only eigenvalues and the $x_{\nu} = s_{\nu\rho}$, $\lambda = w_{\rho}$ are essentially the only solutions.

This is very important, because not only does the knowledge of S, H determine all the solutions of the eigenvalue problem, but conversely, we can also determine S, H as soon as we have solved the eigenvalue problem completely. For example, for H : The w_{μ} are plainly all solutions λ , and each such λ appears in the series $w_1, w_2, \ldots$ as often as there are linearly independent solutions $x_1, x_2, \ldots$ [28] -- hence the $w_1, w_2, \ldots$ are already

---

[28]The S columns, $s_{1\rho}, s_{2\rho}, \ldots$ of the ρ with $w_{\rho} = \lambda$ form a complete set of solutions, and as columns of a matrix which possesses an inverse, they must be linearly independent.

determined except for their order.[29]

The fundamental problem of the matrix theory is then the solution of the eigenvalue equation

$$E_1 \qquad \sum_{\nu} h_{\mu\nu} x_\nu = \lambda \cdot x_\mu \qquad (\mu = 1, 2, \ldots)$$

Let us now go on to the wave theory. The fundamental equation of this theory is the "wave equation"

$$E_2 \quad H\phi(q_1, \ldots, q_k) = \lambda\phi(q_1, \ldots, q_k)$$

in which H is the differential operator already discussed -- we seek all solutions $\phi(q_1, \ldots, q_k)$ and λ, with the exclusion of the trivial $\phi(q_1, \ldots, q_k) \equiv 0$, λ arbitrary. This is analogous to what was required of E_1: the sequence $x_1, x_2, \ldots,$ which we can also regard as a function x_ν of the "discontinuous" variable ν (with the variable values $1, 2, \ldots$) corresponds to the function $\phi(q_1, \ldots, q_k)$ with the "continuous" variables $q_1, \ldots, q_k$; λ has the same role each time. However, the linear transformation

$$x_\mu \longrightarrow \sum_{\nu} h_{\mu\nu} x_\nu$$

shows little similarity to the other

$$\phi(q_1, \ldots, q_k) \longrightarrow H\phi(q_1, \ldots, q_k)$$

How should such an analogy be obtained here?

We have regarded the index ν as variable, and placed it in correspondence with the k variables $q_1, \ldots, q_k$, i.e., a positive integer with the general point

[29]Since an arbitrary permutation of the columns of S, together with the corresponding permutation of the rows of S^{-1}, permutes the diagonal elements in H in the same way, the order of the $w_1, w_2, \ldots$ is in fact indeterminate.

of the k-dimensional configuration space (which may be called Ω from now on). Therefore we should not expect that Σ_ν can be carried over as a sum into Ω. We should rather expect the integral $\int\limits_\Omega \cdots \int\limits_\Omega \ldots dq_1 \ldots dq_k$ (or more briefly, $\int\limits_\Omega \ldots dv$, dv is the volume element $dq_1 \ldots dq_k$ in Ω) to be the correct analog. To the matrix element $h_{\mu\nu}$ which depends on two variables of the type of the index ν, there corresponds a function

$$h(q_1, \ldots, q_k;\ q_1', \ldots, q_k')$$

in which the $q_1, \ldots, q_k$ and the $q_1', \ldots, q_k'$ run through the whole Ω domain independently. The transformation

$$x_\mu \longrightarrow \sum_\nu h_{\mu\nu} x_\nu \quad \text{or} \quad x_\nu \longrightarrow \sum_{\nu'} h_{\nu\nu'} x_{\nu'}$$

then becomes

$$\phi(q_1, \ldots, q_k) \longrightarrow \underbrace{\int \cdots \int}_{\Omega} h(q_1, \ldots, q_k;\ q_1', \ldots, q_k') \phi(q_1', \ldots, q_k') dq_1' \ldots dq_k'$$

and the eigenvalue problem E_1, which we can also write as

$$E_1 \qquad \sum_{\nu'} h_{\nu\nu'} x_{\nu'} = \lambda \cdot x_\nu$$

becomes

$$E_3 \quad \underbrace{\int \cdots \int}_{\Omega} h(q_1, \ldots, q_k;\ q_1', \ldots, q_k') \phi(q_1', \ldots, q_k') dq_1' \ldots dq_k' = \lambda \cdot \phi(q_1, \ldots, q_k)$$

Eigenvalue problems of the type E_3 have been investigated extensively in mathematics, and can in fact be handled in far reaching analogy to the problem E_1. They are known as "integral equations."[30]

But, unfortunately, E_2 does not have this form, or, rather, it can only be brought into this form if a function $h(q_1, \ldots, q_k; q_1', \ldots q_k')$ can be found for the differential operator

$$H = H\left(q_1, \ldots, q_k, \frac{h}{2\pi i}\frac{\partial}{\partial q_1}, \ldots, \frac{h}{2\pi i}\frac{\partial}{\partial q_k}\right)$$

such that

$$H\phi(q_1, \ldots, q_k) = \underbrace{\int \cdots \int}_{\Omega} h(q_1, \ldots, q_k; q_1', \ldots, q_k')\phi(q_1', \ldots, q_k')dq_1' \ldots dq_k'$$

identically (i.e., for all $\phi(q_1, \ldots, q_k)$. This $h(q_1, \ldots, q_k; q_1', \ldots, q_k')$, if it exists, is called the "kernel" of the functional operator H, and H itself is then called an "integral operator."

Now such a transformation is generally impossible, i.e., differential operators H are never integral operators. Even the simplest functional operator, which transforms each ϕ into itself -- this operator is called 1 -- is not one. Let us convince ourselves of this, and for simplicity, take $k = 1$. Then let it be required that

$$\Delta_1 \qquad \phi(q) \equiv \int_{-\infty}^{\infty} h(q_1, q')\phi(q')dq'$$

We replace $\phi(q)$ by $\phi(q + q_0)$, set $q = 0$, and introduce the integration variable $q'' = q' + q_0$. Then

---

[30]The theory of integral equations has received its definitive form through the work of Fredholm and Hilbert. An exhaustive treatment, complete with references, is found in the book by Courant and Hilbert, Methoden der Mathematischen Physik, Berlin, 1931.

$$\phi(q_0) = \int_{-\infty}^{\infty} h(0, q'' - q_0)\phi(q'')dq''.$$

If we replace q_0, q'' by q, q', then we see that $h(0, q' - q)$ along with $h(q, q')$ solves our problem, -- hence we may assume that $h(q, q')$ is only dependent on $q' - q$. Then the requirement becomes

$$\Delta_2 \quad \phi(q) \equiv \int_{-\infty}^{\infty} h(q' - q)\phi(q')dq' \qquad (h(q, q') = h(q' - q))$$

Replacing again $\phi(q + q_0)$ for $\phi(q)$, it suffices to consider $q = 0$, i.e.,

$$\Delta_3 \qquad \phi(0) \equiv \int_{-\infty}^{\infty} h(q)\phi(q)dq$$

Replacing $\phi(q)$ by $\phi(-q)$ shows that $h(-q)$ is also a solution along with $h(q)$, therefore

$$h_1(q) = \tfrac{1}{2}[h(q) + h(-q)]$$

is also, so that $h(q)$ may be considered as an even function of q.

It is clear that these conditions are impossible of fulfillment: If we choose $\phi(q) = h(q)$ for $q \gtrless 0$, $\phi(0) = 0$, then it follows from Δ_3. that $h(q) = 0$ for $q \gtrless 0$.[31] But if we choose $\phi(q) \equiv 1$, then we obtain

$$\int_{-\infty}^{\infty} h(q)dq = 1$$

-- while

---

[31] More precisely, if we take as a basis the Lebesgue concept of the integral, then for $q \gtrless 0$, $h(q) = 0$ except for a set of measure 0 -- i.e., except for such a set, $h(q) = 0$ identically.

$$\int_{-\infty}^{\infty} h(q)dq = 0$$

follows directly from the above.

Dirac nevertheless assumes the existence of such a function

$$\Delta_4\ \delta(q) = 0 \quad \text{for} \quad q \lessgtr 0\ ,\ \delta(q) = \delta(-q)\ ,\ \int_{-\infty}^{\infty} \delta(q)dq = 1$$

This would satisfy Δ_3.:

$$\int_{-\infty}^{\infty} \delta(q)\phi(q)dq = \phi(0)\int_{-\infty}^{\infty} \delta(q)dq + \int_{-\infty}^{\infty} \delta(q)[\phi(q) - \phi(0)]dq$$

$$= \phi(0).1 + \int_{-\infty}^{\infty} 0.dq = \phi(0)$$

therefore Δ_1., Δ_2. also. We should thus think of this function as vanishing everywhere except at the origin, and of its being so strongly infinite there that the total integral still comes out to be 1 for $\delta(q)$.[32]

If we have once accepted this fiction, it is possible to represent the most varied differential operators as integral operators -- provided that, in addition to $\delta(q)$, we also introduce its derivatives. Then we have

$$\frac{d^n}{dq^n}\phi(q) = \frac{d^n}{dq^n}\int_{-\infty}^{\infty} \delta(q-q')\phi(q')dq' = \int_{-\infty}^{\infty} \frac{\partial^n}{\partial q^n}\delta(q-q')\phi(q')dq'$$

$$= \int_{-\infty}^{\infty} \delta^{(n)}(q-q')\phi(q')dq'\ ,$$

$$q^n\phi(q) = \int_{-\infty}^{\infty} \delta(q-q')q^n.\phi(q')dq'$$

---

[32]The area under the curve of $\delta(q)$ is then to be thought of as infinitely thin and infinitely high, for the point

i.e., $\frac{d^n}{dq^n}$ and q^n ... have the kernels $\delta^{(n)}(q - q')$ and $\delta(q - q')q^n$, respectively. According to this same scheme, we can investigate the kernels of rather complicated differential operators. For several variables $q_1, \ldots, q_k$, the delta functions lead to the result

$$\underbrace{\int \cdots \int}_{\Omega} \delta(q_1-q_1')\delta(q_2-q_2')\ldots\delta(q_k-q_k')\phi(q_1'\ldots q_k')dq_1'\ldots dq_k'$$

$$= \int_{-\infty}^{\infty}\left[\cdots\left[\int_{-\infty}^{\infty}\left[\int_{-\infty}^{\infty}\phi(q_1',q_2',\ldots,q_k')\delta(q_1-q_1')dq_1'\right]\right.\right.$$

$$\left.\left.\delta(q_2-q_2')dq_2'\right]\cdots\right]\delta(q_k-q_k')dq_k'$$

$$= \int_{-\infty}^{\infty}\left[\cdots\left[\int_{-\infty}^{\infty}\phi(q_1,q_2',\ldots,q_k')\delta(q_2-q_2')dq_2'\right]\cdots\right]\delta(q_k-q_k')dq_k'$$

$$= \cdots = \phi(q_1,q_2,\ldots,q_k) ,$$

$$\underbrace{\int \cdots \int}_{\Omega} \delta'(q_1-q_1')\delta(q_2-q_2')\ldots\delta(q_k-q_k')\phi(q_1'\ldots q_k')dq_1'\ldots dq_k'$$

$$= \frac{d}{dq_1}\underbrace{\int \cdots \int}_{\Omega} \delta(q_1-q_1')\delta(q_2-q_2')\ldots\delta(q_k-q_k')\phi(q_1'\ldots q_k')dq_1'\ldots dq_k'$$

$$= \frac{d}{dq_1}\phi(q_1,\ldots,q_k) , \text{ etc.}$$

Hence the integral representation I. can be achieved in practice for all operators.

As soon as we have this representation, the analogy of problems E_1. and E_3. is complete. We have only to replace ν; ν'; $\sum_\nu$; x by

---

situated at $q = 0$, and of area unity. This may be viewed as the limiting behavior for the function $\sqrt{\frac{a}{\pi}}\, e^{-aq^2}$ as $a \longrightarrow +\infty$ but it is nevertheless impossible.

$$q_1,\ldots,q_k;\ q'_1,\ldots,q'_k;\ \underbrace{\int\cdots\int}_{\Omega}\ldots dq'_1\ldots dq'_k;\ \phi.$$

As the vectors x_ν correspond to the functions $\phi(q_1,\ldots,q_k)$, the kernels $h(q_1,\ldots,q_k;\ q'_1,\ldots,q'_k)$ must correspond to the matrices $h_{\nu\nu'}$; however, it is useful to regard the kernels themselves as matrices, and consequently to interpret the $q_1,\ldots,q_k$ as row- and the $q'_1,\ldots,q'_k$ as column-indices, corresponding to ν and ν' respectively. We then have, in addition to the ordinary matrices $\{h_{\nu\nu'}\}$ with discrete row and column domains enumerated by the numbers $1,2,\ldots$, others,

$$\{h(q_1,\ldots,q_k;\ q'_1,\ldots,q'_k)\}$$

(the kernels), for which each domain is characterized by k variables, varying continuously throughout the entire Ω .

This analogy may seem entirely formal, but in reality this is not so. The indices ν and ν' can also be regarded as coordinates in a state space, that is, if we interpret them as quantum numbers (in the sense of the Bohr theory: as numbers of the possible orbits in phase space which are then discrete because of the restrictions of the quantum conditions).

We do not desire to follow any further here this train of thought which was shaped by Dirac and Jordan into a unified theory of the quantum processes. The "improper" functions (such as $\delta(x)$, $\delta'(x)$) play a decisive role in this development -- they lie beyond the scope of mathematical methods generally used, and we desire to describe quantum mechanics with the help of these latter methods. We therefore pass over to the other (Schrödinger) method of unification of the two theories.

4. THE EQUIVALENCE OF THE TWO THEORIES: HILBERT SPACE

The method sketched in I.3 resulted in an analogy between the "discrete" space of index values $Z = (1,2,\dots)$ and the continuous state space Ω of the mechanical system (Ω is k-dimensional, where k is the number of classical mechanical degrees of freedom). That this cannot be achieved without some violence to the formalism and to mathematics is not surprising. The spaces Z and Ω are in reality very different, and every attempt to relate the two must run into great difficulties.[33]

What we do have, however, is not a relation of Z to Ω, but only a relation between the functions in these two spaces, i.e., between the sequences $x_1, x_2, \dots$ which are the functions in Z, and the wave functions $\phi(q_1,\dots,q_k)$ which are the functions in Ω. These functions, furthermore, are the entities which enter most essentially into the problems of quantum mechanics.

In the Schrödinger theory, the integral

$$\underbrace{\int\cdots\int}_{\Omega} |\phi(q_1,\dots,q_k)|^2 dq_1\dots dq_k$$

plays an important role -- it must $= 1$, in order that ϕ can be given a physical interpretation (cf. I.2.). In matrix theory, on the other hand (cf. the problem E_1. in I.3.), the vector $x_1, x_2, \dots$ plays the decisive role. The condition of the finiteness of $\Sigma_\nu |x_\nu|^2$ in the sense of the Hilbert theory of such eigenvalue problems (cf. reference in Note 30), is always imposed on this vector. It

---

[33]Such a unification was undertaken, long before quantum mechanics, by E. H. Moore, the originator of the so-called "general analysis." Cf. the article on this subject by Hellinger-Toeplitz in Mth. Enzyklopädie, vol. II, C, 13. Leipzig, 1927.

is also customary, having excluded the trivial solution $x_\nu \equiv 0$, to set up the normalization $\Sigma_\nu |x_\nu|^2 = 1$. It is plain in Z or Ω this limits the field of admissible functions to those with finite

$$\sum_\nu |x_\nu|^2 \quad \text{or} \quad \underbrace{\int \cdots \int}_{\Omega} |\phi(q_1, \ldots, q_k)|^2 dq_1 \ldots dq_k$$

because only with such functions can the above Σ_ν or $\underbrace{\int \ldots \int}_{\Omega}$, be made equal to 1 by multiplication with a constant factor -- i.e., can be normalized in the usual sense.[34] We call the totality of such functions F_Z and F_Ω, respectively.

Now the following theorem holds: F_Z and F_Ω are isomorphic (Fischer and F. Riesz[35]). To be precise,

---

[34]It is a repeatedly observed fact in the Schrödinger theory that only the finiteness of

$$\underbrace{\int \cdots \int}_{\Omega} |\phi(q_1, \ldots, q_k)|^2 dq_1 \ldots dq_k$$

is required in the case of the wave functions ϕ. So, for example, ϕ may be singular, and perhaps become infinite, if only the above integral remains finite. An instructive example for this case is the hydrogen atom in the relativistic theory of Dirac, cf. Proc. Roy. Soc. 117 (1928); also, W. Gordon, Z. Physik 48 (1928).

[35]In the course of our discussion of Hilbert space, a proof of this theorem will be given (cf. II.2., 3., especially THEOREM 5 in II.2.). It is worth mentioning that the part of this theorem sufficient for many purposes, and easier to prove is the isomorphism between F_Ω and an appropriate part of F_Z; this is due to Hilbert (Gött. Nachr. 1906). Thus Schrödinger's original equivalence proof (cf. Note 7) corresponds to just this part of the theorem.

this means the following: It is possible to set up a one-to-one correspondence between F_Z and F_Ω, i.e., to each sequence $x_1, x_2, \ldots$ with finite $\Sigma_\nu |x_\nu|^2$ a function $\phi(q_1, \ldots, q_k)$ with finite $\underset{\Omega}{\int \cdots \int} |\phi(q_1, \ldots, q_k)|^2 dq_1 \ldots dq_k$ can be assigned, and conversely in such a manner that this correspondence is linear and isometric. By "linearity" this is meant: if $x_1, x_2, \ldots$ corresponds to $\phi(q_1, \ldots, q_k)$ and $y_1, y_2, \ldots$ to $\psi(q_1, \ldots, q_k)$, then $ax_1, ax_2, \ldots$ and $x_1 + y_1, x_2 + y_2, \ldots$ correspond respectively to $a\phi(q_1, \ldots, q_k)$ and $\phi(q_1, \ldots, q_k) + \psi(q_1, \ldots, q_k)$; by "isometry" this is meant: if $x_1, x_2, \ldots$ and $\phi(q_1, \ldots, q_k)$ correspond to one another, then

$$\sum_\nu |x_\nu|^2 = \underbrace{\int \cdots \int}_{\Omega} |\phi(q_1, \ldots, q_k)|^2 dq_1 \ldots dq_k$$

(The word isometric has the connotation that it is customary to regard the $x_1, x_2, \ldots$ and $\phi(q_1, \ldots, q_k)$ as vectors, and to consider

$$\sqrt{\sum_\nu |x_\nu|^2}$$

and

$$\sqrt{\underbrace{\int \cdots \int}_{\Omega} |\phi(q_1, \ldots, q_k)|^2 dq_1 \ldots dq_k}$$

as their respective "lengths.") In addition, if $x_1, x_2, \ldots$ and $y_1, y_2, \ldots$ correspond respectively to $\phi(q_1, \ldots, q_k)$ and $\psi(q_1, \ldots, q_k)$, then

$$\sum_\nu x_\nu \bar{y}_\nu = \underbrace{\int \cdots \int}_{\Omega} \phi(q_1, \ldots, q_k) \overline{\psi(q_1, \ldots, q_k)} dq_1 \ldots dq_k$$

(and both sides are absolutely convergent). On this latter point, it should be observed that one might have preferred to have quite generally

$$\sum_{\nu} x_{\nu} = \underbrace{\int \cdots \int}_{\Omega} \phi(q_1, \ldots, q_k) dq_1 \cdots dq_k$$

or something similar, i.e., a complete analogy between addition on the one hand and integration on the other -- but a closer examination shows that the addition Σ_{ν} and the integration

$$\underbrace{\int \cdots \int}_{\Omega} \cdots dq_1 \cdots dq_k$$

are employed in quantum mechanics only in expressions such as $x_{\nu}\bar{y}_{\nu}$ or $\phi(q_1, \ldots, q_k)\overline{\psi(q_1, \ldots, q_k)}$, respectively.

We do not intend to pursue any investigation at this point as to how this correspondence is to be established, since this will be of great concern to us in the next chapter. But we should emphasize what its existence means: Z and Ω are very different, and to set up a direct relation between them must lead to great mathematical difficulties. On the other hand, F_Z and F_{Ω} are isomorphic, i.e., identical in their intrinsic structure (they realize the same abstract properties in different mathematical forms) -- and since they (and not Z and Ω themselves!) are the real analytical substrata of the matrix and wave theories, this isomorphism means that the two theories must always yield the same numerical results. That is, this is the case whenever the isomorphism lets the matrix

$$\overline{H} = H(\overline{Q}_1, \ldots, \overline{Q}_k; \overline{P}_1, \ldots, \overline{P}_k)$$

and the operator

$$H = H\left(q_1, \ldots, q_k; \frac{h}{2\pi i}\frac{\partial}{\partial q_1}, \ldots, \frac{h}{2\pi i}\frac{\partial}{\partial q_k}\right)$$

correspond to one another. Since both are obtained by the

same algebraic operations from the matrices $\overline{Q}_l, \overline{P}_l$ $(l = 1,\ldots,k)$ and the functional operators

$$q_l \ \ldots, \ \frac{h}{2\pi i}\frac{\partial}{\partial q_l} \ \ldots \ (l = 1,\ldots,k)$$

respectively, it suffices to show that $q_l \ \ldots$ corresponds to the matrix $\overline{Q}_l$ and $\frac{h}{2\pi i}\frac{\partial}{\partial q_l} \ \ldots$ to the matrix $\overline{P}_l$. Now nothing further was required of the $\overline{Q}_l, \overline{P}_l$ $(l = 1,\ldots,k)$ than that they satisfy the commutation rules mentioned in I.2.:

$$\left.\begin{array}{l} \overline{Q}_m\overline{Q}_n - \overline{Q}_n\overline{Q}_m = 0 \qquad \overline{P}_m\overline{P}_n - \overline{P}_n\overline{P}_m = 0 \\ Q_mP_n - P_nQ_m \begin{cases} = 0 & \text{for } m \neq n \\ = \frac{h}{2\pi i} 1 & \text{for } m = n \end{cases} \end{array}\right\} (m,n = 1,2,\ldots)$$

But the matrices corresponding to the $q_l \ \ldots, \ \frac{h}{2\pi i}\frac{\partial}{\partial q_l}$ will certainly do this, because the functional operators $q_l,\ldots,\frac{h}{2\pi i}\frac{\partial}{\partial q_l},\ldots$ possess the properties mentioned,[36] and these are not lost in the isomorphic transformation to F_Z.

Since the systems F_Z and F_Ω are isomorphic,

---

[36]We have

$$q_m\cdot q_n\cdot\phi(q_1,\ldots,q_k) = q_n\cdot q_m\cdot\phi(q_1,\ldots,q_k),$$

$$\frac{\partial}{\partial q_m}\frac{\partial}{\partial q_n}\phi(q_1,\ldots,q_k) = \frac{\partial}{\partial q_n}\frac{\partial}{\partial q_m}\phi(q_1,\ldots,q_k),$$

$$\frac{\partial}{\partial q_m}q_n\cdot\phi(q_1,\ldots,q_k) - q_n\cdot\frac{\partial}{\partial q_m}\phi(q_1,\ldots,q_k) \begin{cases} = 0, & \text{for } m \neq n \\ = \phi(q_1,\ldots,q_k), & \text{for } m = 0 \end{cases}$$

from which the desired operator relations follow directly.

and since the theories of quantum mechanics constructed on them are mathematically equivalent, it is to be expected that a unified theory, independent of the accidents of the formal framework selected at the time, and exhibiting only the really essential elements of quantum mechanics, will then be achieved, if we do this: Investigate the intrinsic properties (common to F_Z and F_Ω) of these systems of functions, and choose these properties as a starting point.

The system F_Z is generally known as "Hilbert space." Therefore, our first problem is to investigate the fundamental properties of Hilbert space, independent of the special form of F_Z or F_Ω. The mathematical structure which is described by these properties (which in any specific special case are equivalently represented by calculations within F_Z or F_Ω, but for general purposes are easier to handle directly than by such calculations), is called "abstract Hilbert space."

We wish then to describe the abstract Hilbert space, and then to prove rigorously the following points:

1. That the abstract Hilbert space is characterized uniquely by the properties specified, i.e., that it admits of no essentially different realizations.

2. That its properties belong to F_Z as well as F_Ω. (In this case the properties discussed only qualitatively in I.4 will be analyzed rigorously). When this is accomplished, we shall employ the mathematical equipment thus obtained to shape the structure of quantum mechanics.

CHAPTER II

ABSTRACT HILBERT SPACE

1. THE DEFINITION OF HILBERT SPACE

We must now carry out the program outlined at the end of I.4.: to define Hilbert space, which furnishes the mathematical basis for the treatment of quantum mechanics in terms of those concepts which are subsequently needed in quantum mechanics, and which have accordingly the same meaning in the "discrete" function space F_Z of the sequences x_ν $(\nu = 1,2,\ldots)$ and in the "continuous" F_Ω of the wavefunctions $\phi(q_1,\ldots,q_k)$ ($q_1,\ldots,q_k$ run through the entire state space Ω). These concepts are the following ones, as we have already indicated:

α) The "scalar product," i.e., the product of a (complex) number a with an element f of Hilbert space: af. In F_Z, ax_ν is obtained from x_ν, while in F_Ω, $a\phi(q_1,\ldots,q_k)$ is obtained from $\phi(q_1,\ldots,q_k)$.

β) The addition and subtraction of two elements f, g of Hilbert space: $f \pm g$. In F_Z, $x_\nu \pm y_\nu$ results from x_ν and y_ν; in F_Ω, $\phi(q_1,\ldots,q_k) \pm \psi(q_1,\ldots,q_k)$ results from $\phi(q_1,\ldots,q_k)$ and $\psi(q_1,\ldots,q_k)$.

γ) The "inner product" of two elements f, g of Hilbert space. Unlike α), β), this operation produces a complex number, and not an element of Hilbert space: (f, g).

In F_Z, $\Sigma_\nu x_\nu \overline{y}_\nu$ is obtained from x_ν and y_ν, while in F_Ω,

$$\underbrace{\int \cdots \int}_{\Omega} \phi(q_1, \ldots, q_k)\overline{\psi(q_1, \ldots, q_k)}dq_1, \ldots, dq_k$$

is obtained from $\phi(q_1,\ldots,q_k)$ and $\psi(q_1,\ldots,q_k)$. (The definitions in F_Z and F_Ω are still to be completed by the appropriate convergence proofs. We shall give these proofs in II.3.)

In the following we shall denote the points of Hilbert space by $f, g, \ldots, \phi, \psi, \ldots,$ complex numbers by $a, b, \ldots, x, y, \ldots,$ and positive integers by $k, l, m, \ldots, \mu, \nu, \ldots$. We shall also refer to Hilbert space as $\mathfrak{R}_\infty$ whenever necessary (as an abbreviation for " ∞-dimensional Euclidean space," analogous to the customary designation $\mathfrak{R}_n$ for the "n-dimensional Euclidean space" $[n = 1,2,\ldots]$).

The noteworthy feature of the operations af, $f \pm g$, (f, g) is that they are exactly the basic operations of the vector calculus: those which make possible the establishment of length and angle calculations in Euclidean geometry or the calculations with force and work in the mechanics of particles. The analogy becomes very clear in the case of F_Z, if, in place of the $x_1, x_2, \ldots$ in $\mathfrak{R}_\infty$, we consider the ordinary points $x_1, \ldots, x_n$ of an $\mathfrak{R}_n$ (for which the operations α), β), γ) can be defined in the same way. In particular, for $n = 3$ we have the conditions of ordinary space. Under certain circumstances, it is more appropriate to regard the complexes $x_1, \ldots, x_n$ not as points, but as vectors, directed from the point $0, \ldots, 0$ to the points $x_1, \ldots, x_n$.

In order to define abstract Hilbert space, we then take as a basis the fundamental vector operations af, $f \pm g$, (f, g). We shall consider all $\mathfrak{R}_n$ simultaneously with $\mathfrak{R}_\infty$, as will appear in the discussion to which we now proceed. Therefore, where we do not wish to distinguish between $\mathfrak{R}_\infty$ and the $\mathfrak{R}_n$, we shall use $\mathfrak{R}$ as the common term for the space.

First of all we postulate for $\mathfrak{R}$ the typical vector properties[37]:

A . $\mathfrak{R}$ is a linear space.

That is: an addition $f + g$ and a "scalar" multiplication af are defined in $\mathfrak{R}$ (f, g elements of $\mathfrak{R}$, a a complex number -- $f + g$, af belong to $\mathfrak{R}$), and $\mathfrak{R}$ has a null element.[38] The well known rules of calculation for vector algebra then hold for this space:

$$f + g = g + f \qquad \text{(commutative law of addition),}$$

$$(f + g) + h = f + (g + h) \qquad \text{(associative law of addition),}$$

$$\left.\begin{array}{l}(a + b)f = af + bf, \\ a(f + g) = af + ag\end{array}\right\} \qquad \text{(distributive law of multiplication),}$$

$$(ab)f = a(bf) \qquad \text{(associative law of multiplication),}$$

$$0f = 0, \quad 1f = f \qquad \text{(role of 0 and 1).}$$

The rules of calculation not mentioned here follow directly from these postulates. For example, the role of the null vector in addition:

$$f + 0 = 1.f + 0.f = (1 + 0).f = 1.f = f$$

Or the uniqueness of a substraction: we define

---

[37]The characterization of $\mathfrak{R}_n$ by **A**, **B**, $\mathbf{C}^{(n)}$ originated with Weyl [cf. Raum, Zeit, Materie, Berlin, (1921)]. If we desire to obtain $\mathfrak{R}_\infty$ instead of $\mathfrak{R}_n$, then $\mathbf{C}^{(\infty)}$ must naturally be replaced by $\mathbf{C}^{(n)}$ It is in this case only that **D**, **E** become necessary, cf. the discussion later in the text.

[38]Besides the origin, or the null vector of $\mathfrak{R}$, there is also the number 0, so that the same symbol is used for two things. The relations are such, however, that no confusion should arise.

$$-f = (-1).f , \qquad f - g = f + (-g)$$

then

$$\left.\begin{aligned}(f - g) + g &= (f + (-g)) + g \\ &= f + ((-g) + g) , \\ (f + g) - g &= (f + g) + (-g) \\ &= (f + (g + (-g)) ,\end{aligned}\right\} \begin{aligned}&= f + ((-1).g + 1.g) \\ &= f + ((-1) + 1).g \\ &= f + 0.g = f + 0 = f .\end{aligned}$$

Or the distributive laws of multiplication with subtraction:

$$\begin{aligned}a.(f - g) &= a.f + a.(-g) = af + a((-1).g) = af + (a.(-1)).g \\ &= af + ((-1) \cdot a).g = af + (-1).(ag) = af + (-ag) \\ &= af - ag ,\end{aligned}$$

$$\begin{aligned}(a - b)f &= a.f + (-b).f = af + ((-1)b).f = af + (-1).(bf) \\ &= af + (-bf) = af - bf .\end{aligned}$$

It is not worth while to pursue these matters any further, it ought to be clear that all the rules of the linear vector calculus are valid here.

We can therefore introduce the concept of linear independence for elements $f_1, \ldots, f_k$ of $\mathfrak{R}$ in the same way as this is done for vectors.

DEFINITION 1. The elements $f_1, \ldots, f_k$, are linearly independent if it follows from $a_1 f_1 + \ldots + a_k f_k = 0$ ($a_1, \ldots, a_k$ complex numbers) that $a_1 = \ldots = a_k = 0$.

We further define the analog of the linear entities occurring in the vector calculus (the line, plane, etc., passing through the origin), the linear manifold.

DEFINITION 2. A subset $\mathfrak{M}$ of $\mathfrak{R}$ is called a linear manifold if it contains all the linear combinations $a_1f_1 + \dots + a_kf_k$ of any k (= 1,2,...) of its elements $f_1,\dots,f_k$.[39] If $\mathfrak{A}$ is an arbitrary subset of $\mathfrak{R}$, then the set of all $a_1f_1 + \dots + a_kf_k$ ($k = 1,2,\dots$; $a_1,\dots,a_k$ arbitrary complex numbers; $f_1,\dots,f_k$ arbitrary elements of $\mathfrak{A}$) is a linear manifold, which evidently contains $\mathfrak{A}$. It is clear that it is also a subset of every other linear manifold containing $\mathfrak{A}$. It is called "the linear manifold spanned by $\mathfrak{A}$," and is symbolized by $\{\mathfrak{A}\}$.

Before we develop this concept any further, let us formulate the next basic principle of the vector calculus, the existence of the inner product.

B An Hermitian inner product is defined in $\mathfrak{R}$.

That is: (f, g) is defined (f, g in $\mathfrak{R}$, (f, g) a complex number), and it has the following properties:

$(f' + f'', g) = (f', g) + (f'', g)$ (distributive law of the first factor),

$(a.f, g) = a.(f, g)$ (associative law for the first factor),

[39] It would be sufficient to require: if f belong to $\mathfrak{M}$, then af also; if f, g, then $f + g$ also. Then if the $f_1,\dots,f_k$ belongs to $\mathfrak{M}$, the $a_1f_1,\dots,a_kf_k$ do also, and hence successively the $a_1f_1 + a_2f_2$, $a_1f_1 + a_2f_2 + a_3f_3,\dots,a_1f_1 + \dots + a_kf_k$, too.

$(f, g) = \overline{(g, f)}$ (Hermitian symmetry),

$(f, f) \geq 0$, and $= 0$ only if $f = 0$.[40] (definite form).

In addition, the corresponding relations for the second factor follow from the two properties of the first factor, because of the Hermitian symmetry (we exchange f and g, and take the complex conjugate of both sides):

$$(f, g' + g'') = (f, g') + (f, g'')$$

$$(f, a \cdot g) = \bar{a} \cdot (f, g) .$$

This inner product is of great importance, because it makes possible the definition of length. In Euclidean space, the magnitude of a vector f is defined by $\|f\| = \sqrt{(f, f)}$,[41] and the distance between two points f, g is defined by $\|f - g\|$. We shall start from this point.

DEFINITION 3. The "magnitude" of an element f of $\mathfrak{R}$ is $\|f\| = \sqrt{(f, f)}$, the distance between f, g is $\|f - g\|$.[42]

---

[40] (f, f) is a real number because of the Hermitian symmetry: Indeed, for $f = g$, this gives $(f, f) = \overline{(f, f)}$.

[41] If f has the components $x_1, \ldots, x_n$, then by the observations made in γ), II.1. (if we restrict ourselves to a finite number of components),

$$\sqrt{(f, f)} = \sqrt{\sum_{\nu=1}^{n} |x_\nu|^2}$$

i.e., the ordinary Euclidean length.

[42] Since (f, f) is real and ≥ 0, $\|f\|$ is real, and the square root is chosen ≥ 0. The same holds for $\|f - g\|$.

We shall see that this concept possesses all the properties of distance. For this purpose, we prove the following:

THEOREM 1. $|(f, g)| \leq ||f|| \cdot ||g||$

PROOF. First, we write

$$||f||^2 + ||g||^2 - 2 \operatorname{Re}(f, g)$$

$$= (f, f) + (g, g) - (f, g) - (g, f)$$

$$= (f - g, f - g) \geq 0 ,$$

$$\operatorname{Re}(f, g) \leq \tfrac{1}{2}(||f||^2 + ||g||^2)$$

(if $z = u + iv$ is a complex number -- u, v real --, then $\operatorname{Re} z$, $\operatorname{Im} z$ represent respectively the real and imaginary parts of z , i.e., $\operatorname{Re} z = u$, $\operatorname{Im} z = v$) . If we replace f, g , by af , $(1/a)g$ (a real, > 0) then the left side is not changed, as can easily be seen. But on the right we obtain

$$\tfrac{1}{2}(a^2||f||^2 + \frac{1}{a^2}||g||^2)$$

Since this expression is $\geq \operatorname{Re}(f, g)$, the inequality holds in particular for its minimum, which amounts to $||f|| \cdot ||g||$. (This value is taken on for $f, g \neq 0$, at

$$a = \sqrt{\frac{||g||}{||f||}}$$

and for $f = 0$ or $g = 0$, for $a \longrightarrow +\infty$ or for $a \longrightarrow +0$ respectively.) Therefore

$$\operatorname{Re}(f, g) \leq ||f|| \cdot ||g||$$

If we replace f, g, in this by $e^{i\alpha}f, g$ (α real), then the right side of the equation does not change (because of

$$(af, af) = a\bar{a}(f, f) = |a|^2(f, f)$$

we have

$$||af|| = |a|\cdot||f||,$$

therefore, for $|a| = 1$, $||af|| = ||f||$), while the left side goes over into

$$\text{Re }(e^{i\alpha}(f, g)) = \cos\alpha \text{ Re }(f, g) - \sin\alpha \text{ Im }(f, g)$$

This clearly has the maximum

$$\sqrt{(\text{Re }(f, g))^2 + (\text{Im }(f, g))^2} = |(f, g)|$$

from which the proposition follows:

$$|(f, g)| \leq ||f||\cdot||g||$$

COROLLARY. For the equality to hold, f, g must be identical except for a constant (complex) factor.

PROOF. For the equality to hold in the relation

$$\text{Re }(f, g) \leq \tfrac{1}{2}(||f||^2 + ||g||^2)$$

$(f - g, f - g)$ must be zero, i.e., $f = g$. In the transition from this expression to $|(f, g)| \leq ||f||\cdot||g||$, f, g are replaced by $e^{i\alpha}af, (1/a)g$ (a, α real, $a > 0$), whenever neither f nor $g = 0$. In order that the equality hold in this case we must therefore have

$$e^{i\alpha}af = \tfrac{1}{a}g, \qquad g = a^2e^{i\alpha}f = cf \quad (c \neq 0)$$

Conversely, for f or $g = 0$, or $g = cf$ $(c \neq 0)$, the equality clearly holds.

THEOREM 2. $||f|| \geq 0$ and $= 0$ only if $f = 0$. Also,

$$||a.f|| = |a|\cdot||f||,\ ||f + g|| \leq ||f|| + ||g||$$

always, the equality holding only if f, g are identical except for a constant, real factor, ≥ 0.

PROOF. We have already seen above that the first two propositions are correct. We prove the inequality of the third in the following manner:

$$\begin{aligned}(f + g, f + g) &= (f, f) + (g, g) + (f, g) + (g, f)\\ &= ||f||^2 + ||g||^2 + 2 \operatorname{Re}(f, g)\\ &\leq ||f||^2 + ||g||^2 + 2||f||\cdot||g||\\ &= (||f|| + ||g||)^2,\end{aligned}$$

$$||f + g|| \leq ||f|| + ||g||$$

In order that the equality hold, $\operatorname{Re}(f, g)$ must be equal to $||f||\cdot||g||$, which requires f or $g = 0$, or $g = a^2f = cf$ (c real, > 0) by reason of the observations made in the proof of the above corollary. Conversely, it is clear in this case that the equality holds.

From Theorem 2 it follows immediately that the distance $||f - g||$ has the following properties: f, g have the distance 0 for $f = g$, and never otherwise. The distance between g, f is the same as between f, g. The distance of f, h is less than or equal to the sum of the distances of f, g and g, h. The equality exists only if $g = af + (1 - a)h$ (a real, $0 \leq a \leq 1$).[43] The distance of af, ag is $|a|$ times the distance of f, g.

Now these are the very same properties of the concept of length which make it possible in geometry (and topology) to base the concepts continuity, limit, limit point, etc., on the concept of length. We wish to make use of this, and define:

A function $F(f)$ in $\mathfrak{R}$ (i.e., for which f is defined in $\mathfrak{R}$, and which has for values either always points of $\mathfrak{R}$ or always complex numbers) is continuous at the point f_0 (in $\mathfrak{R}$), if for each $\epsilon > 0$ there exists a $\delta > 0$, such that $\|f - f_0\| < \delta$ implies $\|F(f) - F(f_0)\| < \epsilon$ or $|F(f) - F(f_0)| < \epsilon$ (according to whether the F values are points in $\mathfrak{R}$ or complex numbers). This function is said to be bounded in $\mathfrak{R}$ or in a given subset of $\mathfrak{R}$, if always, $\|F(f)\| \leq C$ or $|F(f)| \leq C$ (C a constant, suitably chosen, but fixed). Analogous definitions hold for several variables. A sequence $f_1, f_2, \ldots$ converges to f, or has the limit f, if the numbers $\|f_1 - f\|, \|f_2 - f\|, \ldots$ converge to zero. A point is a limit point of a set $\mathfrak{A}$ (subset of

---

[43] By THEOREM 2. (which is applied here to $f - g$, $g - h$), $f - g = 0$, i.e., $g = f$, or $g - h = 0$, i.e., $g = h$ or $g - h = c(f - g)$ (c real, > 0), i.e.,

$$g = \frac{c}{c+1} f + \frac{1}{c+1} h$$

i.e., $g = af + (1 - a)h$ with a respectively equal to

$$1, \; 0 \; \frac{c}{c+1}.$$

Geometrically, this means that the point g is collinear with f, h.

$\mathfrak{R}$!) if it is a limit of a sequence from $\mathfrak{A}$.[44] In particular, $\mathfrak{A}$ is said to be closed if it contains all its limit points; and it is said to be everywhere dense if its limit points encompass all $\mathfrak{R}$.

We have yet to prove that af, $f + g$, (f, g) are continuous in all their variables. Since

$$||af - af'|| = |a| \cdot ||f - f'||,$$

$$||(f + g) - (f' + g')|| = ||(f - f') + (g - g')||$$

$$\leq ||f - f'|| + ||g - g'||$$

the first two propositions are clearly true. Furthermore, from

$$||f - f'|| < \epsilon, \qquad ||g - g'|| < \epsilon$$

if we substitute $f' - f = \phi$, $g' - g = \psi$, it follows that

$$|(f, g) - (f', g')| = |(f, g) - (f + \phi, g + \psi)|$$

$$= |(\phi, g) + (f, \psi) + (\phi, \psi)|$$

$$\leq |(\phi, g)| + |(f, \psi)| + |(\phi, \psi)|$$

$$\leq ||\phi||\cdot||g|| + ||f||\cdot||\psi|| + ||\phi||\cdot||\psi||$$

$$\leq \epsilon(||f|| + ||g|| + \epsilon)$$

As $\epsilon \longrightarrow 0$, this expression approaches zero, and can be

[44] The following definition of the limit point is also useful: for each $\epsilon > 0$ let there be an f' of $\mathfrak{A}$ with $||f - f'|| < \epsilon$. The equivalence of the two definitions can be shown exactly as in ordinary analysis.

made smaller than any $\delta > 0$.

The properties **A**., **B**. permit us, as we see, to state a great deal about $\mathfrak{R}$, yet they are not sufficient to enable us to distinguish the $\mathfrak{R}_n$ from each other and from $\mathfrak{R}_\infty$. No mention has been made so far of the number of dimensions. This concept is clearly associated with the maximum number of linearly independent vectors. If $n = 0, 1, 2, \ldots$ is such a maximum, then we may state for this n:

C $^{(n)}$. There are exactly n linearly independent vectors. That is, it is possible to specify n such vectors, but not $n + 1$.

If there exists no maximum number, then we have:

C $^{(\infty)}$. There are arbitrarily many linearly independent vectors.

That is, for each $k = 1, 2, \ldots,$ we can specify k such vectors.

C. is then not an essentially new postulate. If **A**., **B**. hold, then either **C** $^{(n)}$. or **C** $^{(\infty)}$. must hold. We obtain a different space $\mathfrak{R}$, depending on which we decide upon. We shall see that it follows from **C** $^{(n)}$. that $\mathfrak{R}$ has all the properties of the n-dimensional (complex) Euclidean space. **C** $^{(\infty)}$. on the other hand is not sufficient to guarantee the essential identity of $\mathfrak{R}$ with the Hilbert space $\mathfrak{R}_\infty$. Rather we need two additional postulates **D**., **E**. More precisely, the situation is the following: We shall show that an $\mathfrak{R}$ with **A**., **B**., **C** . has all the properties of the $\mathfrak{R}_n$, in particular the **D**., **E**., which will be formulated (and which therefore follow from **A**., **B**., **C** $^{(\infty)}$.). Furthermore, we shall show that an $\mathfrak{R}$ with **A**., **B**., **C**., **D**., **E**. has all the properties of $\mathfrak{R}_\infty$, but that in this case, **D**., **E**. are essential (i.e., they do not follow from **A**., **B**., **C** $^{(\infty)}$.). We therefore proceed to the formulation of **D**., **E**., but the proof that all $\mathfrak{R}_n$, $\mathfrak{R}_\infty$ possess these properties will only be given later (cf. II.3).

D. $\mathfrak{R}$ is complete.[45]

That is, if a sequence $f_1, f_2, \ldots$ in $\mathfrak{R}$ satisfies the Cauchy convergence criterion (for each $\epsilon > 0$, there exists an $N = N(\epsilon)$, such that $\|f_m - f_n\| < \epsilon$ for all $m, n \geq N$), then it is convergent, i.e., it possesses a limit f (cf. the definition of this concept given above).

E. $\mathfrak{R}$ is separable.[45]

That is, there is a sequence $f_1, f_2, \ldots$ in $\mathfrak{R}$ which is everywhere dense in $\mathfrak{R}$.

In II.2, we shall, as we have said, develop the "geometry" of $\mathfrak{R}$ from these basic assumptions, and shall find its identity with that one of $\mathfrak{R}_n$ and $\mathfrak{R}_\infty$.

2. THE GEOMETRY OF HILBERT SPACE

We begin with two definitions. The first contains as much of trigonometry as is necessary for our purposes: the concept of the right angle -- orthogonality.

DEFINITION 4. Two f, g of $\mathfrak{R}$ are orthogonal if $(f, g) = 0$. Two linear manifolds $\mathfrak{M}$, $\mathfrak{N}$ are orthogonal if each element of $\mathfrak{M}$ is orthogonal to each element of $\mathfrak{N}$. A set $\mathfrak{O}$ is called an orthonormal set if for all f, g of $\mathfrak{O}$,

$$(f, g) = \begin{cases} 1 & \text{for } f = g \\ 0 & \text{for } f \neq g \end{cases}$$

[45] We use the topological term for brevity (cf. Hausdorff, Mengenlehre, Berlin, 1927), it is explained further below in the text.

(i.e., each pair of elements are orthogonal and each element has the magnitude 1[46]). Furthermore, $\mathfrak{O}$ is complete if it is not a subset of any other normal set that contains additional elements.[47]

We observe further: That the orthonormal set is complete means obviously that no f exists, with $||f|| = 1$, which is orthogonal to the whole $\mathfrak{O}$ (cf. Note 46). But if f were merely different from zero, and orthogonal to the whole set $\mathfrak{O}$, then all of the above would be satisfied for

$$f' = \frac{1}{||f||} \cdot f$$

(of course, $||f|| > 0$):

$$||f'|| = \frac{1}{||f||} ||f|| = 1 ,$$

f' orthogonal to $\mathfrak{O}$. Therefore the completeness of $\mathfrak{O}$ means that each f orthogonal to the entire set must vanish.

The second definition is such that it is important only in $\mathfrak{R}_\infty$, since in $\mathfrak{R}_n$ every linear manifold is of the type described by it (cf. the end of II.3). Therefore we cannot give an intuitive-geometrical picture of its meaning.

DEFINITION 5. A linear manifold

---

[46]Indeed,

$$||f|| = \sqrt{(f, f)} = 1 .$$

[47]As we see, complete orthonormal sets correspond to the cartesian coordinate systems (i.e., the unit vectors pointing in the directions of the axes) in $\mathfrak{R}_n$.

which is also closed is called a closed linear manifold. If $\mathfrak{A}$ is any set in $\mathfrak{R}$, and we add to $\{\mathfrak{A}\}$ (the linear manifold spanned by $\mathfrak{A}$) all its limit points, we obtain a closed linear manifold which contains $\mathfrak{A}$. It is also a subset of every other closed linear manifold which contains $\mathfrak{A}$.[48] We call it the closed linear manifold spanned by $\mathfrak{A}$, and symbolize it by $[\mathfrak{A}]$.

We now go on to the more detailed analysis of $\mathfrak{R}$, in particular of complete orthonormal sets. For theorems which require $\mathbf{C}^{(n)}$. or $\mathbf{C}^{(\infty)}$., **D**., **E**. in addition to **A**., **B**., we add the index $^{(n)}$ or $^{(\infty)}$ respectively. Such indices are omitted for those theorems which are common to both cases.

THEOREM $3^{(n)}$. Every orthonormal set has $\leq n$ elements, and is complete if and only if it has n elements.

NOTE. It follows from the first proposition that there exists a maximum value for the numbers of elements of orthonormal sets; those orthonormal sets for which this maximum value is reached are by definition complete. By virtue of this theorem complete n orthonormal sets exist in the case $\mathbf{C}^{(n)}$ and every such set has n elements:

---

[48]As a linear manifold, this must contain $\{\mathfrak{A}\}$, and since it is closed, also the limit points of $\{\mathfrak{A}\}$.

PROOF. Each orthonormal set is (if it is finite) linearly independent. If the elements are $\phi_1, \phi_2, \ldots, \phi_m$, it follows from

$$a_1\phi_1 + \ldots + a_m\phi_m = 0$$

by forming the inner product with ϕ_μ $(\mu = 1, 2, \ldots, m)$ that $a_\mu = 0$. Consequently, by $\mathbf{C}^{(n)}$., the set cannot have $n + 1$ elements. An arbitrary orthonormal set therefore can have no subsets with $n + 1$ elements. Therefore it is finite and has $\leq n$ elements.

A set with n elements permits no extension, and is therefore complete. But one with $m < n$ elements, $\phi_1, \ldots, \phi_m$, is not complete. Indeed, among the linear combinations $a_1\phi_1 + \ldots + a_m\phi_m$ there cannot be $n > m$ linearly independent ones. Hence there must exist, by $\mathbf{C}^{(n)}$., an element f which differs from all $a_1\phi_1 + \ldots + a_m\phi_m$, i.e., for which

$$\psi = f - a_1\phi_1 - \ldots - a_m\phi_m$$

is always different from zero. Now $(\psi, \phi_\mu) = 0$ means that $a_\mu = (f, \phi_\mu)$ $(\mu = 1, 2, \ldots, m)$. Therefore this condition can be satisfied for all $\mu = 1, 2, \ldots, m$ simultaneously, thus furnishing a ψ which shows that the set $\phi_1, \ldots, \phi_m$ is incomplete.

THEOREM $3^{(\infty)}$. Each orthonormal set is a finite or a countably infinite set; if it is complete, then it is certainly infinite.

NOTE. We can therefore write all orthonormal sets as sequences: $\phi_1, \phi_2, \ldots$ (perhaps being terminated, i.e., finite), which we shall actually do. It should be observed that the infinite number of elements of the set is necessary for its completeness, but, unlike in the case $\mathbf{C}^{(n)}$., it is not sufficient.[49]

PROOF. Let $\mathfrak{O}$ be an orthonormal set, f, g two different elements belonging to it. Then

$$(f - g, f - g) = (f, f) + (g, g) - (f, g) - (g, f) = 2 ,$$

$$\|f - g\| = \sqrt{2} .$$

Now let $f_1, f_2, \ldots$ be the sequence which is everywhere dense in $\mathfrak{R}$. This sequence exists by postulate $\mathbf{E}$. For each f of $\mathfrak{O}$ there exists an f_m of the sequence for which $\|f - f_m\| < (1/2)\sqrt{2}$. The corresponding f_m, f_n for f, g must be different, because it would follow from $f_m = f_n$ that

$$\|f - g\| = \|(f - f_m) - (g - f_m)\|$$

$$\leq \|f - f_m\| + \|g - f_m\| < \tfrac{1}{2}\sqrt{2} + \tfrac{1}{2}\sqrt{2} = \sqrt{2} .$$

Therefore, to each f of $\mathfrak{O}$ there corresponds an f_m of the sequence $f_1, f_2, \ldots$ with different f_m for different f. Therefore $\mathfrak{O}$ is finite or is a sequence.

As in the proof of Theorem $3^{(n)}$ we show the following: if there are $> m$ linearly independent elements in $\mathfrak{R}$, a set $\phi_1, \phi_2, \ldots, \phi_m$ cannot be complete. But since by $\mathbf{C}^{(\infty)}$., this holds for all m, a complete set must be infinite.

The theorems which now follow, insofar as they are concerned with convergence, apply only to $\mathbf{C}^{(\infty)}$., but

---

[49]Let $\phi_1, \phi_2, \ldots$ be complete. Then $\phi_2, \phi_3, \ldots$ is not complete, but it is still infinite!

it is more desirable to formulate them generally, because of their other implications.

THEOREM 4. Let $\phi_1, \phi_2, \ldots$ be an orthonormal set. Then all series $\Sigma_\nu (f, \phi_\nu)\overline{(g, \phi_\nu)}$, insofar as they have infinitely many terms, are absolutely convergent. In particular, for $f = g$, $\Sigma_\nu |(f, \phi_\nu)|^2 \leq |f|^2$.

PROOF. Let $a_\nu = (f, \phi_\nu)$, $\nu = 1, 2, \ldots$. Then $f - \sum_{\nu=1}^{N} a_\nu \phi_\nu = \psi$ is orthogonal to all ϕ_ν, $\nu = 1, 2, \ldots, N$ (cf. the proof of THEOREM 3[(n)].). Since $f = \sum_{\nu=1}^{N} a_\nu \phi_\nu + \psi$ then

$$(f, f) = \sum_{\substack{\mu=1\\ \nu=1}}^{N} a_\mu \overline{a}_\nu (\phi_\mu, \phi_\nu) + \sum_{\nu=1}^{N} a_\nu (\phi_\nu, \psi) + \sum_{\nu=1}^{N} \overline{a}_\nu (\psi, \phi_\nu)$$

$$+ (\psi, \psi) = \sum_{\nu=1}^{N} |a_\nu|^2 + (\psi, \psi) \geq \sum_{\nu=1}^{N} |a_\nu|^2$$

i.e., $\sum_{\nu=1}^{N} |a_\nu|^2 \leq ||f||^2$. If the set $\phi_1, \phi_2, \ldots$ is finite, then it follows directly that $\Sigma_\nu |a_\nu|^2 = ||f||^2$; if it is infinite, then $N \longrightarrow \infty$ results in the absolute convergence of $\Sigma_\nu |a_\nu|^2$, as well as in the fact that it is $\leq ||f||^2$. This establishes the second proposition.

Because of

$$|(f, \phi_\nu)\overline{(g, \phi_\nu)}| \leq \tfrac{1}{2}\{|(f, \phi_\nu)|^2 + |g, \phi_\nu|^2\}$$

the more general convergence statement of the first proposition follows from the fact of convergence just established.

THEOREM 5. Let $\phi_1, \phi_2, \ldots$ be an infinite orthonormal set. Then the series $\sum_{\nu=1}^{\infty} x_\nu \phi_\nu$ converges if and only if $\sum_{\nu=1}^{\infty} |x_\nu|^2$ does (the latter series has as its terms real, non-negative numbers, and is therefore convergent or else diverges to $+\infty$).

PROOF. Since this proposition has significance only for $\mathbf{C}^{(\infty)}$., we may then use **D**., the Cauchy criterion of convergence. The sum $\sum_{\nu=1}^{\infty} x_\nu \phi_\nu$ then converges, i.e., the sequence of the partial sums $\sum_{\nu=1}^{N} x_\nu \phi_\nu$ converges as $N \longrightarrow \infty$, if for each $\epsilon > 0$ there exists an $N = N(\epsilon)$ such that for $L, M \geq N$, $\left\| \sum_{\nu=1}^{L} x_\nu \phi_\nu - \sum_{\nu=1}^{M} x_\nu \phi_\nu \right\| < \epsilon$. We assume $L > M \geq N$, then

$$\left\| \sum_{\nu=1}^{L} x_\nu \phi_\nu - \sum_{\nu=1}^{M} x_\nu \phi_\nu \right\| = \left\| \sum_{\nu=M+1}^{L} x_\nu \phi_\nu \right\| < \epsilon ,$$

$$\left\| \sum_{\nu=M+1}^{L} x_\nu \phi_\nu \right\|^2 = \left(\sum_{\nu=M+1}^{L} x_\nu \phi_\nu , \sum_{\nu=M+1}^{L} x_\nu \phi_\nu \right)$$

$$= \sum_{\substack{\mu,\nu \\ =M+1}}^{L} x_\mu \overline{x}_\nu (\phi_\mu, \phi_\nu) = \sum_{\nu=M+1}^{L} |x_\nu|^2$$

$$= \sum_{\nu=1}^{L} |x_\nu|^2 - \sum_{\nu=1}^{M} |x_\nu|^2$$

therefore

$$0 \leq \sum_{\nu=1}^{L} |x_\nu|^2 - \sum_{\nu=1}^{M} |x_\nu|^2 < \epsilon^2 .$$

But this is exactly the Cauchy convergence condition for the sequence

$$\sum_{\nu=1}^{N} |x_\nu|^2 , \; N \longrightarrow \infty$$

i.e., for the series

$$\sum_{\nu=1}^{\infty} |x_\nu|^2$$

COROLLARY. For $f = \Sigma_\nu x_\nu \phi_\nu$, $(f, \phi_\nu) = x_\nu$ (regardless of whether the orthogonal set is finite or infinite -- in the latter case, of course, convergence is assumed.)

PROOF. For $N \geq \nu$, we have

$$\left(\sum_{\mu=1}^{N} x_\mu \phi_\mu, \; \phi_\nu \right) = \sum_{\mu=1}^{N} x_\mu (\phi_\mu, \; \phi_\nu) = x_\nu$$

For a finite set $\phi_1, \phi_2, \dots$, we can set N equal to the highest index; for infinite sets $\phi_1, \phi_2, \dots$, we can let $N \longrightarrow \infty$, because of the continuity of the inner product. In either case, $(f, \phi_\mu) = x_\mu$ results.

THEOREM 6. Let $\phi_1, \phi_2, \dots$ be an orthonormal set, f arbitrary. Then $f' = \Sigma_\nu x_\nu \phi_\nu$, $x_\nu = (f, \phi_\nu)$ $(\nu = 1, 2, \dots)$, is always convergent if the series is infinite. The expression $f - f'$ is orthogonal to $\phi_1, \phi_2, \dots$.

PROOF. The convergence follows from THEOREMS 4., 5., and according to the corollary of THEOREM 5.,

$$(f', \phi_\nu) = x_\nu = (f, \phi_\nu) , \; (f - f', \phi_\nu) = 0$$

After these preparations, we can give the general criteria, i.e., even for $\mathbf{C}^{(\infty)}$., for the completeness of an orthonormal set.

THEOREM 7. Let $\phi_1, \phi_2, \cdots$ be an orthonormal set. For completeness, each one of the following conditions is necessary and sufficient:

α) The closed linear manifold $[\phi_1, \phi_2, \cdots]$ spanned by $\phi_1, \phi_2, \cdots$ is equal to $\mathfrak{R}$.

β) It is always true that $f = \Sigma_\nu x_\nu \phi_\nu$, $x_\nu = (f, \phi_\nu)$ ($\nu = 1, 2, \ldots,$ convergence by THEOREM 6.).

γ) It is always true that

$$(f, g) = \sum_\nu (f, \phi_\nu)\overline{(g, \phi_\nu)}$$

(absolute convergence by THEOREM 4.).

PROOF. If $\phi_1, \phi_2, \cdots$ is complete, then $f - \Sigma_\nu x_\nu \phi_\nu$ is equal to zero $(x_\nu = (f, \phi_\nu), \nu = 1, 2, \ldots)$, since it is orthogonal to $\phi_1, \phi_2, \cdots$ by THEOREM 6. Then **β**) is satisfied. If **β**) holds, then each f is the limit of its partial sums

$$\sum_{\nu=1}^{N} x_\nu \phi_\nu$$

$N \longrightarrow \infty$ (if $\phi_1, \phi_2, \cdots$ is at all infinite) and therefore belongs to $[\phi_1, \phi_2, \cdots]$. Therefore $[\phi_1, \phi_2, \cdots] = \mathfrak{R}$ i.e., **α**) is satisfied. If **α**) holds, then we may argue as follows: If f is orthogonal to all $\phi_1, \phi_2, \cdots$, then it is also orthogonal to their linear combinations, and by reason of continuity also to their limit points, i.e., to all $[\phi_1, \phi_2, \cdots]$. Therefore it is orthogonal to all $\mathfrak{R}$, and hence to itself: $(f, f) = 0$, $f = 0$. Consequently, $\phi_1, \phi_2, \cdots$ is complete.

We then have the logical scheme:

$$\text{Completeness} \longrightarrow \beta) \longrightarrow \alpha) \longrightarrow \text{completeness}$$

i.e., α), β) have been shown to be necessary and sufficient conditions.

From γ) it follows that if f is orthogonal to all $\phi_1, \phi_2, \cdots$, and if we set $f = g$, then we obtain $(f, f) = \Sigma_\nu\, 0.0 = 0$, $f = 0$, i.e., $\phi_1, \phi_2, \cdots$ is complete. On the other hand, from β) (which is now equivalent to completeness),

$$(f, g) = \lim_{N\to\infty} \left(\sum_{\nu=1}^{N} (f, \phi_\nu)\cdot\phi_\nu, \sum_{\nu=1}^{N} (g, \phi_\nu)\cdot\phi_\nu \right)$$

$$= \lim_{N\to\infty} \sum_{\mu,\nu=1}^{N} (f, \phi_\mu)\overline{(g, \phi_\nu)}\cdot(\phi_\mu, \phi_\nu)$$

$$= \lim_{N\to\infty} \sum_{\nu=1}^{N} (f, \phi_\nu)\overline{(g, \phi_\nu)} = \sum_{\nu=1}^{\infty} (f, \phi_\nu)\overline{(g, \phi_\nu)}$$

(if the set is finite, then the limit process is unnecessary), i.e. γ). Therefore γ) is also a necessary and sufficient condition.

THEOREM 8. To each set $f_1, f_2, \cdots$, there corresponds an orthonormal set $\phi_1, \phi_2, \cdots$, which spans the same linear manifold as the former set (both sets can be finite).

PROOF. First we replace $f_1, f_2, \cdots$ by a subset $g_1, g_2, \cdots$ which spans the same linear manifold and which consists of linearly independent elements. This may be done as follows. Let g_1 be the first f_n which is different from zero; g_2 the first f_n which is different from all $a_1 g_1$; g_3 the first f_n which is different from all $a_1 g_1 + a_2 g_2$; ... (if for any p there exists no f_n which is different from all $a_1 g_1 + \cdots + a_p g_p$, we terminate the set with g_p .) These $g_1, g_2, \ldots$ obviously furnish the desired result.

We now form

$$\gamma_1 = g_1 , \qquad \phi_1 = \frac{1}{\|\gamma_1\|} \cdot \gamma_1 ,$$

$$\gamma_2 = g_2 - (g_2, \phi_1)\cdot\phi_1 , \qquad \phi_2 = \frac{1}{\|\gamma_2\|} \cdot \gamma_2 ,$$

$$\gamma_3 = g_3 - (g_3, \phi_1)\cdot\phi_1 - (g_3, \phi_2)\cdot\phi_2 , \quad \phi_3 = \frac{1}{\|\gamma_3\|} \cdot \gamma_3 ,$$

(this is the well known Schmidt orthonormalization process). Each ϕ_p construction is actually possible, i.e., the denominators $\|\gamma_p\|$ are all different from zero. For otherwise, if $\gamma_p = 0$, then g_p would be a linear combination of the $\phi_1, \ldots, \phi_{p-1}$, i.e. of the $g_1, \ldots, g_{p-1}$, which is contrary to the hypothesis. Furthermore, it is clear that g_p is a linear combination of the $\phi_1, \ldots, \phi_p$, and ϕ_p is a linear combination of the $g_1, \ldots, g_p$ -- therefore $g_1, g_2, \ldots$ and $\phi_1, \phi_2, \ldots$ determine the same linear manifold.

Finally, by the construction $\|\phi_p\| = 1$, and for $q < p$, $(\gamma_p, \phi_q) = 0$, therefore $(\gamma_p, \gamma_q) = 0$. Since we can interchange p, q , the latter statement holds for $p \neq q$. Therefore $\phi_1, \phi_2, \ldots$ is an orthonormal set.

THEOREM 9. Corresponding to each closed linear manifold $\mathfrak{M}$ there is an orthonormal set which spans the same $\mathfrak{M}$ as closed linear manifold.

PROOF. In the case $\mathbf{C}^{(n)}$., this theorem is

immediate: Because $\Re$ satisfies **A**., **B**., $\mathbf{C}^{(n)}$., each linear manifold $\mathfrak{M}$ in $\Re$ satisfies **A**., **B**., $\mathbf{C}^{(m)}$. with an $m \leq n$, so that the note on THEOREM $3^{(n)}$ is applicable to $\mathfrak{M}$: There is an orthonormal set $\phi_1, \ldots, \phi_m$ which is complete in $\mathfrak{M}$. Because of THEOREM 7., α, this is exactly the proposition to be proved. (As can be seen, the premise of the closed nature of $\mathfrak{M}$ is itself unnecessary, since it is actually proved. In this case, compare the statements on DEFINITION 5.)

In the case $\mathbf{C}^{(\infty)}$., we recall that $\Re$ is separable according to **E**. We want to show that $\mathfrak{M}$ is also separable -- in general, that each subset of $\Re$ is separable. For this purpose, we form the sequence $f_1, f_2, \ldots,$ everywhere dense in $\Re$ (cf. **E**. in II.1.), and for each f_n and $m = 1, 2, \ldots,$ we form the sphere $\mathfrak{K}_{n,m}$ consisting of all f with $\|f - f_n\| < \frac{1}{m}$. For each $\mathfrak{K}_{n,m}$ which contains points of $\mathfrak{M}$, we select one such point: $g_{n,m}$. For some n, m, this $g_{n,m}$ may be undefined, but the defined points form a sequence in $\mathfrak{M}$.[50] Now let f be any point of $\mathfrak{M}$ and $\epsilon > 0$. Then there exists an m with $1/m < \epsilon/2$, and an f_n with $\|f_n - f\| < \frac{1}{m}$. Since $\mathfrak{K}_{n,m}$ then contains a point of $\mathfrak{M}$ (namely f), g_{nm} is defined, and $\|f_n - g_{nm}\| < \frac{1}{m}$, therefore $\|f - g_{nm}\| < \frac{2}{m} < \epsilon$. Consequently, f is the limit point of the g_{nm} thus defined; hence this sequence yields the desired result.

We shall denote by $f_1, f_2, \ldots$ the sequence from $\mathfrak{M}$, everywhere dense in $\mathfrak{M}$. The closed linear manifold determined by it, $[f_1, f_2, \ldots]$, contains all its limit points, and hence all $\mathfrak{M}$; but, since $\mathfrak{M}$ is a closed linear manifold, and $f_1, f_2, \ldots$ belong to it, therefore $[f_1, f_2, \ldots]$ is a part of $\mathfrak{M}$ -- therefore it is equal to

---

[50] It should be recalled that a double sequence g_{nm} $(n, m = 1, 2, \ldots)$ can also be written as a simple sequence: $g_{11}, g_{12}, g_{21}, g_{13}, g_{22}, g_{31}, \ldots$.

$\mathfrak{M}$. We now choose the orthonormal set $\phi_1, \phi_2, \ldots$ by THEOREM 8. Then $\{\phi_1, \phi_2, \ldots\} = \{f_1, f_2, \ldots\}$, and if we add the limit points to both sides, we obtain $[\phi_1, \phi_2, \ldots] = [f_1, f_2, \ldots] = \mathfrak{M}$. But this was our proposition.

We now need only put $\mathfrak{M} = \mathfrak{R}$ in THEOREM 9., and we have by THEOREM 7. α a complete orthonormal set $\phi_1, \phi_2, \ldots$. So we see: There are complete orthonormal sets. On the basis of this we can now show that $\mathfrak{R}$ is an $\mathfrak{R}_n$ or an $\mathfrak{R}_\infty$ (according to whether $\mathbf{C}^{(n)}$. or $\mathbf{C}^{(\infty)}$. holds), i.e., all its properties are completely determined.

It is only necessary to show that $\mathfrak{R}$ allows a one to one mapping on the set of all $\{x_1, \ldots, x_n\}$ or of all $\{x_1, x_2, \ldots\}$ ($\sum_{\nu=1}^{\infty} |x_\nu|^2$ finite) respectively, in such a way that

1. $af \longleftrightarrow \{ax_1, ax_2, \ldots\}$

follows from $f \longleftrightarrow \{x_1, x_2, \ldots\}$.

2. $f + g \longleftrightarrow \{x_1 + y_1, x_2 + y_2, \ldots\}$

follows from $\left\{ \begin{array}{l} f \longleftrightarrow \{x_1, x_2, \ldots\} \\ g \longleftrightarrow \{y_1, y_2, \ldots\} \end{array} \right\}$

3. $(f, g) = \sum_{\nu=1}^{n \text{ or } \infty} x_\nu \bar{y}_\nu$

follows from $\left\{ \begin{array}{l} f \longleftrightarrow \{x_1, x_2, \ldots\} \\ g \longleftrightarrow \{y_1, y_2, \ldots\} \end{array} \right\}$

(In the infinite case in 3., the absolute convergence must be shown.) We now specify the mapping $f \longleftrightarrow \{x_1, x_2, \ldots\}$.

Let $\phi_1, \phi_2, \ldots$ be a complete orthonormal set; in case $\mathbf{C}^{(n)}$., it terminates with ϕ_n , in case $\mathbf{C}^{(\infty)}$., it is infinite (THEOREMS $3^{(n)}$., $3^{(\infty)}$). We set

$$f = \sum_{\nu=1}^{n \text{ or } \infty} x_\nu \phi_\nu$$

By THEOREM 5., this series converges even in the infinite case (since

$$\sum_{\nu=1}^{\infty} |x_\nu|^2$$

is finite), i.e., the elements of either $\mathfrak{R}_n$ or $\mathfrak{R}_\infty$ are exhausted. By THEOREM 7., $\boldsymbol{\beta}$, and because

$$\sum_{\nu=1}^{n \text{ or } \infty} |(f, \phi_\nu)|^2$$

is finite, (THEOREM 4.) the elements of $\mathfrak{R}$ are also exhausted ($x_\nu = (f, \phi_\nu)$ is to be substituted). It is clear that only one f corresponds to each $\{x_1, x_2, \cdots\}$, while the converse follows from the corollary to THEOREM 5.

Statements 1., 2. are obviously satisfied, while 3. follows from THEOREM 7., $\boldsymbol{\gamma}$.

3. DIGRESSION ON THE CONDITIONS **A.** – **E.**[51]

We must still verify the proposition 2. at the end of I.4.: That F_Z, F_Ω actually satisfy the conditions **A.** – **E.** For this purpose, it is sufficient to consider F_Ω, because we have already shown in II.2. that an $\mathfrak{R}$ with **A.** – **E.** must be identical in all properties with $\mathfrak{R}_\infty$, i.e., F_Z, so that **A.** – **E.** must be valid for F_Z also. Moreover, we shall show the independence of the conditions **D.**, **E.** from **A.** – **C**$^{(n)}$., mentioned in II.2., as well as the fact that they follow from **A.** – **C**$^{(n)}$., i.e. that they hold in $\mathfrak{R}_n$. These three purely mathematical questions form the subject matter of this chapter.

We begin with the verification of **A.** – **E.** in

---

[51]This section is not necessary for the understanding of the later portions of the text.

F_Ω. For this, we must rely upon the Lebesque concept of the integral, for whose foundations, reference should be made to special works on the subject.[52] (The Lebesque integral is of importance to us only upon this occasion, and a knowledge of it is not necessary for the later chapters).

In I.4., we had introduced Ω as the k-dimensional space of the $q_1, \ldots, q_k$ and F_Ω as the totality of all functions $f(q_1, \ldots, q_k)$ with finite

$$\underbrace{\int \cdots \int}_{\Omega} |f(q_1, \ldots, q_k)|^2 dq_1 \ldots dq_k$$

We now allow all the $q_1, \ldots, q_k$ to vary from $-\infty$ to $+\infty$. All our deductions would of course remain valid, and even the proofs would be carried over for the most part verbatim, if we were to limit the range of variation of the $q_1, \ldots, q_k$ (so that Ω would be, for example, a half space, or the inside of a cube, or the inside of a sphere or the outside of these figures, etc.) -- indeed even if we were to choose Ω as a curved surface (e.g., as the surface of a sphere, etc.). But in order not to become lost in unnecessary complications (whose discussion can be carried out without difficulty by the reader himself, with the aid of our typical proof) we limit ourselves to the simplest case first mentioned. We shall now go through **A.** - **E.** consecutively:

For A. We must show: If f, g belong to F_Ω, then af, f $\pm$ g also belong to it, i.e., if

$$\int_\Omega |f|^2 , \quad \int_\Omega |g|^2$$

(we abbreviate

---

[52]For example, Carathéodory, <u>Vorlesungen über reelle Funktionen</u>, Leipzig, 1927, in particular, pp. 237-274; Kamke, <u>Das Lebesguesche Integral</u>, Leipzig, 1925.

$$\underbrace{\int \cdots \int}_{\Omega} |f(q_1, \ldots, q_k)|^2 dq_1 \ldots dq_k ,$$

$$\underbrace{\int \cdots \int}_{\Omega} |g(q_1, \ldots, q_k)|^2 dq_1 \ldots dq_k$$

since no confusion can result) are finite, then

$$\int_{\Omega} |af|^2 = |a|^2 \int_{\Omega} |f|^2 , \quad \int_{\Omega} |f \pm g|^2$$

are also finite. The first case is trivial, while the second is established, because of $|f \pm g|^2 = |f|^2 + |g|^2 \pm 2R_e(f\bar{g})$,[53] as soon as the finite nature of

$$\int_{\Omega} |f \cdot \bar{g}| = \int_{\Omega} |f||g|$$

is ascertained. But since $|f||g| \leq \frac{1}{2}(|f|^2 + |g|)^2$, this follows directly from the hypothesis.

For B . We define (f, g) as $\int_{\Omega} f\bar{g}$. This integral is, as we have already seen, absolutely convergent. All properties postulated in **B**. are apparent except the last: That $(f, f) = 0$ implies $f \equiv 0$. Now $(f, f) = 0$ means that $\int_{\Omega} |f|^2 = 0$, so that the set of points for which $|f|^2 > 0$, i.e. $f(q_1, \ldots, q_k) \neq 0$, must have the Lebesgue measure zero. If we now consider two functions f, g , for which $f \neq g$ (i.e. $f(q_1, \ldots, q_k) \neq g(q_1, \ldots, q_k)$) holds only in a $q_1, \ldots, q_k$ set of Lebesgue measure zero, as being not essentially different,[54] then we can assert that $f \equiv 0$.

---

[53]In general,

$$|x + y|^2 = (x + y)(\bar{x} + \bar{y}) = x\bar{x} + y\bar{y} + (x\bar{y} + \bar{x}y)$$
$$= |x|^2 + |y|^2 + 2 \operatorname{Re} (x\bar{y})$$

[54]This is customary in the theory of the Lebesgue integral.

For C . Let $O_1, \ldots, O_m$ be n domains in Ω, no two of which have a point in common, and let the Lebesgue measure of all be greater than zero, but finite. Let $f_l(q_1, \ldots, q_k)$ be 1 in O_l, and zero elsewhere. Since $\int_\Omega |f_l|^2$ is equal to the measure of O_l, it belongs to F_Ω $(l = 1, \ldots, n)$. These $f_1, \ldots, f_n$ are linearly independent. For $a_1 f_1 + \cdots + a_n f_n \equiv 0$ means, that the function on the left fails to vanish only in a set of measure zero. It therefore has roots in each O_l, but since it is constant $= a_l$ in O_l, then $a_l = 0$; $l = 1, \ldots, m$. This construction holds for all n, so that $\mathbf{C}^{(\infty)}$. holds.

For D . Let the sequence $f_1, f_2, \ldots$ satisfy the Cauchy convergence criterion, i.e., for each $\epsilon > 0$ there exists an $N = N(\epsilon)$ such that $\int_\Omega |f_m - f_n|^2 < \epsilon$ whenever $m, n \geq N$. We choose $n_1 = N(\frac{1}{8})$; $n_2 \geq n_1, N(\frac{1}{8^2})$; $n_3 \geq n_1, n_2, N(\frac{1}{8^3})$; . . . . Then $n_1 \leq n_2 \leq \cdots$; $n_\nu, n_{\nu+1} \geq N(\frac{1}{8^\nu})$; hence

$$\int_\Omega |f_{n_{\nu+1}} - f_{n_\nu}|^2 < \frac{1}{8^\nu}$$

Let us now consider the set $P^{(\nu)}$ of all points for which

$$|f_{n_{\nu+1}} - f_{n_\nu}| > \frac{1}{2^\nu}$$

If its Lebesgue measure is $\mu^{(\nu)}$, then

$$\int_\Omega |f_{n_{\nu+1}} - f_{n_\nu}|^2 \geq \mu^{(\nu)} \left(\frac{1}{2^\nu}\right)^2 = \frac{\mu^{(\nu)}}{4^\nu}, \quad \frac{\mu^{(\nu)}}{4^\nu} < \frac{1}{8^\nu}, \quad \mu^{(\nu)} < \frac{1}{2^\nu}$$

Let us also consider the set $Q^{(\nu)}$, which consists of the union of $P^{(\nu)}, P^{(\nu+1)}, P^{(\nu+2)}, \ldots$. Its Lebesgue measure is

$$\leq \mu^{(\nu)} + \mu^{(\nu+1)} + \mu^{(\nu+2)} + \cdots < \frac{1}{2^\nu} + \frac{1}{2^{\nu+1}} + \frac{1}{2^{\nu+2}} + \cdots = \frac{1}{2^{\nu-1}}$$

Outside of $Q^{(\nu)}$ it is true that

$$|f_{n_{\nu+1}} - f_{n_\nu}| < \frac{1}{2^\nu}, \quad |f_{n_{\nu+2}} - f_{n_{\nu+1}}| < \frac{1}{2^{\nu+1}},$$
$$|f_{n_{\nu+3}} - f_{n_{\nu+2}}| < \frac{1}{2^{\nu+2}}, \ \ldots,$$

therefore, in general, for $\nu \leq \nu' \leq \nu''$

$$|f_{n_{\nu''}} - f_{n_{\nu'}}| \leq |f_{n_{\nu'+1}} - f_{n_{\nu'}}|$$
$$+ |f_{n_{\nu'+2}} - f_{n_{\nu'+1}}| + \ldots + |f_{n_{\nu''}} - f_{n_{\nu''-1}}|$$
$$< \frac{1}{2^{\nu'}} + \frac{1}{2^{\nu'+1}} + \ldots + \frac{1}{2^{\nu''-1}} < \frac{1}{2^{\nu'-1}}$$

As $\nu' \longrightarrow \infty$, this approaches zero independently of ν'', i.e., the sequence $f_{n_1}, f_{n_2}, \ldots$ fulfills the Cauchy condition, in the case that the $q_1, \ldots, q_k$ do not lie in $Q^{(\nu)}$. Since we are dealing with numbers (for fixed $q_1, \ldots, q_k$), this sequence also converges. Therefore we can say conversely: If the $f_{n_1}, f_{n_2}, \ldots$ sequence does not converge for a certain $q_1, \ldots, q_k$, then this lies in $Q^{(\nu)}$. Let the set of all $q_1, \ldots, q_k$ for which convergences do not occur be Q. Then Q is a subset of $Q^{(\nu)}$, its measure is therefore not larger than that of $Q^{(\nu)}$, i.e. $< \frac{1}{2^{\nu-1}}$. This must be true for all ν, although Q is defined independently of ν. Therefore Q has the Lebesgue measure zero. Consequently, nothing is changed if we, for example, set all f_n in Q equal to zero (cf. Note 54). But then $f_{n_1}, f_{n_2}, \ldots$ converges also in Q, and hence everywhere.

We have thus specified a subsequence $f_{n_1}, f_{n_2}, \ldots$ of $f_1, f_2, \ldots$ which converges at all points $q_1, \ldots, q_k$ (this need not be the case for $f_1, f_2, \ldots$). Let the limit of the $f_{n_1}, f_{n_2}, \ldots$ be $f = f(q_1, \ldots, q_k)$. We must then prove: 1. f belongs to F_Ω, i.e., $\int_\Omega |f|^2$ is finite; 2. f is the limit of the $f_{n_1}, f_{n_2}, \ldots$ not only in the sense of convergence for each $q_1, \ldots, q_k$, but also in the sense of "length convergence" of Hilbert space, i.e., $\|f - f_{n_\nu}\| \longrightarrow 0$ or $\int_\Omega |f - f_{n_\nu}|^2 \longrightarrow 0$; 3. In this sense it is also a limit of the entire sequence $f_1, f_2, \ldots$, i.e., $\|f - f_n\| \longrightarrow 0$ or $\int_\Omega |f - f_n|^2 \longrightarrow 0$.

Let $\epsilon > 0$, and let ν_0 be chosen with $n_{\nu_0} \geq N(\epsilon)$ (for example, $\frac{1}{8^{\nu_0}} \leq \epsilon$), and $\nu \geq \nu_0$, $n \geq N(\epsilon)$. Then $\int_\Omega |f_{n_\nu} - f_n|^2 < \epsilon$. If we let $\nu \longrightarrow \infty$, then the integrand approaches $|f - f_n|^2$, therefore $\int_\Omega |f - f_n|^2 \leq \epsilon$ (according to a convergence theorem of Lebesgue integrals. See Note 52). Consequently, first, $\int_\Omega |f - f_n|^2$ is finite, i.e., $f - f_n$ in F_Ω; also, since f_n belongs to F_Ω, f does likewise; 1. is then proved. Second, it follows from the above inequality that $\int_\Omega |f - f_n|^2 \longrightarrow 0$ as $n \longrightarrow \infty$ i.e., 2. and 3. are proved.

For E. We must specify a function sequence $f_1, f_2, \ldots$ everywhere dense in F_Ω.

Let $\Omega_1, \Omega_2, \ldots$ be a sequence of regions in Ω, each of which has a finite measure, and which cover the entire Ω. (For example, let Ω_N be a sphere of radius N about the origin.) Let $f = f(q_1, \ldots, q_k)$ be any element of F_Ω. We define an $f_N = f_N(q_1, \ldots, q_k)$ for each $N = 1, 2, \ldots$:

$$f_N(q_1, \ldots, q_k) = \begin{cases} f(q_1, \ldots, q_k) & \left\{ \begin{array}{l} \text{if } q_1, \ldots, q_k \text{ are in } \Omega_N \\ \text{and } |f(q_1, \ldots q_k)| \leq N \end{array} \right. \\ 0 & \left\{ \text{otherwise} \right. \end{cases}$$

As $N \longrightarrow \infty$, $f_N(q_1, \ldots, q_k) \longrightarrow f(q_1, \ldots, q_k)$ (from a certain N on, equality is obtained), therefore $|f - f_N|^2 \longrightarrow 0$. Furthermore, $f - f_N = 0$, or f, therefore $|f - f_N|^2 \leq f^2$. The integrals

$$\int_\Omega |f - f_N|^2$$

are therefore dominated by $\int_\Omega |f|^2$ (finite). Since the integrands approach zero, the integrals do likewise (cf. the convergence theorem cited above).

Let the class of all functions $g = g(q_1, \ldots, q_k)$, for which the set of all points with $g \neq 0$ has finite measure, and which satisfies an inequality $|g| \leq C$ throughout all space, with arbitrary but fixed C, be called G. The above f_N all belong to G. Therefore G is everywhere dense (in F_Ω).

Let g belong to G, $\epsilon > 0$. Let the measure of the $g \neq 0$ set be M, and the upper bound for $|g|$ be C. We choose a series of rational numbers $-C < \rho_1 < \rho_2 < \cdots < \rho_t < C$ such that

$$\rho_1 < -C + \epsilon, \rho_2 < \rho_1 + \epsilon, \ldots, \rho_t < \rho_{t-1} + \epsilon, \ C < \rho_1 + \epsilon$$

which can easily be done. We now change each $\operatorname{Re} g(q_1, \ldots, q_k)$ value into the nearest ρ_s $(s = 1, 2, \ldots, t)$, only we let zero remain zero. Then a new function $h_1(q_1, \ldots, q_k)$ is obtained which differs from $\operatorname{Re} g$ everywhere by less than ϵ. In the same way, we construct an $h_2(q_1, \ldots, q_k)$ for $\operatorname{Im} g$. Then for $h = h_1 + ih_2$,

$$\int_\Omega |g - h|^2 = \int_\Omega |\operatorname{Re} g - h_1|^2 + \int_\Omega |\operatorname{Im} g - h_2|^2$$

$$\leq M\epsilon^2 + M\epsilon^2 = 2M\epsilon^2,$$

$$\|g - h\| \leq \sqrt{2M}\,\epsilon$$

If $\delta > 0$ is given, then we set $\epsilon < \delta/\sqrt{2M}$ and then $\|g - h\| < \delta$.

Let the class of all functions $h = h(q_1, \ldots, q_k)$ which take on only a finite number of different values, actually only those of the form $\rho + i\sigma$, ρ, σ rational, and each such value, except zero, only on sets of finite measure, be called H. The above h belong to H, therefore H is everywhere dense in G, and therefore in F_Ω also.

Let Π be a set of finite Lebesgue measure. We define a function $f_\Pi = f_\Pi(q_1, \ldots, q_k)$:

$$f_\Pi(q_1, \ldots, q_k) = \begin{cases} 1 & \text{in } \Pi \\ 0 & \text{elsewhere} \end{cases}$$

The class H obviously consists of all

$$\sum_{s=1}^{t} (\rho_s + i\sigma_s) f_{\Pi_s} \quad (t = 1, 2, \ldots ;\ \rho_s, \sigma_s \text{ rational}).$$

We now seek a Π-set sequence $\Pi^{(1)}, \Pi^{(2)}, \ldots$ with the following property: for each Π-set and for each $\epsilon > 0$ there exists a $\Pi^{(n)}$ such that the measure of the set of all points which belong to Π but not to $\Pi^{(n)}$, or to $\Pi^{(n)}$ but not to Π is $< \epsilon$ (this set is known as the difference set of $\Pi, \Pi^{(n)}$). If we have such a sequence, then the

$$\sum_{s=1}^{t} (\rho_s + i\sigma_s) f_{\Pi^{(n_s)}}$$

$(t = 1, 2, \ldots ;\ \rho_s, \sigma_s$ rational, $n_s = 1, 2, \ldots)$ are everywhere dense in H: because if we choose for each Π_s, the $\Pi^{(n_s)}$ according to the above discussion, then

$$\sqrt{\int_\Omega |\sum_{s=1}^t (\rho_s + i\sigma_s) f_{\Pi_s} - \sum_{s=1}^t (\rho_s + i\sigma_s) f_{\Pi}(n_s)|^2}$$

$$\leq \sum_{s=1}^t \sqrt{\int_\Omega |(\rho_s + i\sigma_s) f_{\Pi_s} - (\rho_s + i\sigma_s) f_{\Pi}(n_s)|^2}$$

$$= \sum_{s=1}^t \sqrt{(\rho_s{}^2 + \sigma_s{}^2) \int_\Omega |f_{\Pi_s} - f_{\Pi}(n_s)|^2}$$

$$= \sum_{s=1}^t \sqrt{(\rho_s{}^2 + \sigma_s{}^2) . \text{measure of the difference set} . (\Pi_s, \Pi^{(n_s)})}$$

$$< \sum_{s=1}^t \sqrt{(\rho_s{}^2 + \sigma_s{}^2) . \epsilon} = \left(\sum_{s=1}^t \sqrt{\rho_s{}^2 + \sigma_s{}^2} \right) \sqrt{\epsilon}$$

If a $\delta > 0$ is given, then

$$\epsilon = \delta^2 / \left(\sum_{s=1}^t \sqrt{\rho_s{}^2 + \sigma_s{}^2} \right)^2$$

furnishes the result

$$|| \sum_{s=1}^t (\rho_s + i\sigma_s) f_{\Pi_s} - \sum_{s=1}^t (\rho_s + i\sigma_s) f_{\Pi}(n_s) || < \delta$$

But the

$$\sum_{s=1}^t (\rho_s + i\sigma_s) f_{\Pi}(n_s)$$

form a sequence, if we order them appropriately. This can be done in the following way. Let the common denominator of all $\rho_1, \sigma_1, \ldots, \rho_t, \sigma_t$ be τ, and the new numerators $\rho_1', \sigma_1', \ldots, \rho_t', \sigma_t'$, then the relation becomes

$$\frac{1}{\tau}\sum_{s=1}^{t}(\rho_s' + i\sigma_s')f_{\Pi}(n_s)$$

in which we have $t, \tau = 1,2,\ldots$; $\rho_s', \sigma_s' = 0, \pm 1, \pm 2, \ldots$, $n_s = 1,2,\ldots$ for $s = 1,\ldots,t$. To order these functions as a sequence is the identical problem as doing the same thing for the integers $t,\tau,\rho_1',\sigma_1',\ldots,\rho_t',\sigma_t',n_1,\ldots,n_t$. Among these complexes of numbers group those together for which the positive integer

$$I = t + \tau + |\rho_1'| + |\sigma_1'| + \cdots + |\rho_t'| + |\sigma_t'| + n_1 + \cdots + n_t$$

has the same value. Then arrange these groups in the order of increase of their indices I . Each one of these groups (with fixed I) consists obviously of a finite number of the complexes in question. If we now arrange each one of these finite sets in any order, we have in fact obtained a simple sequence containing all the complexes.

In order to be able to specify the set sequence $\Pi^{(1)},\Pi^{(2)},\ldots$ mentioned, we make use of the fact that for each set Π with finite Lebesgue measure M , and for each $\delta > 0$, there exists an open point set Π' , which covers Π , but whose measure exceeds it by $< \delta$ (cf. the references of Note 52 and Note 45, where the concept "open point set" has been defined.) For each open Π' and a $\delta > 0$, there obviously exists a set Π'' consisting of a finite number of cubes, which is contained in Π' , and whose measure is less than that of Π' by $< \delta$. Clearly the lengths of the edges of these cubes and their center coordinates can all be chosen rational. We now easily recognize that the "difference set" of Π, Π'' , as defined above, has a measure $< \delta + \delta = 2\delta$ and therefore for $\delta = \frac{\epsilon}{2}$ a measure $< \epsilon$. We have then accomplished our purpose, if we can order in a sequence the sets of cubes of the type just described.

These sets of cubes are now characterized by the

number of their cubes, $n = 1, 2, \ldots$, together with the lengths of their edges $\kappa^{(\nu)}$ and the coordinates of their center points $\xi_1^{(\nu)}, \ldots, \xi_k^{(\nu)}$ $(\nu = 1, \ldots, n)$. The $\kappa^{(\nu)}$, $\xi^{(\nu)}, \ldots, \xi_k^{(\nu)}$ are rational. Let their common denominator (for all $\nu = 1, \ldots n$) be $\eta = 1, 2, \ldots$, their numerators

$$\kappa'^{(\nu)} = 1, 2, \ldots \; ; \; \xi_1'^{(\nu)}, \ldots, \xi_k'^{(\nu)} = 0, \pm 1, \pm 2, \ldots$$

Then our sets of cubes are characterized by the complexes of numbers

$$n, \eta, \kappa'^{(1)}, \xi_1'^{(1)}, \ldots, \xi_k'^{(1)}, \ldots, \kappa'^{(n)}, \xi_1'^{(n)}, \ldots, \xi_k'^{(n)}$$

If we arrange these in the order of increase of the positive integers,

$$n + \eta + \kappa'^{(1)} + |\xi_1'^{(1)}| + \ldots + |\xi_k'^{(1)}| + \ldots + \kappa'^{(n)} + |\xi_1'^{(n)}| + \ldots + |\xi_\kappa'^{(n)}|$$

then we obtain a simple sequence, exactly as in the earlier analogous case of the linear combinations of functions.

Before we continue, let us answer the following question: Given an $\mathfrak{R}$ satisfying an **A.** - **E.** (with **C**$^{(\infty)}$.), in which subsets $\mathfrak{M}$ of $\mathfrak{R}$, are **A.** - **E.** again satisfied? (with unchanged definitions of af, $f \pm g$ as well as (f, g) ?

In order that **A.** hold, $\mathfrak{M}$ must be a linear manifold. **B** . is valid of itself. We postpone **C** . momentarily; in any case, a **C**$^{(n)}$. or a **C**$^{(\infty)}$. holds. **D**. means: if a sequence in $\mathfrak{M}$ satisfies the Cauchy convergence criterion, then it has a limit in $\mathfrak{M}$. Since such a sequence will certainly possess a limit in $\mathfrak{R}$, **D**. means simply that this limit also belongs to $\mathfrak{M}$. I.e., $\mathfrak{M}$ must be closed. The condition **E**. always holds, as we saw in the proof of THEOREM 9. Therefore, we may summarize thus: $\mathfrak{M}$ must be a closed linear manifold. We call the orthonormal set which spans $\mathfrak{M}$ (THEOREM 9.) $\phi_1, \phi_2, \ldots$. If it is infinite, then **C**$^{(\infty)}$. obviously holds, and $\mathfrak{M}$ is isomorphic to $\mathfrak{R}_\infty$ therefore to $\mathfrak{R}$ itself; if it terminates at ϕ_n, then **C**$^{(n)}$. holds (e.g., because of THEOREM $3^{(n)}$.), i.e., $\mathfrak{M}$ is isomorphic to $\mathfrak{R}_n$.

But since **D**., **E**. are valid in $\mathfrak{M}$ in any case, they are valid in each $\mathfrak{R}_n$. Therefore they also follow from **A**. - **C**$^{(n)}$.

As we see, we have avoided the direct verification of **A**. - **E**. (with **C**$^{(n)}$. or **C**$^{(\infty)}$.) in $\mathfrak{R}_n$ or $\mathfrak{R}_\infty$. This was achieved by indirect, logical devices. However, a direct, analytical demonstration causes no essential difficulties either. It may be left to the reader for proof.

It still remains to show that **D**. and **E**. are independent of **A**. - **C**$^{(\infty)}$. As we have seen previously, every linear manifold in $\mathfrak{R}_\infty$ satisfies **A**., **B**., **E**., as well as **C**$^{(n)}$. or **C**$^{(\infty)}$., if it is not closed, then **D**. is not fulfilled. In this case **C**$^{(\infty)}$. must hold in it, because **D**. follows from **C**$^{(n)}$. Now it is not difficult to exhibit such a non-closed, linear manifold. Let $\phi_1, \phi_2, \ldots$ be an orthonormal set, then the

$$\sum_{\nu=1}^{N} x_\nu \phi_\nu$$

($N = 1, 2, \ldots$; $x_1, \ldots, x_N$ arbitrary) form a linear manifold, but one which is not closed, because

$$\sum_{\nu=1}^{\infty} \frac{1}{\nu} \phi_\nu$$

($\sum_1^\infty (\frac{1}{\nu})^2$ is finite!) is a limiting point, but not an element of the manifold

$$\left(\sum_{\nu=1}^{N} \frac{1}{\nu} \phi_\nu \longrightarrow \sum_{\nu=1}^{\infty} \frac{1}{\nu} \phi_\nu \quad \text{as} \quad N \longrightarrow \infty \right)$$

Consequently **D.** is independent of **A.** - **C**$^{(\infty)}$., **E.**

Let us consider next all complex functions $x(\alpha)$ whose parameter α is continuous: $-\infty < \alpha < +\infty$. Moreover, suppose that it is possible to write the $x(\alpha) \neq 0$ in a series, such that the sum $\Sigma_\alpha |x(\alpha)|^2$ extended over these terms is finite.[55] All these functions $x(\alpha)$ form a space $\mathfrak{R}_{\text{cont}}$. Since for any two points $x(\alpha)$, $y(\alpha)$ of the latter space, $x(\alpha)$ or $y(\alpha) \neq 0$ only for two α-sequences, and since we can join these two sequences into a single one, $x(\alpha) = y(\alpha) = 0$ except for a certain α-sequence $\alpha_1, \alpha_2, \ldots$. Therefore we need discuss only the values $x_n = x(\alpha_n)$, $y_n = y(\alpha_n)$ for all $n = 1,2,\ldots$. These all behave the same as in $\mathfrak{R}_\infty$, as long as only two $\mathfrak{R}_{\text{cont}}$ points appear. Hence **A.**, **B.** hold in $\mathfrak{R}_{\text{cont}}$ exactly as in $\mathfrak{R}_\infty$.[56] The same follows for k $(= 1,2,\ldots)$ $\mathfrak{R}_{\text{cont}}$ points, therefore **C**$^{(\infty)}$. holds also. Moreover this is even true for a sequence of $\mathfrak{R}_{\text{cont}}$ points. Consider $x_1(\alpha), x_2(\alpha), \ldots$, the α with $x_n(\alpha) \neq 0$ form a sequence for each $n = 1,2,\ldots$: $\alpha_1^{(n)}, \alpha_2^{(n)}, \ldots$. These sequences together form a double sequence $\alpha_m^{(n)}$ $(n, m = 1,2,\ldots)$ which can be written also as a simple sequence $\alpha_1^{(1)}, \alpha_2^{(1)}, \alpha_1^{(2)}, \alpha_3^{(1)}, \alpha_2^{(2)}, \alpha_1^{(3)}, \ldots$. Consequently **D.** holds in $\mathfrak{R}_{\text{cont}}$ as well as in $\mathfrak{R}_\infty$. It is otherwise with **E.** In that case, all points of $\mathfrak{R}$ play a role (all must be limit points of an appropriate sequence), therefore, we

---

[55] Although α varies continuously, this is a sum and not an integral, since only a sequence of the α appears in the sum!

[56] We naturally define $(x(\alpha), y(\alpha))$ as $\sum_\alpha x(\alpha)\overline{y(\alpha)}$.

cannot reason from $\mathfrak{R}_\infty$ to $\mathfrak{R}_{\text{cont}}$. Also, the condition is actually not satisfied, because one deduction from it is invalid: there exists an orthonormal set which cannot be written as a sequence (contrary to THEOREM $3^{(\infty)}$.).

Let

$$x_\beta(\alpha) \begin{cases} = 1 & \text{for} \quad \alpha = \beta \\ = 0 & \text{for} \quad \alpha \neq \beta \end{cases}$$

for each β, $x_\beta(\alpha)$ is an element of $\mathfrak{R}_{\text{cont}}$, and the $x_\beta(\alpha)$ form an orthonormal set. But they could be written as a sequence only if this were possible for all $\beta > -\infty, < +\infty$, which is well known not to be the case.[57] Therefore **E.** is also independent of **A.** - $\mathbf{C}^{(\infty)}$., **D.**

(In addition, the fundamental difference between the function space of the $f(x)$ with finite

$$\int_{-\infty}^{\infty} |f(x)|^2 dx$$

and that of the $x(\alpha)$ with finite $\Sigma_\alpha |x(\alpha)|^2$ should be noted. We could just as well characterize the former as the space of all $x(\alpha)$ with finite

$$\int_{-\infty}^{\infty} |x(\alpha)|^2 d\alpha \,!$$

The entire difference is the replacing of $\int_{-\infty}^{\infty} \ldots d\alpha$ by $\Sigma_\alpha \ldots$, and yet the first named space is $\mathbf{F}_\Omega$, therefore satisfies **A.** - **E.**, and is isomorphic to $\mathfrak{R}_\infty$, while the latter, $\mathfrak{R}_{\text{cont}}$, violates **E.**, and is essentially different from $\mathfrak{R}_\infty$. Nevertheless the two spaces are identical

---

[57] This is the set theoretical theorem on the "Non-Denumerability of the Continuum." Cf. for example, the book of Hausdorff mentioned in Note 45.

except for their differing definitions of magnitude!)

4. CLOSED LINEAR MANIFOLDS

The § II.2. is of importance for us not only because of the proof of isomorphism, but also because several theorems on orthonormal sets were proved therein. We now desire to go further into the geometric analysis of Hilbert space, and to investigate in detail the closed linear manifolds which play in the $\mathfrak{R}_\infty$ a role analogous to that played by straight lines, planes, etc. in the $\mathfrak{R}_n$ (i.e., the $\mathfrak{R}_m$, $m \leq n$).

We first recall the notations introduced in the DEFINITIONS 2., 5.: if $\mathfrak{A}$ is any set in $\mathfrak{R}$, then $\{\mathfrak{A}\}$ or $[\mathfrak{A}]$ are the linear manifold spanned by $\mathfrak{A}$ or the closed linear manifold respectively, i.e., the smallest representant of either type which contains $\mathfrak{A}$.

We now extend this notation so that by $\{\mathfrak{A}, \mathfrak{B}, \ldots, f, g, \ldots\}$ or $[\mathfrak{A}, \mathfrak{B}, \ldots, f, g, \ldots]$ (if $\mathfrak{A}, \mathfrak{B}, \ldots$ are any subsets and $f, g, \ldots$ elements of $\mathfrak{R}$) we understand respectively the linear manifold, or the closed linear manifold spanned by that set which results from the combination of the $\mathfrak{A}, \mathfrak{B}, \ldots$ and the $f, g, \ldots$.

If, in particular, $\mathfrak{M}, \mathfrak{N}, \ldots$ (finite or infinite in number) are closed linear manifolds, then we designate the closed linear manifold $[\mathfrak{M}, \mathfrak{N}, \ldots]$ by $\mathfrak{M} + \mathfrak{N} + \ldots$. The linear manifold $\{\mathfrak{M}, \mathfrak{N}, \ldots\}$ clearly consists of all sums $f + g + \ldots$ (f running through $\mathfrak{M}$, g running through $\mathfrak{N}, \ldots$), while $[\mathfrak{M}, \mathfrak{N}, \ldots] = \mathfrak{M} + \mathfrak{N} + \ldots$ is obtained from this by the addition of the limit points. If only a finite number of sets $\mathfrak{M}, \mathfrak{N}$ are present, and each element of one is orthogonal to all elements of the others, then, as we shall soon see, these two representations are equal to each other, which is not necessarily the case in general.

If $\mathfrak{M}$ is a subset of $\mathfrak{R}$, then we consider the totality of elements of $\mathfrak{R}$ which are orthogonal to all elements of $\mathfrak{M}$. This also is obviously a closed linear manifold, which may be called $\mathfrak{R} - \mathfrak{M}$. THEOREM 14. will clarify the reason for denoting this as subtraction. The set $\mathfrak{R} - \mathfrak{M}$ of all f orthogonal to the entire $\mathfrak{M}$ is of special importance. This is called the closed linear manifold complementary to $\mathfrak{M}$.

Finally, we select three particularly simple closed linear manifolds: first, $\mathfrak{R}$ itself; second, the set $\{0\} = [0]$ consisting of zero alone; and third, the set of all af (f a given element of $\mathfrak{R}$, a variable), which is clearly a closed linear manifold, and therefore simultaneously, $= \{f\} = [f]$.

We now introduce the concept of "projection," one which is completely analogous to that term in Euclidean geometry:

THEOREM 10. Let $\mathfrak{M}$ be a closed linear manifold. Then each f can be resolved in one and only one way into two components, $f = g + h$, g from $\mathfrak{M}$, h from $\mathfrak{R} - \mathfrak{M}$.

NOTE. We call g the projection of f in $\mathfrak{M}$, h (which is orthogonal to all $\mathfrak{M}$) the normal from f onto $\mathfrak{M}$. We introduce the notation $P_{\mathfrak{M}} f$ for g.

PROOF. Let $\phi_1, \phi_2, \cdots$ be the orthonormal set, existing by reason of THEOREM 9., spanning the closed linear manifold $\mathfrak{M}$. We write $g = \sum_n (f, \phi_n) \cdot \phi_n$. By THEOREM 6., this series converges (if it is infinite at all), its sum g obviously belonging to $\mathfrak{M}$. Furthermore, by THEOREM 6., $h = f - g$ is orthogonal to all $\phi_1, \phi_2, \cdots$, but since the vectors orthogonal to h form a closed linear manifold, along with $\phi_1, \phi_2, \ldots$ all $\mathfrak{M}$ is also orthogonal to h, i.e., h belongs to $\mathfrak{R} - \mathfrak{M}$.

If there were still another resolution $f = g' + h'$, g' from $\mathfrak{M}$, h' from $\mathfrak{R} - \mathfrak{M}$, then $g + h = g' + h'$, $g - g' = h' - h = j$. The j would have to belong simultaneously to $\mathfrak{M}$ and $\mathfrak{R} - \mathfrak{M}$, and would therefore be orthogonal to itself. Therefore $(j, j) = 0$, $j = 0$ and consequently $g = g'$, $h = h'$.

The operation $P_{\mathfrak{M}} f$ is therefore one which assigns to each f of $\mathfrak{R}$ its projection in $\mathfrak{M}$, $P_{\mathfrak{M}} f$. In the next section we shall define: an operator R is a function defined in a subset of $\mathfrak{R}$ with values from $\mathfrak{R}$, i.e., a correspondence which assigns to certain f of $\mathfrak{R}$ certain $\mathfrak{R}f$ of $\mathfrak{R}$. (Not necessarily for all f. For other f of $\mathfrak{R}$, the operation may be undefined, i.e., "meaningless.") $P_{\mathfrak{M}}$ is then an operator defined everywhere in $\mathfrak{R}$ and is known as the projection operator of $\mathfrak{M}$, or merely the projection of $\mathfrak{M}$.

THEOREM 11. The operator $P_{\mathfrak{M}}$ has the following properties:

$$P_{\mathfrak{M}} (a_1 f_1 + \ldots + a_n f_n) = a_1 P_{\mathfrak{M}} f_1 + \ldots + a_n P_{\mathfrak{M}} f_n ,$$

$$(P_{\mathfrak{M}} f, g) = (f, P_{\mathfrak{M}} g)$$

$$P_{\mathfrak{M}} (P_{\mathfrak{M}} f) = P_{\mathfrak{M}} f .$$

$\mathfrak{M}$ is the set of all values of $P_{\mathfrak{M}}$, i.e., the set of all $P_{\mathfrak{M}} f$; but it can also be characterized as the set of all solutions of $P_{\mathfrak{M}} f = f$, while $\mathfrak{R} - \mathfrak{M}$ is the set of all solutions of $P_{\mathfrak{M}} f = 0$.

NOTE. In the succeeding sections, we shall see that the first property determines the so-called linear operators, and the second the so-called Hermitian operators. The third expresses the following: double application of the operator $P_{\mathfrak{M}}$ has the same effect as a single application. The customary symbolic representation of this is

$$P_{\mathfrak{M}} P_{\mathfrak{M}} = P_{\mathfrak{M}} \quad \text{or} \quad P_{\mathfrak{M}}^2 = P_{\mathfrak{M}}$$

PROOF. From

$$f_1 = g_1 + h_1, \ldots, f_n = g_n + h_n$$

$$(g_1, \ldots, g_n \text{ from } \mathfrak{M}, \; h_1, \ldots, h_n \text{ from } \mathfrak{R} - \mathfrak{M})$$

it follows that

$$a_1 f_1 + \cdots + a_n f_n = (a_1 g_1 + \cdots + a_n g_n) + (a_1 h_1 + \cdots + a_n h_n)$$

$$(a_1 g_1 + \cdots + a_n g_n \text{ from } \mathfrak{M}, \; a_1 h_1 + \cdots + a_n h_n \text{ from } \mathfrak{R} - \mathfrak{M})$$

therefore

$$P_{\mathfrak{M}} (a_1 f_1 + \cdots + a_n f_n) = a_1 g_1 + \cdots + a_n g_n$$

$$= a_1 P_{\mathfrak{M}} f_1 + \cdots + a_n P_{\mathfrak{M}} f_n$$

This is the first assertion.

In the second case, let

$$f = g' + h', \; g = g'' + h'' \quad (g', g'' \text{ from } \mathfrak{M}, \; h', h'' \text{ from } \mathfrak{R} - \mathfrak{M})$$

then g', g'' are orthogonal to h', h'', therefore

$$(g', g) = (g', g'' + h'') = (g', g'') = (g' + h', g'') = (f, g'')$$

i.e., $(P_{\mathfrak{M}} f, g) = (f, P_{\mathfrak{M}} g)$. This is the second assertion.

Finally, $P_{\mathfrak{M}} f$ belongs to $\mathfrak{M}$, therefore $P_{\mathfrak{M}} f = P_{\mathfrak{M}} f + 0$ is the resolution into components guaranteed by THEOREM 10. for $P_{\mathfrak{M}} f$, i.e., $P_{\mathfrak{M}} (P_{\mathfrak{M}} f) = P_{\mathfrak{M}} f$. This is the third assertion.

The relations $P_{\mathfrak{M}} f = f$ or 0 signifies that in the resolution $f = g + h$, g from $\mathfrak{M}$, h from $\mathfrak{R} - \mathfrak{M}$ (THEOREM 10.), either $f = g$, $h = 0$, or $g = 0$, $f = h$; i.e., that f belongs either to $\mathfrak{M}$ or to $\mathfrak{R} - \mathfrak{M}$. These are the fifth and sixth assertions. All $P_{\mathfrak{M}} f$ belong to $\mathfrak{M}$ by definition, and each f' of $\mathfrak{M}$ is equal to a $P_{\mathfrak{M}} f$: e.g., according to the statements just made, to $P_{\mathfrak{M}} f$. This is the fourth assertion.—

We observe next, that the second and third assertions imply this:

$$(P_{\mathfrak{M}} f, P_{\mathfrak{M}} g) = (f, P_{\mathfrak{M}} P_{\mathfrak{M}} g) = (f, P_{\mathfrak{M}} g) = (P_{\mathfrak{M}} f, g)$$

We now want to characterize the projection operators independently of the $\mathfrak{M}$.

THEOREM 12. An operator E, defined everywhere (cf. the discussion preceding THEOREM 11.) is a projection, i.e., $E = P_{\mathfrak{M}}$ for a closed linear manifold $\mathfrak{M}$, if and only if it has the following properties:

$$(Ef, g) = (f, Eg), \quad E^2 = E$$

(cf. the note on THEOREM 11.) In this case, $\mathfrak{M}$ is then uniquely determined by E (according to THEOREM 11.).

PROOF. The necessity of this condition as well as the determination of $\mathfrak{M}$ by E is obvious from THEOREM 11. We then have only to show that if E possesses the above properties, then there is a closed linear manifold $\mathfrak{M}$ with $E = P_{\mathfrak{M}}$.

Let $\mathfrak{M}$ be the closed manifold spanned by all Ef. Then $g - Eg$ is orthogonal to all Ef:

$$(Ef, g - Eg) = (Ef, g) - (Ef, Eg) = (Ef, g) - (E^2f, g) = 0$$

The elements orthogonal to $g - Eg$ of $\mathfrak{R}$ form a closed linear manifold; therefore they include $\mathfrak{M}$ along with Ef -- then $g - Eg$ belongs to $\mathfrak{R} - \mathfrak{M}$. The resolution of g for $\mathfrak{M}$, in the sense of THEOREM 10. is then $g = Eg + (g - Eg)$, hence $P_{\mathfrak{M}} g = Eg$, where g is arbitrary. Therefore the entire theorem has been proved.—

If $\mathfrak{M} = \mathfrak{R}$ or $= [0]$, then $\mathfrak{R} - \mathfrak{M} = [0]$ or $\mathfrak{R}$ respectively, therefore $f = f + 0$ or $= 0 + f$ is the resolution, by THEOREM 11.; hence $P_{\mathfrak{M}} f = f$ or $= 0$ respectively. We call 1 the operator defined (everywhere!) by $Rf = f$, and 0 the operator defined by $Rf = 0$. Hence $P_{\mathfrak{R}} = 1$, $P_{[0]} = 0$. Furthermore, it is clear that the resolution $f = g + h$ (g from $\mathfrak{M}$, h from $\mathfrak{R} - \mathfrak{M}$) belonging to $\mathfrak{M}$ is also useful for $\mathfrak{R} - \mathfrak{M}$ in the form $f = h + g$ (h from $\mathfrak{R} - \mathfrak{M}$, g from $\mathfrak{M}$). (For, since g belongs to $\mathfrak{M}$, it is orthogonal to each element of $\mathfrak{R} - \mathfrak{M}$ therefore it belongs to $\mathfrak{R} - (\mathfrak{R} - \mathfrak{M})$.) Therefore $P_{\mathfrak{M}} f = g$, $P_{\mathfrak{R} - \mathfrak{M}} f = h = f - g$, i.e., $P_{\mathfrak{R} - \mathfrak{M}} f = f - P_{\mathfrak{M}} f$. This fact, $P_{\mathfrak{R} - \mathfrak{M}} f = 1f - P_{\mathfrak{M}} f$, we express symbolically as $P_{\mathfrak{R} - \mathfrak{M}} = 1 - P_{\mathfrak{M}}$ (for addition, subtraction and multiplication of operators, cf. the discussion in THEOREM 14.).

The following should be noted: A short time ago we easily recognized that $\mathfrak{M}$ is a subset of $\mathfrak{R} - (\mathfrak{R} - \mathfrak{M})$. It is difficult to prove directly that both sets are equal. But this equality follows immediately from

$$P_{\mathfrak{R} - (\mathfrak{R} - \mathfrak{M})} = 1 - P_{\mathfrak{R} - \mathfrak{M}} = 1 - (1 - P_{\mathfrak{M}}) = P_{\mathfrak{M}}$$

Moreover, it follows from the above that if E is a projection, $1 - E$ is also a projection, and because $1 - (1 - E) = E$, the converse is also true.

THEOREM 13. It is always true that

$$\|Ef\|^2 = (Ef, f), \quad \|Ef\| \leq \|f\|,$$

$\|Ef\| = 0$ or $= \|f\|$ is characteristic for the f of $\mathfrak{R} - \mathfrak{M}$ and $\mathfrak{M}$ respectively.

NOTE. In particular, therefore,

$$\|Ef - Eg\| = \|E(f - g)\| \leq \|f - g\|$$

i.e., the operator E is continuous (cf. the discussion after THEOREM 2. in II.1.).

PROOF. We have (cf. the discussion after THEOREM 11.)

$$\|Ef\|^2 = (Ef, Ef) = (Ef, f)$$

Since $1 - E$ is also a projection,

$$\|Ef\|^2 + \|f - Ef\|^2 = \|Ef\|^2 + \|(1 - E)f\|^2$$

$$= (Ef, f) + ((1 - E)f, f) = (f, f) = \|f\|^2$$

Since both components are ≥ 0, they are also $\leq \|f\|^2$, in particular $\|Ef\|^2 \leq \|f\|^2$, $\|Ef\| \leq \|f\|$. That

$||Ef|| = 0$, $Ef = 0$ expresses the fact that f belongs to $\mathfrak{R} - \mathfrak{M}$, we know from THEOREM 11. Because of the above relation, $||Ef|| = ||f||$ means that $||f - Ef|| = 0$, $Ef = f$, therefore, by THEOREM 11., that f belongs to $\mathfrak{M}$.

If R, S are two operators, then we understand by $R \pm S$, aR (a a complex number), RS , the operators defined by

$$(R \pm S)f = Rf \pm Sf \ , \quad (aR)f = a \cdot Rf \ , \quad (RS)f = R(Sf) \ ,$$

and we use the then natural notation

$$R^0 = 1, R^1 = R, R^2 = RR, R^3 = RRR, \ldots$$

The rules of calculation which are valid here can be discussed rather easily. For $R \pm S$, aR , we can verify without difficulty all elementary laws of calculation valid for numbers, but such is not the case for RS . The distributive law holds, as can easily be verified: $(R \pm S)T = RT \pm ST$ and $R(S \pm T) = RS \pm RT$ (for the latter, the linearity of R is of course necessary; cf. the note on THEOREM 11., and the discussion in the following paragraph). The associative law also holds: $(RS)T = R(ST) = RST$, but the commutative law $RS = SR$ is not generally valid. [$(RS)f = R(Sf)$ and $(SR)f = S(Rf)$ need not be equal to each other!] If this law does hold for two particular R, S , they are said to commute. Hence, for example, 0 and 1 commute with all R which are defined everywhere:

$$R0 = 0R = 0 \ , \quad R1 = 1R = R$$

Also, R^m, R^n commute, since $R^m R^n = R^{m+n}$, and therefore does not depend on the order of m, n .

THEOREM 14. Let E, F be projections, those of the closed linear manifolds $\mathfrak{M}, \mathfrak{N}$. Then EF is also a projection if and only if E, F commute, i.e., if $EF = FE$. Also, EF belongs to the closed linear manifold $\mathfrak{P}$ which consists of the elements common to $\mathfrak{M}, \mathfrak{N}$. The operator $E + F$ is a projection if and only if $EF = 0$ (or equally: if $FE = 0$). This means that all $\mathfrak{M}$ is orthogonal to all $\mathfrak{N}$; $E + F$ then belongs to $\mathfrak{M} + \mathfrak{N} = [\mathfrak{M}, \mathfrak{N}]$, which in this case $= \{\mathfrak{M}, \mathfrak{N}\}$. The operator $E - F$ is a projection if and only if $EF = F$ (or equally: if $FE = F$). This means that $\mathfrak{N}$ is a subset of $\mathfrak{M}$, and $E - F$ belongs to $\mathfrak{M} - \mathfrak{N}$.

PROOF. For EF, we must re-examine the two conditions of THEOREM 12.:

$$(EFf, g) = (f, EFg), \quad (EF)^2 = EF$$

Because $(EFf, g) = (Ff, Eg) = (f, FEg)$, the first signifies that

$$(f, EFg) = (f, FEg), \quad (f, (EF - FE)g) = 0$$

Since this holds for all f, $(EF - FE)g = 0$, and since this holds for all g, $EF - FE = 0$, $EF = FE$. The commutativity is therefore necessary and sufficient for the first condition, but it also has the second one as a consequence:

$$(EF)^2 = EFEF = EEFF = E^2F^2 = EF$$

Since $E + F$ always satisfies the first condition

$((E + F)f, g) = (f, (E + F)g)$ (because E, F do so) only the second condition $(E + F)^2 = E + F$ remains to be proved. Since

$$(E + F)^2 = E^2 + F^2 + EF + FE = (E + F) + (EF + FE)$$

this means simply that $EF + FE = 0$. Now for $EF = 0$, EF is a projection. Therefore, by the above proof, $EF = FE$, therefore $EF + FE = 0$. Conversely, from $EF + FE = 0$, it follows that

$$E(EF + FE) = E^2F + EFE = EF + EFE = 0 ,$$

$$E(EF + FE)E = E^2FE + EFE^2 = EFE + EFE = 2 \cdot EFE = 0$$

therefore $EFE = 0$, and therefore $EF = 0$. Consequently, $EF = 0$ is necessary and sufficient or, since E, F play the same roles, $FE = 0$, too, is necessary and sufficient.

$E - F$ is a projection if and only if $1 - (E - F) = (1 - E) + F$ is one, and since $1 - E, F$ are projections, by the same proof, $(1 - E)F = 0$, $F - EF = 0$, $EF = F$ are characteristic of this, or equally, $F(1 - E) = 0$, $F - FE = 0$, $FE = F$.

We have yet to prove the propositions on $\mathfrak{M}, \mathfrak{N}$ $(E = P_{\mathfrak{M}}, F = P_{\mathfrak{N}})$. First, let $EF = FE$. Then each $EFf = FEf$ belongs to $\mathfrak{M}$ and to $\mathfrak{N}$, therefore to $\mathfrak{P}$, and for each g of $\mathfrak{P}$, $Eg = Fg = g$, therefore $EFg = Eg = g$, i.e., it has the form EFf. Consequently, $\mathfrak{P}$ is the totality of values of EF, hence by THEOREM 11. $EF = P_{\mathfrak{P}}$. Second, let $EF = 0$ (therefore $FE = 0$ also). Each $(E + F)f = Ef + Ff$ belongs to $\{\mathfrak{M}, \mathfrak{N}\}$ and each g of $\{\mathfrak{M}, \mathfrak{N}\}$ is equal to $h + j$, h from $\mathfrak{M}$, j from $\mathfrak{N}$, therefore $Eh = h$, $Fh = FEh = 0$, $Fj = j$, $Ej = EFj = 0$. Therefore

$$(E + F)(h + j) = Eh + Fh + Ej + Fj = h + j , \quad (E + F)g = g$$

Then g has the form $(E + F)f$. Consequently $\{\mathfrak{M}, \mathfrak{N}\}$ is the totality of values of $E + F$, but since $E + F$ is a projection, $\{\mathfrak{M}, \mathfrak{N}\}$ is the corresponding closed linear manifold (THEOREM 11.). Since $\{\mathfrak{M}, \mathfrak{N}\}$ is closed, it $= [\mathfrak{M}, \mathfrak{N}] = \mathfrak{M} + \mathfrak{N}$. Third, let $EF = F$ (therefore $FE = F$ also). Then $E = P_{\mathfrak{M}}$, $1 - F = P_{\mathfrak{R} - \mathfrak{N}}$, therefore $E - F = E - EF = E(1 - F)$, equalling $P_{\mathfrak{P}}$, where $\mathfrak{P}$ is the intersection of $\mathfrak{M}$ and $\mathfrak{R} - \mathfrak{N}$, i.e., $\mathfrak{M} - \mathfrak{N}$.

Finally, $EF = 0$ means that $(EFf, g) = 0$ always, i.e., $(Ff, Eg) = 0$, i.e., that the entire $\mathfrak{M}$ is orthogonal to the entire $\mathfrak{N}$. And $EF = F$ means $F(1 - E) = 0$, i.e., all $\mathfrak{N}$ is orthogonal to $\mathfrak{R} - \mathfrak{M}$, or equally: $\mathfrak{N}$ is a subset of $\mathfrak{R} - (\mathfrak{R} - \mathfrak{M}) = \mathfrak{M}$.

If $\mathfrak{N}$ is a subset of $\mathfrak{M}$, then we also want to say for $F = P_{\mathfrak{N}}$, $E = P_{\mathfrak{M}}$ that F is part of E, symbolically, $E \geq F$ or $F \leq E$. (This then means that $EF = F$, or also $FE = F$, and has the commutativity as a consequence. This can be seen either by observation of $\mathfrak{M}$, $\mathfrak{N}$ or by direct calculation. It is always true that $0 \leq E \leq 1$. From $E \leq F$, $F \leq E$ it follows that $E = F$. From $E \leq F$, $F \leq G$ it follows that $E \leq G$. This possesses the characteristics of an ordering according to magnitude. It should further be observed that $E \leq F$, $1 - E \geq 1 - F$ and E orthogonal to $1 - F$ are all equivalent. Furthermore, the orthogonality of E', F' follows from that of E, F if $E' \leq E$, $F' \leq F$.) If $\mathfrak{M}$, $\mathfrak{N}$ are orthogonal, we say that E, F are also orthogonal. (Hence this means that $EF = 0$ or also $FE = 0$.) Conversely, if E, F commute, we say that their $\mathfrak{M}$, $\mathfrak{N}$ also commute.

THEOREM 15. The statement $E \leq F$ is equivalent to the general validity of $\|Ef\| \leq \|Ff\|$.

PROOF. From $E \leq F$ it follows that $E = EF$, therefore $\|Ef\| = \|EFf\| \leq \|Ff\|$ (cf. THEOREM 13.).

Conversely, this relation has the following consequence: If $Ff = 0$, then $||Ef|| \leq ||Ff|| = 0$, $Ef = 0$. Now, because of $F(1 - F)f = (F - F^2)f = 0$, we have $E(1 - F)f = 0$ identically, i.e., $E(1 - F) = E - EF = 0$, $E = EF$, therefore $E \leq F$.

THEOREM 16. Let $E_1, \ldots, E_k$ be projections. Then $E_1 + \ldots + E_k$ is a projection if and only if all E_m, E_l ($m, l = 1, \ldots, k$; $m \neq l$) are mutually orthogonal. Another necessary and sufficient condition is that

$$||E_1 f||^2 + \ldots + ||E_k f||^2 \leq ||f||^2$$

(for all f). Moreover $E_1 + \ldots + E_k$ ($E_1 = P_{\mathfrak{M}_1}, \ldots, E_k = P_{\mathfrak{M}_k}$) is then the projection of $\mathfrak{M}_1 + \ldots + \mathfrak{M}_k = [\mathfrak{M}_1, \ldots \mathfrak{M}_k]$ which in this case $= \{\mathfrak{M}_1, \ldots, \mathfrak{M}_k\}$.

PROOF. The last proposition obtains by repeated application of THEOREM 14. And so does the sufficiency of the first criterion. If the second criterion is satisfied, then the first one is, too. For $m \neq l$, $E_m f = f$,

$$||f||^2 = ||E_l f||^2 = ||E_m f||^2 = ||E_l f||^2 \leq ||E_1 f||^2 + \ldots + ||E_k f||^2 \leq ||f||^2 ,$$

$$||E_l f||^2 = 0 , \; E_l f = 0$$

But since $E_m(E_m f) = E_m f$ holds identically, $E_l(E_m f) = 0$, i.e., $E_l E_m = 0$. Finally, the second condition is necessary: If $E_1 + \ldots + E_k$ is a projection, then (THEOREM 13.)

$$||E_1f||^2 + \dots + ||E_kf||^2 = (E_1f, f) + \dots + (E_kf, f)$$

$$= ((E_1 + \dots + E_k)f, f) = ||(E_1 + \dots + E_k)f||^2 \leq ||f||^2$$

We have therefore the following logical scheme:
$E_1 + \dots + E_k$ is a projection $\longrightarrow$ second criterion $\longrightarrow$ first criterion $\longrightarrow E_1 + \dots + E_k$ is a projection. Therefore all three are equivalent.

In conclusion, we prove a theorem on the convergence of the projections:

THEOREM 17. Let $E_1, E_2, \dots$ be an increasing or decreasing sequence of projections: $E_1 \leq E_2 \leq \dots$ or $E_1 \geq E_2 \geq \dots$. These converge to a projection E in the sense that for all f , $E_nf \longrightarrow Ef$; also, all $E_n \leq E$ or all $E_n \geq E$ respectively.

PROOF. It suffices to investigate the second case, since the first can be reduced to it by substitution of $1 - E_1, 1 - E_2, \dots, 1 - E$ for $E_1, E_2, E_3, \dots, E$. Let therefore $E_1 \geq E_2 \geq \dots$.

By THEOREM 15., $||E_1f||^2 \geq ||E_2f||^2 \geq \dots \geq 0$ therefore $\lim_{m \to \infty} ||E_mf||^2$ exists. Then for each $\epsilon > 0$ there exists an $N = N(\epsilon)$, such that for $m , l \geq N$

$$||E_mf||^2 - ||E_lf||^2 < \epsilon$$

Now for $m \leq l$, $E_m \geq E_l$, $E_m - E_l$ is a projection, therefore

$$||E_mf||^2 - ||E_lf||^2 = (E_mf, f) - (E_lf, f) = ((E_m - E_l)f, f)$$

$$= ||(E_m - E_l)f||^2 = ||E_mf - E_lf||^2$$

from which it follows that $||E_m f - E_l f|| < \sqrt{\epsilon}$. The sequence $E_1 f, E_2 f, \ldots$ therefore satisfies the Cauchy convergence criterion and has a limit f^* (D. from II.1.!). $Ef = f^*$ therefore defines an operator which has meaning everywhere.

From $(E_n f, g) = (f, E_n g)$ it follows by transition to the limit that $(Ef, g) = (f, Eg)$, from $(E_n f, E_n g) = (E_n f, g)$ that $(Ef, Eg) = (Ef, g)$ -- therefore $(E^2 f, g) = (Ef, g)$, $E^2 = E$. Consequently E is a projection. For $l \geq m$, $||E_m f|| \geq ||E_l f||$, and as $l \longrightarrow \infty$ we obtain $||E_m f|| \geq ||Ef||$, therefore $E_m \geq E$ (THEOREM 15.).

If $E_1, E_2, \ldots$ are projections, each pair of which are mutually orthogonal, then

$$E_1, E_1 + E_2, E_1 + E_2 + E_3, \ldots$$

are also projections, and form an increasing sequence. By THEOREM 17., they then converge to a projection which is $\geq$ all of them, and which we can denote by $E_1 + E_2 + \ldots$. Let $E_1 = P_{\mathfrak{M}_1}, E_2 = P_{\mathfrak{M}_2}, \ldots, E_1 + E_2 + \ldots = P_{\mathfrak{M}}$. Since all $E_m \leq E$, $\mathfrak{M}_m$ is a subset of $\mathfrak{M}$, therefore $\mathfrak{M}$ also includes $[\mathfrak{M}_1, \mathfrak{M}_2, \ldots] = \mathfrak{M}_1 + \mathfrak{M}_2 + \ldots = \mathfrak{M}'$. Conversely, all $\mathfrak{M}_m$ are subsets of $\mathfrak{M}'$; therefore $E_m \leq P_{\mathfrak{M}'} = E'$. Consequently, by reasons of continuity (cf. the treatment in the proof above), $E \leq E'$. Therefore $\mathfrak{M}$ is a subset of $\mathfrak{M}'$ and hence $\mathfrak{M} = \mathfrak{M}'$, $E = E'$ i.e., $\mathfrak{M} = \mathfrak{M}_1 + \mathfrak{M}_2 + \ldots$ or, written in another way,

$$P_{\mathfrak{M}_1 + \mathfrak{M}_2 + \ldots} = P_{\mathfrak{M}_1} + P_{\mathfrak{M}_2} + \ldots$$

With this we conclude our study of projection operators.

5. OPERATORS IN HILBERT SPACE

We have now given sufficient consideration to the geometric relations of the (Hilbert) space $\mathfrak{R}_\infty$ of infinitely many dimensions, that we may turn our attention to its linear operators -- i.e., to the linear mappings of $\mathfrak{R}_\infty$ on itself. For this purpose, we must introduce several concepts which were actually anticipated, to a certain extent, in the last few sections.

In these sections, we concerned ourselves with operators, which we define as follows. (in accord with the statements made preceding THEOREM 11.).

> DEFINITION 6. An operator R is a function defined in a subset of $\mathfrak{R}$ with values from $\mathfrak{R}$, i.e., a relation which established a correspondence between certain elements f of $\mathfrak{R}$ and elements Rf of $\mathfrak{R}$.

(We have admitted the $\mathfrak{R}_n$ here in addition to $\mathfrak{R}_\infty$. It should be observed that if $\mathfrak{R}_\infty$ is an F_Ω, then the operator R is defined for the elements of F_Ω, i.e., ordinary configuration space functions, and its values are defined likewise. The operators are then so-called "function-functions" or "functionals." Cf. the examples of I.2., 4.) The class of the f for which Rf is defined, the domain of R, need not encompass the entire $\mathfrak{R}$, but if it does so, R is said to be defined everywhere. In addition, it is not necessary that the set of all the Rf, the range of R (the mapping of its domain mediated by R), be contained in the domain of R, i.e., if Rf has meaning, it does not necessarily follow that $R(Rf) = R^2f$ is defined.[58]

---

[58] For example, let $\mathfrak{R}_\infty$ be an F_Ω, where Ω is the space

We have already given the meaning of $R \pm S$, aR, RS, R^m (R, S, operators, a a complex number, $m = 0,1,2,\ldots$) in the preceding section.

$$(R \pm S)f = Rf \pm Sf\ ,\quad (aR)f = a.Rf\ ,\quad (RS)f = R(Sf)\ ,$$

$$R^0 = 1, R^1 = R, R^2 = RR, R^3 = RRR, \ldots$$

In determining the domains of these operators, it should be observed that the left sides (i.e., the operators $R \pm S$, aR, RS) are defined only if the right sides are also defined. Therefore, for example, $R \pm S$ is defined only in the common part of the domains of R and S, etc. If Rf takes on each of its values only once,

---

of all real x, $-\infty < x < \infty$. $\frac{d}{dx}$ is a function-function, i.e., an operator, but defined in our sense only for such $f(x)$ which, in the first place, are differentiable, and which, in the second place, have finite

$$\int_{-\infty}^{\infty} \left|\frac{d}{dx} f(x)\right|^2 dx$$

(cf. II. 8., where this is discussed in more detail). Naturally, in general,

$$\frac{d^2}{dx^2} f(x)$$

will not exist, and

$$\int_{-\infty}^{\infty} \left|\frac{d^2}{dx^2} f(x)\right|^2 dx$$

is not necessarily finite. For example,

$$f(x) = |x|^{\frac{3}{2}} e^{-x^2}$$

behaves in this manner.

then it has an inverse R^{-1} : R^{-1} is defined if $Rg = f$ has a solution g, and its value is then this g. The discussion in the preceding sections was on the laws of calculation valid for $R \pm S$, aR, RS, here we will only add the following with regard to their domains. The operators there designated as equal also have identical domains, while operator equations such as $0 \cdot R = 0$ do not hold for the domains. $0f$ always has meaning, while $(0 \cdot R)f$, by definition, only has meaning if Rf is defined (but if both are defined, then both $= 0$). On the other hand, $1 \cdot R = R \cdot 1 = R$ hold, and also $R^m \cdot R^l = R^{m+l}$, and the same is true for their domains.

If R, S have inverses, then RS also possesses an inverse, which is, as can easily be seen, $(RS)^{-1} = S^{-1}R^{-1}$. Furthermore, for $a \neq 0$, $(aR)^{-1} = \frac{1}{a} R^{-1}$. If R^{-1} exists, we can also form the other negative powers or R:

$$R^{-2} = R^{-1}R^{-1}, R^{-3} = R^{-1}R^{-1}R^{-1}, \ldots$$

After this general development, we proceed to the investigation in more detail of those special classes of operators which will be of particular importance to use.

DEFINITION 7. An operator A is said to be linear if its domain is a linear manifold, i.e., if it contains $a_1 f_1 + \cdots + a_k f_k$ along with $f_1, \ldots, f_k$, and if

$$A(a_1 f_1 + \cdots + a_k f_k) = a_1 A f_1 + \cdots + a_k A f_k$$

In the following, we shall consider only linear operators, and indeed only those whose domains are everywhere dense.

The latter remark provides a sufficient substitute, for many purposes, for the requirement that operators be defined everywhere, which we must abandon in quantum mechanics. This circumstance is sufficiently important for us to consider it in more detail. For example, let us consider the configuration space in Schrödinger's wave mechanics which, for simplicity, we shall take as one dimensional: $-\infty < q < \infty$. The wave functions are the $\phi(q)$ with finite

$$\int_{-\infty}^{+\infty} |\phi(q)|^2 dq$$

These form a Hilbert space (cf. II.3.). We also consider the operators $q \ldots$ and $\frac{h}{2\pi i}\frac{d}{dq} \ldots$. They are evidently linear operators, but their domain is not the entire Hilbert space. It is not so for $q \ldots$ because

$$\int_{-\infty}^{+\infty} |q\phi(q)|^2 dq = \int_{-\infty}^{+\infty} q^2 |\phi(q)|^2 dq$$

can very well become infinite, even if

$$\int_{-\infty}^{+\infty} |\phi(q)|^2 dq$$

is finite, so that $q\phi(q)$ no longer lies in the Hilbert space; and it is not so for $\frac{h}{2\pi i}\frac{d}{dq}$ also, because there are non-differentiable functions, as well as those for which

$$\int_{-\infty}^{+\infty} |\phi(q)|^2 dq$$

but not

$$\int_{-\infty}^{\infty} \left|\frac{h}{2\pi i}\frac{d}{dq}\phi(q)\right|^2 dq = \frac{h^2}{4\pi^2}\int_{-\infty}^{\infty} \left|\frac{d}{dq}\phi(q)\right|^2 dq$$

is finite (for example, $|q|^{\frac{1}{2}} e^{-q^2}$ or $e^{-q^2}\sin(e^{q^2})$).

But the domains are everywhere dense. Both operators are certainly applicable to each $\phi(q)$ which $\neq 0$ only in a finite interval $-c \leq q \leq c$, and which is everywhere continuously differentiable; and this set of functions is everywhere dense.[59]

We further define:

DEFINITION 8. Two operators A, A^* are said to be adjoint if they have the same domain, and in this domain

---

[59]According to the development of II.3. (in the discussion of the condition **E**.), it is sufficient if we can approximate all linear combinations of the following function arbitrarily well: $f(x) = 1$ in a set consisting of a finite number of intervals, $= 0$ elsewhere. This is possible if we can approximate each one of these functions separately; and this in turn is possible if the same can be done for functions with a single 1-interval (the other functions are sums of such). For example, let the interval be $a < x < b$. The function

$$f(x) = 0 \qquad \text{for} \quad x < a - \epsilon \quad \text{or} \quad x > b + \epsilon$$

$$f(x) = \cos^2 \frac{\pi}{2} \frac{a - x}{\epsilon} \qquad \text{for} \quad a - \epsilon \leq x \leq a$$

$$f(x) = \cos^2 \frac{\pi}{2} \frac{x - b}{\epsilon} \qquad \text{for} \quad b \leq x \leq b + \epsilon$$

$$f(x) = 1 \qquad \text{for} \quad a < x < b$$

actually satisfies our requirements of regularity and approximates the given function arbitrarily well for sufficiently small ϵ.

$$(Af, g) = (f, A^*g) \ , \ (A^*f, g) = (f, Ag)$$

(By exchanging f, g, and taking the complex conjugate of both sides, each of these relations follows from the other. Furthermore, it is clear that the relation A, A^* is symmetric, i.e., that A^*, A are also adjoint, so that $A^{**} = A$.)

We note further that for A, only one adjoint A^* can be given, i.e., if A is adjoint to A_1^* and to A_2^*, then $A_1^* = A_2^*$. In fact, for all g with Ag defined,

$$(A_1^*f, g) = (f, Ag) = (A_2^*f, g)$$

and since these g are everywhere dense, $A_1^*f = A_2^*f$. Since this holds in general, $A_1^* = A_2^*$. Consequently, A determines A^* uniquely, just as A^* does A.

The following can be seen immediately: 0, 1, and, in general, all projections E are self-adjoint (cf. THEOREM 12.), i.e., $0^*, 1^*, E^*$ exist and are respectively equal to $0, 1, E$. Furthermore, $(aA)^* = \bar{a}A^*$, and, so far as $A \pm B$ can be formed in general (i.e., their domain is everywhere dense), $(A \pm B)^* = A^* \pm B^*$. Finally, with limitations on the domain which can easily be ascertained, $(AB)^* = B^*A^*$ (namely $(ABf, g) = (Bf, A^*g) = (f, BA^*g)$), as well as $(A^{-1})^* = A^{*-1}$ (here, $(A^{-1}f, g) = (A^{-1}f, A^*A^{*-1}g) = (AA^{-1}f, A^{*-1}g) = (f, A^{*-1}g)$).

In particular, for the case of the Schrödinger wave mechanics (which we considered previously, but here a k-dimensional configuration space will be assumed), where the Hilbert space consists of the $\phi(q_1, \ldots, q_k)$ with finite

$$\int_{-\infty}^{\infty} \cdots \int_{-\infty}^{\infty} |\phi(q_1, \ldots, q_k)|^2 dq_1 \ldots dq_k$$

we have for the operators $q_1 \ldots$ and $\frac{h}{2\pi i} \frac{\partial}{\partial q_1}$,

$$(q_l)^* = q_l \ , \quad \left(\frac{h}{2\pi i}\frac{\partial}{\partial q_l}\right)^* = \frac{h}{2\pi i}\frac{\partial}{\partial q_l} \ .$$

The former is clear since

$$\int_{-\infty}^{\infty}\cdots\int_{-\infty}^{\infty} q_l\cdot\phi(q_1,\ldots,q_k)\cdot\overline{\psi(q_1,\ldots,q_k)}\cdot dq_1\ldots dq_k$$

$$= \int_{-\infty}^{\infty}\cdots\int_{-\infty}^{\infty}\phi(q_1,\ldots,q_k)\cdot\overline{q_l\cdot\psi(q_1,\ldots,q_k)}\cdot dq_1\ldots dq_k$$

The latter implies that

$$\int_{-\infty}^{\infty}\cdots\int_{-\infty}^{\infty}\frac{h}{2\pi i}\frac{\partial}{\partial q_l}\phi(q_1,\ldots,q_k)\cdot\overline{\psi(q_1,\ldots,q_k)}\cdot dq_1\ldots dq_k$$

$$= \int_{-\infty}^{\infty}\cdots\int_{-\infty}^{\infty}\phi(q_1,\ldots,q_k)\cdot\overline{\frac{h}{2\pi i}\frac{\partial}{\partial q_l}\psi(q_1,\ldots,q_k)}\cdot dq_1\ldots dq_k$$

i.e.,

$$\int_{-\infty}^{\infty}\cdots\int_{-\infty}^{\infty}\left\{\frac{\partial}{\partial q_l}\phi(q_1,\ldots,q_k)\cdot\overline{\psi(q_1,\ldots,q_k)}\right.$$

$$\left. + \ \phi(q_1,\ldots,q_k)\cdot\frac{\partial}{\partial q_l}\overline{\psi(q_1,\ldots,q_k)}\right\}\cdot dq_1\ldots dq_k = 0 \ ,$$

$$\lim_{\substack{A\to+\infty\\ B\to+\infty}}\int_{-\infty}^{\infty}\cdots\int_{-\infty}^{\infty}\left[\phi(q_1,\ldots,q_k)\overline{\psi(q_1,\ldots,q_k)}\right]_{q_l=-B}^{q_l=+A}\cdot$$

$$dq_1\ldots dq_{l-1}dq_{l+1}\ldots dq_k = 0$$

The limit must exist because the convergence of all integrals

$$\int_{-\infty}^{\infty}\cdots\int_{-\infty}^{\infty}\ldots dq_1\ldots dq_k$$

is certain (since ϕ , ψ , $\frac{\partial}{\partial q_l}\phi$, $\frac{\partial}{\partial q_l}\psi$ belong to the Hilbert space) so that it is only its vanishing which is of importance. If it were $\neq 0$, then the limit (which certainly exists) would be $\neq 0$, for $q_l \longrightarrow +\infty$ or for $q_l = -\infty$:

$$\int_{-\infty}^{\infty} \cdots \int_{-\infty}^{\infty} \phi(q_1,\ldots,q_k)\overline{\psi(q_1,\ldots,q_k)}dq_1\ldots dq_{l-1}dq_{l+1}\ldots dq_k$$

which is incompatible with the absolute convergence of the integral

$$\int_{-\infty}^{\infty} \cdots \int_{-\infty}^{\infty} \phi(q_1,\ldots,q_k)\psi(q_1,\ldots,q_k)dq_1\ldots dq_{l-1}dq_l dq_{l+1}\ldots dq_k$$

(ϕ, ψ belong to the Hilbert space!).

If A is the integral operator

$$A\ \phi(q_1,\ldots,q_k)$$
$$= \int_{-\infty}^{\infty} \cdots \int_{-\infty}^{\infty} K(q_1,\ldots,q_k;\ q_1',\ldots,q_k')\phi(q_1',\ldots,q_k')dq_1'\ldots dq_k'$$

then the following is obtained directly. A^* is also an integral operator, only its kernel is not

$$K(q_1,\ldots,q_k;\ q_1',\ldots,q_k')$$

but

$$\overline{K(q_1',\ldots,q_k';\ q_1,\ldots,q_k)}$$

Let us now consider the situation in matrix theory, where the Hilbert space consists of all sequences $x_1, x_2, \ldots$ with finite

$$\sum_{\mu=1}^{\infty} |x_\mu|^2$$

A linear operator A transforms $\{x_1, x_2, \ldots\}$ into $\{y_1, y_2, \ldots\}$:

$$A\{x_1, x_2, \dots\} = \{y_1, y_2, \dots\}$$

where, because of the linearity of A, the $y_1, y_2, \dots$ must depend linearly on the $x_1, x_2, \dots$

$$y_\mu = \sum_{\nu=1}^{\infty} a_{\mu\nu} x_\nu \qquad {}^{60}$$

Therefore A is characterized by the matrix $a_{\mu\nu}$. We see immediately that the matrix $\overline{a}_{\nu\mu}$ (the complex-conjugate-transposed matrix!) belongs to A^*.[60]

---

[60] This consideration is not rigorous, since it uses linearity in the case of infinite sums, etc. But it can be perfected as follows: Let $\phi_1, \phi_2, \dots$ be a complete orthonormal set, A, A^* adjoint operators. Let

$$f = \sum_{\nu=1}^{\infty} x_\nu \phi_\nu, \quad Af = \sum_{\nu=1}^{\infty} y_\nu \phi_\nu$$

Then

$$y_\mu = (Af, \phi_\mu) = (f, A^*\phi_\mu) = \sum_{\nu=1}^{\infty} (f, \phi_\nu)\overline{(A^*\phi_\mu, \phi_\nu)}$$

[by THEOREM 7., γ)]

$$= \sum_{\nu=1}^{\infty} x_\nu \overline{(\phi_\mu, A\phi_\nu)} = \sum_{\nu=1}^{\infty} (A\phi_\nu, \phi_\mu) x_\nu$$

If we then set $a_{\mu\nu} = (A\phi_\nu, \phi_\mu)$, we have the formula

$$y_\mu = \sum_{\nu=1}^{\infty} a_{\mu\nu} x_\nu$$

of the text, and absolute convergence is assured.

In the Hilbert space of the sequences $x_1, x_2, \dots$,

The analogy with the situation in matrix theory which has just now been developed suggests introducing the concept of the Hermitian operator, in a manner which we shall now expound. Simultaneously, we shall introduce two other concepts, which will be important for our later purposes.

DEFINITION 9. The operator A is said to be Hermitian if $A^* = A$. It is also said to be definite if it is always true that $(Af, f) \geq 0$.[61] The operator U is said to be unitary if $UU^* = U^*U = 1$.[62]

---

the sequences $\phi_1 = 1, 0, 0, \ldots; \phi_2 = 0, 1, 0, \ldots; \ldots$, form a complete orthonormal set (as can easily be seen). For

$$f = \{x_1, x_2, \ldots\}, \quad f = \sum_{\nu=1}^{\infty} x_\nu \phi_\nu$$

for

$$Af = \{y_1, y_2, \ldots\}, \quad Af = \sum_{\nu=1}^{\infty} y_\nu \phi_\nu$$

In this way complete concordance with the text is reached.

If we form $a^*_{\mu\nu}$ for A^*, then we see that

$$a^*_{\mu\nu} = (A^*\phi_\nu, \phi_\mu) = (\phi_\nu, A\phi_\mu) = \overline{(A\phi_\mu, \phi_\nu)} = \bar{a}_{\nu\mu}$$

[61] (Af, f) is real in any case, since it equals

$$(A^*f, f) = (f, Af) = \overline{(Af, f)}$$

[62] Consequently, U, U^* must be defined everywhere. Furthermore, they are inverse to each other. Therefore they each take on every value once and only once.

For unitary operators, we thus have $U^* = U^{-1}$. By definition,

$$(Uf, Ug) = (U^*Uf, g) = (f, g)$$

therefore, in particular (for $f = g$) , $||Uf|| = ||f||$. Conversely, the unitary nature follows from the latter properties, if U is defined everywhere, and takes on every value (cf. Note 62). We prove this as follows: First, if it is true that

$$||Uf|| = ||f|| \text{ , i.e., } (Uf, Uf) = (f, f) \text{ , } (U^*Uf, f) = (f, f)$$

If we replace f by $\frac{f+g}{2}$ and again by $\frac{f-g}{2}$ and subtract, then we obtain, as may easily be calculated, Re (Uf, Ug) = Re (f, g) . If we substitute $i \cdot f$ here for f , then we get Im in place of Re . Consequently it is true in general that

$$(Uf, Ug) = (f, g) \text{ , i.e., } (U^*Uf, g) = (f, g)$$

For fixed f , this holds for all g . Therefore $U^*Uf = f$. Since this holds for all f , then $U^*U = 1$. We must still show that $UU^* = 1$. For each f there is a g with $Ug = f$, then

$$UU^*f = UU^*.Ug = U.U^*Ug = Ug = f \text{ , therefore } UU^* = 1$$

Since, because of linearity,

$$||Uf - Ug|| = ||U(f - g)|| = ||f - g||$$

each unitary operator is continuous, which is not at all necessary for Hermitian operators. For example, the operators $q\ldots$ and $\frac{h}{2\pi i}\frac{d}{dq}\ldots$, so important for quantum mechanics, are discontinuous .[63]

---

[63]For given

$$\int_{-\infty}^{+\infty} |\phi(q)|^2 dq \text{ ,}$$

From our formal rules of calculation for A^*, it follows immediately that if U, V are unitary, U^{-1}, UV are also. Therefore all powers of U are also unitary. If A, B are Hermitian, then $A \pm B$ are also Hermitian. On the other hand, aA is Hermitian only for real a (except $A = 0$), and AB only if A, B commute, i.e., if $AB = BA$. Furthermore, we know that all projections (in particular, $0, 1$) are Hermitian, also the operators $q_1 \dots$ and $\frac{h}{2\pi i}\frac{\partial}{\partial q_1} \dots$ of the Schrödinger theory. All powers of A are also Hermitian (also A^{-1}, if it exists), and all polynomials with real coefficients. It is noteworthy that for Hermitian A and arbitrary X, XAX^* is also Hermitian:

$$(XAX^*)^* = X^{**}A^*X^* = XAX^*$$

Therefore, for example, all XX^* ($A = 1$) and X^*X (X^* in place of X) are Hermitian. For unitary U, UAU^{-1} is Hermitian because $U^{-1} = U^*$.

$$\int_{-\infty}^{+\infty} q^2 |\phi(q)|^2 dq$$

as well as

$$\int_{-\infty}^{+\infty} |\frac{d}{dq}\,\phi(q)|^2 dq$$

can be made arbitrarily large. Take, for example,

$$\phi(q) = ae^{-bq^2}$$

The three integrals are all finite ($b > 0$!), but are proportional respectively to

$$a^2b^{-\frac{1}{2}},\ a^2b^{-\frac{3}{2}},\ a^2b^{\frac{1}{2}}$$

so that the value of any two of them can be prescribed at will.

The continuity of operators, just as in the case of the numerical functions treated in analysis, is a property of basic importance. We therefore want to state several characteristic conditions for its existence in the case of linear operators.

THEOREM 18. A linear operator R is everywhere continuous if it is continuous at the point $f = 0$. A necessary and sufficient condition for this latter property is the existence of a constant C for which, in general, $\|Rf\| \leq C\cdot\|f\|$. In turn, this condition is equivalent to the general validity of

$$|(Rf, g)| \leq C\cdot\|f\|\cdot\|g\|$$

For Hermitian R, this need be required only for $f = g$: $|(Rf, f)| \leq C\cdot\|f\|^2$, or, since (Rf, f) is real (Note 61):

$$-C\cdot\|f\|^2 \leq (Rf, f) \leq C\cdot\|f\|^2$$

NOTE. The concept of continuity for operators originated with Hilbert.[64] He characterized it as "boundedness," and defined it by the next to the last criterion given above. If only one of the $\leq$ in the last criterion is generally valid, then R is said to be half-bounded: above or below. For example, each definite R is half-bounded below (with $C = 0$).

PROOF. Continuity for $f = 0$ implies that for each $\epsilon > 0$ there exists a $\delta > 0$ such that $\|Rf\| < \epsilon$ follows from $\|f\| < \delta$. Then it follows from

---

[64] Gött. Nachr. 1906.

$||f - f_0|| < \delta$ that

$$||R(f - f_0)|| = ||Rf - Rf_0|| < \epsilon$$

i.e., that R is also continuous for $f = f_0$, and therefore everywhere.

If $||Rf|| \leq C.||f||$ (of course, $C > 0$), then we have continuity: we can set $\delta = \frac{\epsilon}{C}$. Conversely, if continuity exists, we can determine the δ for $\epsilon = 1$, and set $C = \frac{2}{\delta}$. Then

$$||Rf|| \leq C.||f||$$

holds for $f \neq 0$. For $f \neq 0$, $||f|| > 0$, but then let

$$g = \frac{\frac{1}{2}\delta}{||f||} \cdot f$$

In this case, $||g|| = \frac{1}{2}\delta$ and therefore

$$||Rg|| = \frac{\frac{1}{2}\delta}{||f||} . ||Rf|| < 1 , \quad ||Rf|| < \frac{||f||}{\frac{1}{2}\delta} = C.||f||$$

From $||Rf|| \leq C.||f||$ it follows that

$$|(Rf, g)| \leq ||Rf||\cdot||g|| \leq C.||f||\cdot||g||$$

Conversely, from $|(Rf, g)| \leq C.||f||\cdot||g||$ we obtain $||Rf||^2 \leq C.||f||\cdot||Rf||$ if we set $g = Rf$, and therefore $||Rf|| \leq C.||f||$. It still remains to show for Hermitian R that $|(Rf, f)| \leq C.||f||^2$ leads to $|(Rf, g)| \leq C.||f||\cdot||g||$. Substitution of $\frac{f + g}{2}$ and $\frac{f - g}{2}$ for f gives

$$|\mathrm{Re}\ (Rf, g)| = |(R\,\frac{f + g}{2}, \frac{f + g}{2}) - (R\,\frac{f - g}{2}, \frac{f - g}{2})|$$

$$\leq C(||\frac{f + g}{2}||^2 + ||\frac{f - g}{2}||^2) = C\,\frac{||f||^2 + ||g||^2}{2} \quad [65]$$

---

[65] The Hermitian character of R is important in the reduction

We now substitute $af, \frac{1}{a} g$ $(a > 0)$ for f, g, as in the proof of THEOREM 1. Minimizing the right side gives $|\mathrm{Re}\ Rf, g|| \leq C.||f||.||g||$. Then, replacing f by $e^{i\alpha f}$ (α real) gives for the maximum of the left side

$$|(Rf, g)| \leq C.||f||.||g||$$

Of course this is valid only if Rg is defined, but since these g are everywhere dense and Rg no longer enters into the final result, it holds generally by reason of continuity.

We prove still another theorem on definite operators:

THEOREM 19. If R is Hermitian and definite, then

$$\overline{|(Rf, g)|} \leq \sqrt{(Rf, f).(Rg, g)}$$

From $(Rf, f) = 0$, it then follows that $Rf = 0$.

PROOF. The above inequality follows from the general validity of $(Rf, f) \geq 0$ (definiteness!), just as the Schwarz inequality

$$|(f, g)| \leq \sqrt{(f, f).(g, g)}$$

(i.e., $\leq ||f|| \cdot ||g||$) was proved in THEOREM 1. from $(f, f) \geq 0$. If now $(Rf, f) = 0$, then it follows from this inequality that $(Rf, g) = 0$ also, if Rg is defined.

$$(R\frac{f+g}{2}, \frac{f+g}{2}) - (R\frac{f-g}{2}, \frac{f-g}{2}) = \frac{(Rf, g) + (Rg, f)}{2}$$

$$= \frac{(Rf, g) + \overline{(f, Rg)}}{2}$$

(in the third step).

Consequently, it holds for a g-set which is everywhere dense, therefore, by reason of continuity, for all g. Then $Rf = 0$.

Finally we shall make reference to the important concept of the commutativity of two operators R, S, i.e., the relation $RS = SR$.

From $RS = SR$ it follows that

$$S\ldots SSR = S\ldots SRS = S\ldots RSS = \ldots = RS\ldots SS$$

i.e., R, S^n commute $(n = 1,2,\ldots)$. Since $R1 = 1R = R$ and $S^0 = 1$, this is also true for $n = 0$. If S^{-1} exists, then $S^{-1}.SR.S^{-1} = S^{-1}RSS^{-1}$, therefore

$$S^{-1}.SR.S^{-1} = S^{-1}S.RS^{-1} = RS^{-1}, \quad S^{-1}.RS.S^{-1} = S^{-1}R.SS^{-1} = S^{-1}R$$

and hence $RS^{-1} = S^{-1}R$ also. Consequently, $n = -1$, and therefore $n = -2,-3,\ldots$, are also admissible. That is, R commutes with all powers of S. Repeated application shows that each power of R commutes with each power of S. If R commutes with S, T, then it obviously commutes with all aS, and also with $S \pm T, ST$. Together with the above results, it follows from this that if R, S commute, all polynomials of R commute with all polynomials of S. In particular, for $R = S$, all polynomials of R commute with each other.

6. THE EIGENVALUE PROBLEM

We have now come far enough that we can consider, in the abstract Hilbert space that problem which was of central importance in quantum mechanics, in its relationship to the special cases F_Z and F_Ω : the solution of the equations $\mathbf{E}_1$. and $\mathbf{E}_2$. respectively in I.3. We call this the eigenvalue problem, and we must formulate it anew and in a unified fashion.

In I.3., $\mathbf{E}_1$. and $\mathbf{E}_2$. both required the finding of all solutions $\phi \neq 0$ of

$$\textbf{E.} \qquad H\phi = \lambda\phi$$

where H is the Hermitian operator corresponding to the Hamiltonian function (cf. the discussion in I.3.), ϕ an element of the Hilbert space, λ a real number (H given, ϕ, λ to be determined). In connection with this, however, certain requirements were made regarding the number of solutions to be found. It was required to find such a number that

1. in matrix theory, a matrix $S = \{s_{\mu\nu}\}$ could be formed from these solutions

$$\phi_1 = \{s_{11}, s_{21}, \ldots\}, \phi_2 = \{s_{12}, s_{22}, \ldots\}, \ldots$$

(we are in F_Z!) which possesses inverse S^{-1} (cf. I.3.);

2. in the wave theory, each wave function (which need not be a solution) can be developed in a series of the solutions

$$\phi_1 = \phi_1(q_1, \ldots, q_f), \phi_2 = \phi_2(q_1, \ldots, q_f), \ldots$$

($\phi_1, \phi_2, \ldots$ may belong to different λ) :

$$\phi(q_1, \ldots, q_f) = \sum_{n=1}^{\infty} c_n \phi_n(q_1, \ldots, q_f)$$

(There was no mention of this latter circumstance in I.3., but this requirement is indispensable for the further development of the wave theory, in particular for the Schrödinger "perturbation theory."[66])

Now 1. amounts to the same thing as 2., because the matrix S transforms $\{1, 0, 0, \ldots\}, \{0, 1, 0, \ldots\}, \ldots$ respectively into

---

[66] Cf. the fifth paper in the book mentioned in Note 9 [Ann. Phys [4] 80 (1926)].

$$\{s_{11}, s_{21}, s_{31}, \ldots\}, \{s_{12}, s_{22}, s_{32}, \ldots\}, \ldots$$

and therefore the entire Hilbert space $\mathfrak{R}_\infty$ into the closed linear manifold spanned by $\phi_1, \phi_2, \ldots$. In order that S^{-1} exist therefore, the latter must also be equal to $\mathfrak{R}_\infty$. But 2. states the same thing directly: it also requires that each ϕ can be approximated to an arbitrary degree of accuracy by a linear combination of the $\phi_1, \phi_2, \ldots$.[67] Let us make the significance of this condition clear, and also prove once again the properties of the equation **E**. with the formal apparatus now at our command.

First, since $\phi \neq 0$ is required, and since $a\phi$ is a solution if ϕ is, it is sufficient to consider solutions with $\|\phi\| = 1$. Second, we do not need to require the λ to be real, since this follows from $H\phi = \lambda\phi$:

$$(H\phi, \phi) = (\lambda\phi, \phi) = \lambda(\phi, \phi) = \lambda$$

(cf. II.5., Note 61). Third, the solutions ϕ_1, ϕ_2 , which belong to different λ_1, λ_2 are mutually orthogonal:

$$(H\phi_1, \phi_2) = \lambda_1(\phi_1, \phi_2) , \quad (H\phi_1, \phi_2) = (\phi_1, H\phi_2) = \lambda_2(\phi_1, \phi_2)$$

Therefore, $(\phi_1, \phi_2) = 0$ because $\lambda_1(\phi_1, \phi_2) = \lambda_2(\phi_1, \phi_2)$ and $\lambda_1 \neq \lambda_2$.

Now let $\lambda_1, \lambda_2, \ldots$ be the λ , all different from one another, for which **E**. is solvable. (If we choose a solution ϕ_λ of absolute value 1 for each λ with solvable $H\phi = \lambda\phi$, then the ϕ_λ form an orthonormal set, by reason of previous comments. By II.2., THEOREM $3^{(\infty)}$., this set is then a finite or infinite sequence. Therefore,

---

[67] We purposely do not go into the finer questions of convergence; these questions were not treated with exactness in the original forms of the matrix and wave theories; also, we shall settle them later (cf. e.g., II.9.).

we can also write the λ as a sequence, which may or may not terminate). For each $\lambda = \lambda_\rho$ all solutions of $H\phi = \lambda\phi$ form a linear manifold, and indeed a closed one.[68] According to THEOREM 9., therefore, there exists an orthonormal set $\phi_{\rho,1}, \ldots, \phi_{\rho,\nu_\rho}$ of such solutions which spans this closed linear manifold. The number ν_ρ is clearly the maximum number of linearly independent solutions for $\lambda = \lambda_\rho$. This is known as the multiplicity of the eigenvalue λ_ρ. ($\nu = 1, 2, \ldots, \infty$; $\nu = \infty$ can occur. For example, for $H = 1$, $\lambda = 1$.) According to the preceding discussion, the $\phi_{\rho,1}, \ldots, \phi_{\rho,\nu_\rho}$ of two different ρ are also mutually orthogonal. Therefore the totality

$$\phi_{\rho,\nu} \quad (\rho = 1, 2, \ldots \; ; \; \nu = 1, \ldots, \nu_\rho)$$

also forms an orthonormal set. By reason of its origin, we recognize that it spans the same closed linear manifold as all solutions ϕ of **E**.

We number the $\phi_{\rho,\nu}$ in any order by $\psi_1, \psi_2, \ldots$ and let the corresponding λ_ρ be $\lambda^{(1)}, \lambda^{(2)}, \ldots$. The condition previously formulated that all solutions of **E**. could span $\mathfrak{R}_\infty$ as the closed linear manifold then implies that $\psi_1, \psi_2, \ldots$ (a subset of solutions!) must do this by

---

[68]The latter is evident without further discussion only for continuous H everywhere defined, i.e., if $Hf_n \longrightarrow Hf$ follows from $f_n \longrightarrow f$. Moreover, the following, more limited property is also a consequence, as can easily be seen: from $f_n \longrightarrow f$, $Hf_n \longrightarrow f^*$ it follows that $Hf = f^*$ [this is the so-called closure of H , cf. the work of the author in Math. Ann. 102 (1929)]. This is always satisfied with the operators of quantum mechanics, even the discontinuous ones; a Hermitian operator which is not closed can be made (Hermitian and) closed by a unique extension of its domain (which is not the case for the property of continuity, for example), cf. II.9, p. 148 Note.

itself -- therefore, by THEOREM 7., that this orthonormal set is complete.

The solution of the eigenvalue problem in the sense of quantum mechanics would therefore require the finding of a sufficient number of solutions

$$\phi = \psi_1, \psi_2, \ldots \quad \text{and} \quad \lambda = \lambda_1, \lambda_2, \ldots$$

of **E**., so that a complete orthonormal set can be formed from them. But this is not possible in general. For example, in wave theory, we see that for a certain subset of solutions of **E**. (i.e., $\mathbf{E}_2$. in I.3.) -- all of which are needed to develop each wave function by solutions (cf. above) -- there exists no finite value for the integral of the square of the absolute value.[69] Therefore it does not belong to Hilbert space. Hence there is no complete orthonormal set of solutions in Hilbert space (and we consider Hilbert space only in **E**.!).

On the other hand, the Hilbert theory of the eigenvalue problem shows that this phenomenon does not at all represent an exception in the behavior of operators (not even of the continuous ones).[70] We must therefore analyze the situations which result when it does occur. (We shall soon see what this means physically. Cf. III.3.) If it occurs, i.e., if the orthonormal set selected from the solutions of **E**. is not complete, then we say that a "continuous spectrum of H" exists. ($\lambda_1, \lambda_2, \ldots$ form the "point" or "discrete" spectrum of H.)

Our next problem, since **E**. has failed, is then to find a formulation for the eigenvalue problem of the Hermitian operator and to apply this to quantum mechanics.

---

[69] Cf. for example, Schrödinger's treatment of the hydrogen atom, the reference cited in Note 16.

[70] Cf. the reference cited in Note 64.

First, this formulation of the eigenvalue problem, following a pattern set by Hilbert (cf. above, Note 70) should be given and explained.

7. CONTINUATION

The equation

$$H \phi = \lambda \phi$$

as well as the requirement that a complete orthonormal set can be formed from its solutions, originates in an analogy with the case of finite dimensions, the $\mathfrak{R}_n$.

In the $\mathfrak{R}_n$, H is a matrix

$$\{h_{\mu\nu}\}, \ \mu, \ \nu = 1, \ldots, n, \ h_{\mu\nu} = \bar{h}_{\nu\mu}$$

and it is a well known algebraic fact that the solutions of $H \phi = \lambda\phi$, $\phi = \{x_1, x_2, \ldots, x_n\}$, i.e.,

$$\sum_{\nu=1}^{n} h_{\mu\nu} x_\nu = \lambda x_\mu \qquad (\mu = 1, \ldots, n)$$

form a complete orthonormal set.[71]

This property of the $\mathfrak{R}_n$ cannot, as we have seen, carry over by $n \longrightarrow \infty$ to the $\mathfrak{R}_\infty$. Hence the eigenvalue problem in $\mathfrak{R}_\infty$ must be formulated differently. We shall now see that the eigenvalue problem in $\mathfrak{R}_n$ can be transformed in such a way that for this new formulation (which in $\mathfrak{R}_n$ is equivalent to the old), a transition to $\mathfrak{R}_\infty$ is possible. That is, both express the same thing in each $\mathfrak{R}_n$ $(n = 1, 2, \ldots)$ (namely, the possibility of the diagonalization of the Hermitian matrices), but the one can also be carried over into $\mathfrak{R}_\infty$, while the other cannot.

---

[71] Cf. Courant-Hilbert: the reference cited in Note 30.

Let $\{x_{11},\ldots,x_{1n}\},\ldots,\{x_{n1},\ldots,x_{nn}\}$ be the complete orthonormal set of solutions of the eigenvalue equation, and $\lambda_1,\ldots,\lambda_n$ the corresponding λ. The vectors $\{x_{11},\ldots,x_{1n}\},\ldots,\{x_{n1},\ldots,x_{nn}\}$ then form a cartesian coordinate system in $\mathfrak{R}_n$. The transformation formulas of the coordinates $\mathfrak{x}_1,\ldots,\mathfrak{x}_n$ in this coordinate system to such $\xi_1,\ldots,\xi_n$ then run as follows:

$$\{\xi_1,\ldots,\xi_n\} = \mathfrak{x}_1\{x_{11},\ldots,x_{1n}\} + \cdots + \mathfrak{x}_n\{x_{n1},\ldots,x_{nn}\}$$

i.e.,

$$\xi_1 = \sum_{\mu=1}^{n} x_{\mu 1}\mathfrak{x}_\mu, \ldots, \xi_n = \sum_{\mu=1}^{n} x_{\mu n}\mathfrak{x}_\mu$$

and conversely,

$$\mathfrak{x}_1 = \sum_{\mu=1}^{n} \overline{x}_{1\mu}\xi_\mu, \ldots, \mathfrak{x}_n = \sum_{\mu=1}^{n} \overline{x}_{n\mu}\xi_\mu$$

We can write the conditions

$$\sum_{\nu=1}^{n} h_{\mu\nu}x_{\rho\nu} = \lambda_\rho x_{\rho\mu}$$

as follows, with the help of the variables $\mathfrak{x}_1,\ldots,\mathfrak{x}_n$ and a new set of variables $\eta_1,\ldots,\eta_n$ (in addition to the corresponding $\mathfrak{y}_1,\ldots,\mathfrak{y}_n$, by reason of the above formula):

$$\sum_{\rho,\mu=1}^{n}\left(\sum_{\nu=1}^{n} h_{\mu\nu}x_{\rho\nu}\right)\mathfrak{x}_\rho\overline{\eta}_\mu = \sum_{\rho,\mu=1}^{n}\lambda_\rho x_{\rho\mu}\mathfrak{x}_\rho\overline{\eta}_\mu$$

i.e.,

$$(\mathbf{D.})\quad \sum_{\mu,\nu=1}^{n} h_{\mu\nu}\xi_\nu\overline{\eta}_\mu = \sum_{\rho=1}^{n}\lambda_\rho\left(\sum_{\mu=1}^{n}\overline{x}_{\rho\mu}\xi_\mu\right)\overline{\left(\sum_{\mu=1}^{n}\overline{x}_{\rho\mu}\eta_\mu\right)}$$

The cartesian character of our coordinate system may then be expressed as

$$(\mathbf{O.})\quad \sum_{\mu=1}^{n} \xi_\mu \overline{\eta}_\mu = \sum_{\rho=1}^{n} \left(\sum_{\mu=1}^{n} \overline{x}_{\rho\mu} \xi_\mu \right) \overline{\left(\sum_{\mu=1}^{n} \overline{x}_{\rho\mu} \eta_\mu \right)}$$

The finding of a matrix $\{x_{\mu\nu}\}$ with the properties **D.**, **O.** is then equivalent in the $\mathfrak{R}_n$ to the solution of the eigenvalue problem; and in this form the transition to $\mathfrak{R}_\infty$ fails. But this failure is not surprising, for the following reason. The conditions **D.**, **O.** do not determine the unknowns λ_ρ, $x_{\mu\nu}$ completely. Indeed, as the theory of this diagonal transformation shows (see the reference in Note 71), the λ_ρ are determined uniquely except for the order, but the situation is much worse in the case of the $x_{\mu\nu}$. Each row $x_{\rho 1}, \ldots, x_{\rho n}$ can evidently be multiplied by a factor θ_ρ of absolute value 1 -- and if several λ_ρ coincide, even an arbitrary unitary transformation of the corresponding columns $x_{\rho 1}, \ldots, x_{\rho n}$ is possible! To attempt the difficult transition to the limit $n \longrightarrow \infty$ with such quantities, which are not uniquely determined, is hopeless: for how can the process converge if the λ_ρ, $x_{\mu\nu}$ can undergo arbitrarily large fluctuations, which is possible because of the incompleteness of their determination.

But this points out the way in which the problem can be handled correctly: we must first seek to replace the conditions **D.**, **O.**, and the unknowns λ_ρ, $x_{\mu\nu}$ by such which possess the lacking uniqueness property -- it will then be shown that the limiting process causes less difficulty.

If l is any value which one or more λ_ρ assume, then

$$\sum_{\lambda_\rho = l} \left(\sum_{\mu=1}^{n} \overline{x}_{\rho\mu} \xi_\mu \right) \overline{\left(\sum_{\mu=1}^{n} \overline{x}_{\rho\mu} \eta_\mu \right)}$$

is invariant under the variations (compatible with **D.**; **O.**) of the λ_ρ, $x_{\mu\nu}$ mentioned above. If l is different from all λ_ρ, then the sum $= 0$, and is therefore certainly invariant. Consequently the Hermitian form (ξ and η signify $\xi_1, \ldots, \xi_n$ and $\eta_1, \ldots, \eta_n$ respectively)

$$E(l;\ \xi,\ \eta) = \sum_{\lambda_\rho \leq l} \left(\sum_{\mu=1}^{n} \overline{x}_{\rho\mu} \xi_\mu \right) \overline{\left(\sum_{\mu=1}^{n} \overline{x}_{\rho\mu} \eta_\mu \right)}$$

is also invariant (l arbitrary!) If we know the $E(l;\ \xi,\ \eta)$ (i.e., their coefficients), then it is easy to reason back from this point to the λ_ρ, $x_{\mu\nu}$. Then, if we so formulate the eigenvalue problem (i.e., **D.**, **O.**) that only the $E(l;\ \xi,\ \eta)$ appear instead of the λ_ρ, $x_{\mu\nu}$, we have achieved the desired unique formulation.

Therefore, let $E(l)$ be the matrix of the Hermitian form $E(l;\ \xi,\ \eta)$.[72] What do **D.**, **O.** now mean for the family of matrices $E(l)$?

O. means: if l is sufficiently large (namely, larger than all λ_ρ), then $E(l) = 1$ (the unit matrix). From the nature of $E(l)$, it follows that if l is sufficiently small (that is, smaller than all λ_ρ), then $E(l) = 0$, and if l increases from $-\infty$ to $+\infty$, then $E(l)$ is always constant, except for a finite number of points (the different values among the $\lambda_1, \ldots, \lambda_n$, which we call $l_1 < l_2 < \ldots l_m$, $m \leq n$), at which it changes discontinuously. Furthermore, the discontinuity lies to

---

[72] That is,

$$E(l) = (e_{\mu\nu}(l)),\quad E(l;\ \xi,\ \eta) = \sum_{\mu,\nu=1}^{n} e_{\mu\nu}(l) \xi_\nu \overline{\eta}_\mu$$

Consequently,

$$e_{\mu\nu}(l) = \sum_{\lambda_\rho \leq l} x_{\rho\mu} \overline{x}_{\rho\nu}$$

the left of the point in question (because the

$$\sum_{\lambda_\rho \leq l}$$

is continuous to the right as a function of l, while for the case of

$$\sum_{\lambda_\rho < l}$$

it would be just the opposite). Finally, as we shall show, for $l' \leq l''$

$$E(l')E(l'') = E(l'')E(l') = E(l')$$

(matrix product!).

It is more convenient to prove this for $E(l'; \xi, \eta)$, $E(l''; \xi, \eta)$ in the coordinate system of the $\mathfrak{x}_1, \ldots, \mathfrak{x}_n$ and $\mathfrak{y}_1, \ldots, \mathfrak{y}_n$. After introduction of these variables, we obtain from $E(l'; \xi, \eta)$ and $E(l''; \xi, \eta)$

$$\sum_{\lambda_\rho \leq l'} \mathfrak{x}_\rho \overline{\mathfrak{y}}_\rho \quad \text{and} \quad \sum_{\lambda_\rho \leq l''} \mathfrak{x}_\rho \overline{\mathfrak{y}}_\rho$$

The matrices are therefore as follows: 0 except on the diagonal; 1 on the diagonal in the ρ-th field if $\lambda_\rho \leq l'$ or $\leq l''$, otherwise 0 also. For these matrices, the above proposition is evident.

Now let us formulate **D.** again. It clearly means:

$$\sum_{\mu,\nu=1}^{n} h_{\mu\nu} \xi_\nu \overline{\eta}_\mu = \sum_{\tau=1}^{m} l_\tau \{E(l_\tau; \xi, \eta) - E(l_{\tau-1}; \xi, \eta)\}$$

(l_0 is any number $< l_1$). But since $E(l; \xi, \eta)$ is constant in each of the intervals

$$-\infty < l < l_1;\ l_1 \leq l < l_2;\ \ldots;\ l_{m-1} \leq l < l_m;\ l_m \leq l < +\infty$$

then for each set of numbers

$$\Lambda_0 < \Lambda_1 < \Lambda_2 < \dots < \Lambda_k$$

if $l_1, \dots, l_m$ appear among the $\Lambda_1, \dots, \Lambda_k$,

$$\sum_{\mu,\nu=1}^{n} h_{\mu\nu}\xi_\nu\bar{\eta}_\mu = \sum_{\tau=1}^{k} \Lambda_\tau \{E(\Lambda_\tau;\ \xi,\ \eta) - E(\Lambda_{\tau-1};\ \xi,\ \eta)\}$$

By application of the Stieltjes concept of an integral,[73]

---

[73]For the concept of the Stieltjes integral, cf. Perron, Die Lehre von den Kettenbrüchen, Leipzig, 1913, also, for particular consideration of the requirements of the operator theory, Carleman: Équations intégrales singulières, Upsala, 1923. For the reader who is less interested in such things, the following definition will suffice: for a subdivision $\Lambda_0, \Lambda_1, \dots, \Lambda_k$ of the interval a, b

$$a \leq \Lambda_0 < \Lambda_1 < \dots < \Lambda_k \leq b$$

we form the sum

$$\sum_{\tau=1}^{k} f(\Lambda_\tau)\{g(\Lambda_\tau) - g(\Lambda_{\tau-1})\}$$

If this always converges as the subdivisions

$$\Lambda_0, \Lambda_1, \Lambda_2, \dots, \Lambda_k$$

are made smaller and smaller, then the integral

$$\int_a^b f(x)dg(x)$$

exists and is defined to be equal to this limit. (For $g(x) = x$, this goes over into the well-known Riemann integral.)

this can also be written as

$$\sum_{\mu,\nu=1}^{n} h_{\mu\nu}\xi_\nu\overline{\eta}_\mu = \int_{-\infty}^{+\infty} \lambda \; dE(\lambda;\ \xi,\ \eta)$$

($\int_{-\infty}^{+\infty}$ can obviously be replaced by each $\int_a^b$, $a < l_1$, $b > l_m$.) Or, if we consider the coefficients, and write down for the matrices themselves the equation which is valid for all the coefficients:

$$H = \int_{-\infty}^{+\infty} \lambda \; dE(\lambda)$$

in which $H = \{h_{\mu\nu}\}$.

Thus far then, the problem is the following: For a given Hermitian matrix $H = \{h_{\mu\nu}\}$ a family of Hermitian matrices $E(\lambda)$ $(-\infty < \lambda < +\infty)$ with the following properties is sought:

S_1. For sufficiently $\left\{{\text{small} \atop \text{large}}\right\}$ λ, $E(\lambda) = \left\{{0 \atop 1}\right\}$. $E(\lambda)$ is (considered as a function of λ) everywhere constant, with the exception of a finite number of points at which it changes discontinuously. Also, the discontinuity always occurs to the left of the given point.

S_2. It is always true that

In our case therefore, the equation which has been derived means that

$$\int_{-\infty}^{\infty} x \; dE(x;\ \xi,\ \eta)$$

exists (we denoted the variable by λ instead of x) and is equal to

$$\sum_{\mu,\nu=1}^{n} h_{\mu\nu}\xi_\nu\overline{\eta}_\mu$$

$$E(\lambda')E(\lambda'') = E(\mathrm{Min}\ (\lambda', \lambda''))\ ^{74}$$

S₃. We have (using the Stieltjes integral)

$$H = \int_{-\infty}^{+\infty} \lambda\, dE(\lambda)$$

At present we shall not stop to carry out the converse process, i.e., going from **S₁.** - **S₃.** back to the solutions **D.**, **O.** (although this would be simple), because only the present form of the eigenvalue problem will be needed in quantum mechanics. Instead, we shall proceed at once to generalize **S₁.** - **S₃.** from a finite number to an infinite number of variables, i.e., from $\mathfrak{R}_n$ to $\mathfrak{R}_\infty$.

In $\mathfrak{R}_\infty$ we shall have to understand by H and the $E(\lambda)$ clearly Hermitian operators, -- that is, we shall seek to determine for a given H a family $E(\lambda)$ which is related to it in a certain fashion, modeled on **S₁.** - **S₃.** Therefore, it suffices to find the $\mathfrak{R}_\infty$ analog of **S₁.** - **S₃.**

The property **S₂.** remains invariant in the transition, since the number of dimensions of $\mathfrak{R}_n$ plays no role in it. But we want to transform it in some way by use of our results on projections (II.4.). First, the property implies that $E(\lambda)^2 = E(\lambda)$ for $\lambda' = \lambda'' = \lambda$, i.e., that the $E(\lambda)$ must be projections. But then **S₂.** means this (we can limit ourselves to $\lambda' \leq \lambda''$, since corresponding results are obtained for $\lambda' \geq \lambda''$): $\lambda' \leq \lambda''$ implies $E(\lambda') \leq E(\lambda'')$ (cf. THEOREM 14., and the subsequent text in II.4.). Some caution is required in the case of **S₃.** The expression

$$A = \int_{-\infty}^{+\infty} \lambda\, dE(\lambda)$$

---

[74] Min (a, b, ..., e) is the smallest, Max (a, b, ..., e) is the largest, of the finite set of real numbers a, b, ..., e .

is meaningless, since the Stieltjes integral is defined for numbers and not for operators. But it is easy to replace H , $E(\lambda)$ by numbers which again give the operator relation desired. We require

$$(Hf, g) = \int_{-\infty}^{+\infty} \lambda \, d(E(\lambda)f, g)$$

for all f, g of $\mathfrak{R}_\infty$ -- so far as Hf is defined. The relation

$$H = \int_{-\infty}^{+\infty} \lambda \, dE(\lambda)$$

is to be understood symbolically, as an abbreviation for this.

Lastly, the property $\mathbf{S}_1$. is essentially affected by the transition to an infinite number of dimensions. The points beyond which $E(\lambda)$ assumes its terminal values 0 or 1 , or where $E(\lambda)$ executes its discontinuous jumps, are (in $\mathfrak{R}_n$) the eigenvalues of H and intervals of constancy are those free from eigenvalues. If now $n \longrightarrow \infty$, many things can happen. The smallest or largest value can approach $-\infty$ or $+\infty$ respectively, but the others may cluster increasingly densely, since their numbers can increase arbitrarily, and thus the intervals of constancy may gradually contract into points. (This last circumstance is that symptom which, in the Hilbert theory, under certain circumstances indicates the appearance of the so-called continuous spectrum.[75]) We must therefore change $\mathbf{S}_1$. considerably in the transition from the $\mathfrak{R}_n$ to $\mathfrak{R}_\infty$. Allowance must be made for the possibility that the variation of $E(\lambda)$ may no longer show a discrete, discontinuous character.

With this point of view, it is very plausible to

---

[75]See the reference of Note 64, as well as the book of Carleman mentioned in Note 73. We shall have a great deal to do with this "continuous spectrum," cf. II.8.

disregard the requirement of the final assumption of the values 0, 1 by $E(\lambda)$, and then to require only convergence to 0 or 1 (as $\lambda \longrightarrow -\infty$ or $\lambda \longrightarrow +\infty$ respectively). Then, in place of intervals of constancy and points of discontinuity, there emerges the additional possibility of a continuous increase. On the other hand, we may try to maintain the less stringent requirement that at the possible points of discontinuity, the discontinuity should appear only from the left. Consequently we formulate $\mathbf{S}_1$. as follows: for $\lambda \longrightarrow -\infty$, $E(\lambda) \longrightarrow 0$, for $\lambda \longrightarrow +\infty$, $E(\lambda) \longrightarrow 1$, and for $\lambda \longrightarrow \lambda_0$, $\lambda \geq \lambda_0$, $E(\lambda) \longrightarrow E(\lambda_0)$.[76]

Something must still be said regarding $\mathbf{S}_3$. In a space with a finite number of dimensions,

$$A = \sum_{\tau=1}^{m} l_\tau F_\tau$$

if by F_τ we understand the matrix $E(l_\tau) - E(l_{\tau-1})$. Because of $\mathbf{S}_1$., we have for $\sigma \geq \tau$

$$F_\tau E(l_\sigma) = E(l_\tau)E(l_\sigma) - E(l_{\tau-1})E(l_\sigma)$$

$$= E(l_\tau) - E(l_{\tau-1}) = F_\tau$$

while for $\sigma \leq \tau-1$

$$F_\tau E(l_\sigma) = E(l_\tau)E(l_\sigma) - E(l_{\tau-1})E(l_\sigma)$$

$$= E(l_\sigma) - E(l_\sigma) = 0$$

Therefore, because $F_\sigma = E(l_\sigma) - E(l_{\sigma-1})$,

---

[76] By $A(\lambda) \longrightarrow B$, ($A(\lambda)$, B operators in the $\mathfrak{R}_\infty$, λ a parameter), we mean that for all f of $\mathfrak{R}_\infty$, $A(\lambda)f \longrightarrow Bf$ This is then an abbreviation for convergence statements in Hilbert space.

$$F_\tau \cdot F_\sigma = \begin{cases} F_\tau & \text{for } \tau = \sigma \\ 0 & \text{for } \tau \neq \sigma \end{cases}$$

From this,

$$H^2 = \left(\sum_{\tau=1}^{m} 1_\tau F_\tau \right)^2 = \sum_{\tau,\sigma=1}^{m} 1_\tau 1_\sigma F_\tau F_\sigma = \sum_{\tau=1}^{m} 1_\tau^2 F_\tau$$

(In the same way, $H^p = \sum_{\tau=1}^{m} 1_\tau^p F_\tau$). Consequently, this transformation results in

$$H^2 = \int_{-\infty}^{+\infty} \lambda^2 dE(\lambda)$$

which is analogous to the equation that holds for H itself. In $\mathfrak{R}_\infty$ therefore, we assume the symbolic equation which is constructed in analogous fashion. Hence numerically,

$$(H^2 f, g) = \int_{-\infty}^{+\infty} \lambda^2 d(E(\lambda)f, g)$$

(This will actually be confirmed by our subsequent considerations.) For $f = g$ this gives

$$(H^2 f, f) = (Hf, Hf) = ||Hf||^2, \quad (E(\lambda)f, f) = ||E(\lambda)f||^2$$

hence

$$||Hf||^2 = \int_{-\infty}^{+\infty} \lambda^2 d(||E(\lambda)f||^2)$$

This formula, however, causes us to expect that the $E(\lambda)$ not only determine the value Hf , whenever it is defined, but also, whether it is defined for a particular f . For the integral

$$\int_{-\infty}^{+\infty} \lambda^2 d(||E(\lambda)f||^2)$$

has a non-negative integrand $(\lambda^2 \geq 0)$, and a monotonically increasing expression under the differential sign $(\|E(\lambda)f\|^2$, cf. S_2. and THEOREM 15. in II.4.). Therefore it is by its nature convergent, i.e., zero or positive and finite, or is properly divergent, i.e., $+\infty$.[77] This is true independently of the relation to H, i.e., without consideration as to whether Hf is defined or not. It is then to be expected that Hf is defined (i.e., exists in $\mathfrak{R}_\infty$) if and only if the presumed value of $\|Hf\|^2$ is finite, that is, the expression

$$\int_{-\infty}^{+\infty} \lambda^2 d(\|E(\lambda)f\|^2)$$

is defined for all f.

Therefore our new formulation of S_1. - S_3. runs as follows: For the given Hermitian operator H, we seek a family of projections $E(\lambda)$ $(-\infty < \lambda < +\infty)$ with the following properties:

$\overline{S_1}$. For $\lambda \longrightarrow -\infty$ or $\lambda \longrightarrow \infty$, $E(\lambda)f \longrightarrow 0$ or $\longrightarrow f$ respectively. For $\lambda \longrightarrow \lambda_0$, $\lambda \geq \lambda_0$, $E(\lambda)f \longrightarrow E(\lambda_0)f$ (for each f!).

$\overline{S_2}$. From $\lambda' \leq \lambda''$ it follows that $E(\lambda') \leq E(\lambda'')$.

$\overline{S_3}$. The integral

$$\int_{-\infty}^{\infty} \lambda^2 d(\|E(\lambda)f\|^2$$

which by its nature is convergent (zero or positive and finite) or properly divergent $(+\infty)$, determines the domain of H: Hf is defined if and only if the former is the case. In this case, then

[77]This follows from the definition of the Stieltjes integral given in Note 73. For proof, see the reference given there.

$$(Hf, g) = \int_{-\infty}^{+\infty} \lambda (E(\lambda)f, g)$$

(The latter integral is absolutely convergent whenever the former is finite.[78])

The operator H does not enter into the properties $\overline{S_1}$., $\overline{S_2}$. A family of projections $E(\lambda)$ with these two properties is known as a resolution of the identity. A resolution of the identity which has the relation $\overline{S_3}$. to H is said to belong to H.

The eigenvalue problem of the $\mathfrak{R}_\infty$ then runs as follows. Do there always exist, for a given Hermitian operator, H, resolutions of the identity belonging to H, and if so, how many? (The desired answer would be: there always exists exactly one.) Furthermore, we must prove how our definition of the eigenvalue problem is related to the general methods used in quantum mechanics (especially in the wave theory) for the determination of the eigenvalues of the Hermitian operators.

8. INITIAL CONSIDERATIONS CONCERNING THE EIGENVALUE PROBLEM

The first question which arises from our definition of the eigenvalue problem is this: $\overline{S_1}$. - $\overline{S_3}$. sound entirely different from the problem with which we started at the beginning of the last section, and their relation is no longer recognizable. It is true that we gave a derivation of S_1. - S_3. in the $\mathfrak{R}_n$ from those conditions, but in $\mathfrak{R}_\infty$ the relations are essentially modified. The two formulations are here no longer equivalent (they were, as was mentioned at the time, in $\mathfrak{R}_n$). Therefore, the entire question is reopened, and we must ascertain how far the new formulation coincides with the old, i.e., when and

[78]Cf. Math. Ann. 102 (1929).

how our $E(\lambda)$ determine the $\lambda_1, \lambda_2, \ldots$ and the $\phi_1, \phi_2, \ldots$.

If the resolution of the identity $E(\lambda)$ belongs to the Hermitian operator A , when is the equation

$$A\phi = \lambda_0\phi$$

solvable? $A\phi = \lambda_0\phi$ means the same as $(A\phi, g) - \lambda_0(\phi, g) = 0$ for all g , i.e.,

$$0 = \int_{-\infty}^{\infty}\lambda d(E(\lambda)f, g) - \lambda_0(\phi, g) = \int_{-\infty}^{\infty}\lambda d(E(\lambda)f, g)$$

$$- \lambda_0 \int_{-\infty}^{\infty}d(E(\lambda)f, g) = \int_{-\infty}^{\infty}(\lambda - \lambda_0)d(E(\lambda)f, g)$$

We first set $g = E(\lambda_0)f$, then

$$0 = \int_{-\infty}^{\infty}(\lambda - \lambda_0)d(E(\lambda)f, E(\lambda_0)f)$$

$$= \int_{-\infty}^{\infty}(\lambda - \lambda_0)(E(\lambda_0)E(\lambda)f, f)$$

$$= \int_{-\infty}^{\infty}(\lambda - \lambda_0)d(E(\mathrm{Min}\ (\lambda, \lambda_0))f, f)$$

$$= \int_{-\infty}^{\infty}(\lambda - \lambda_0)d(||E(\mathrm{Min}\ (\lambda, \lambda_0))f||^2)$$

We can now break up $\int_{-\infty}^{\infty}$ into $\int_{-\infty}^{\lambda_0} + \int_{\lambda_0}^{\infty}$. In $\int_{-\infty}^{\lambda_0}$ we can replace $\mathrm{Min}\ (\lambda, \lambda_0)$ by λ , in the $\int_{\lambda_0}^{\infty}$ by λ_0 . In the latter integral then, a constant appears behind the differential sign. Therefore the integral vanishes.

For the first integral, there remains

$$\int_{-\infty}^{\lambda_0}(\lambda - \lambda_0)d(||E(\lambda)f||^2) = 0$$

Second, we set $g = f$, then

$$0 = \int_{-\infty}^{\infty}(\lambda - \lambda_0)d(E(\lambda)f, f) = \int_{-\infty}^{\infty}(\lambda - \lambda_0)d(||E(\lambda)f||^2)$$

By subtraction of the first equation from this, we get (if we reverse the sign of the integrand),

$$\int_{\lambda_0}^{\infty}(\lambda_0 - \lambda)d(||E(\lambda)f||^2) = 0$$

Let us now examine

$$\int_{-\infty}^{\lambda_0}(\lambda_0 - \lambda)d(||E(\lambda)f||^2), \quad \int_{\lambda_0}^{\infty}(\lambda - \lambda_0)d(||E(\lambda)f||^2)$$

somewhat more closely. The integrand in each case is ≥ 0, and behind the differential sign there is a monotonically increasing function of λ. Therefore, we have for each $\epsilon > 0$,

$$\int_{-\infty}^{\lambda_0}(\lambda_0 - \lambda)d(||E(\lambda)f||^2) \geq \int_{-\infty}^{\lambda_0-\epsilon}(\lambda_0 - \lambda)d(||E(\lambda)f||^2)$$

$$\geq \int_{-\infty}^{\lambda_0-\epsilon}\epsilon d(||E(\lambda)f||^2) = \epsilon||E(\lambda_0 - \epsilon)f||^2,$$

$$\int_{\lambda_0}^{\infty}(\lambda - \lambda_0)d(||E(\lambda)f||^2) \geq \int_{\lambda_0+\epsilon}^{\infty}(\lambda - \lambda_0)d(||E(\lambda)f||^2)$$

$$\geq \int_{\lambda_0+\epsilon}^{\infty}\epsilon d(||E(\lambda)f||^2) = \epsilon(||f||^2 - ||E(\lambda_0 + \epsilon)f||^2)$$

$$= \epsilon||f - E(\lambda_0 + \epsilon)f||^2$$

The right sides are then ≤ 0, but since they are also ≥ 0, they must vanish. Therefore,

$$E(\lambda_0 - \epsilon)f = 0, \quad E(\lambda_0 + \epsilon)f = f$$

Because of the right hand continuity of $E(\lambda)$, we may carry out $\epsilon \longrightarrow 0$ on the right in the second equation: $E(\lambda_0)f = f$. For $\lambda \geq \lambda_0$ then, because of the second equation $(\epsilon = \lambda - \lambda_0 \geq 0)$ $E(\lambda)f = f$, while for $\lambda < \lambda_0$, because of the first equation $(\epsilon = \lambda_0 - \lambda > 0)$, $E(\lambda)f = 0$. Therefore:

$$E(\lambda)f = \begin{cases} f & \text{for} \quad \lambda \geq \lambda_0 \\ 0 & \text{for} \quad \lambda < \lambda_0 \end{cases}$$

But this necessary condition is also sufficient, because from it, it follows that

$$(Af, g) = \int_{-\infty}^{+\infty} \lambda d(E(\lambda)f, g) = \lambda_0(f, g)$$

(the definition of the Stieltjes integral given in Note 73 should be recalled), therefore $(Af - \lambda_0 f, g) = 0$ for all g; i.e., $Af = \lambda_0 f$.

What does this condition mean? First, it involves a discontinuity of $E(\lambda)$ at the point $\lambda = \lambda_0$. By THEOREM 17. in II.4., $E(\lambda)$ converges to a projection $E^{(1)}(\lambda_0)$ or $E^{(2)}(\lambda_0)$, for $\lambda \longrightarrow \lambda_0$, $\lambda < \lambda_0$ and for $\lambda \longrightarrow \lambda_0$, $\lambda > \lambda_0$,[79] respectively. By $\mathbf{S}_1$., $E^{(2)}(\lambda_0) = E(\lambda_0)$, but in the case of the discontinuity, $E^{(1)}(\lambda_0) \neq E(\lambda_0)$. Further, because $E(\lambda) \leq E(\lambda_0)$ for $\lambda < \lambda_0$ ($\mathbf{S}_2$.) $E^{(1)}(\lambda_0) \leq E(\lambda_0)$. Therefore, $E(\lambda_0) - E^{(1)}(\lambda_0)$ is a

---

[79] This was shown only for λ sequences. However, the limit for all such λ sequences ($\lambda \longrightarrow \lambda_0$ and $\lambda < \lambda_0$ or $\lambda > \lambda_0$) must be the same, because two such sequences can be combined to form one -- and since this has a limit, both constituents must have the same limit. From this it follows that the convergence (to the common limit of all sequences also occurs in the case of continuous variation of the λ.

projection, and it is characteristic for the discontinuity that it $\neq 0$.

$E(\lambda)f = 0$ for all $\lambda < \lambda_0$ has $E^{(1)}(\lambda_0)f = 0$ as a consequence, but (because $E(\lambda) \leq E^{(1)}(\lambda_0)$), it is also a consequence of this. $E(\lambda)f = f$ for all $\lambda \geq \lambda_0$ follows from $E(\lambda_0)f = f$: $E(\lambda_0) \leq E(\lambda)$, $E(\lambda)E(\lambda_0) = E(\lambda_0)$, therefore $E(\lambda)f = E(\lambda)E(\lambda_0)f = E(\lambda_0)f = f$. Hence $E^{(1)}(\lambda_0)f = 0$, $E(\lambda_0)f = f$ is characteristic for $Af = \lambda_0 f$, or (THEOREM 14. in II.4.) $[E(\lambda_0) - E^{(1)}(\lambda_0)]f = f$. That is, if we write $E(\lambda_0) - E^{(1)}(\lambda_0) = P_{\mathfrak{M}_{\lambda_0}}$ then the above implies that f belongs to $\mathfrak{M}_{\lambda_0}$.

Consequently it has been shown that $Af = \lambda f$ is solvable by an $f \neq 0$ only at the discontinuities of $E(\lambda)$, and the solutions f form the closed linear manifold $\mathfrak{M}_{\lambda_0}$ defined above.

The complete orthonormal set sought in II.6., to be selected from these solutions (combining any λ) then exists if and only if the $\mathfrak{M}_{\lambda_0}$ $(-\infty < \lambda_0 < \infty)$ together span the closed linear manifold $\mathfrak{R}_\infty$. [We have discussed in II.6. how the construction of this set would then be accomplished. The mutual orthogonality of the $\mathfrak{M}_{\lambda_0}$ can be seen in another way. From $\lambda_0 < \mu_0$ it follows that

$$P_{\mathfrak{M}_{\lambda_0}} P_{\mathfrak{M}_{\mu_0}} = [E(\lambda_0) - E^{(1)}(\lambda_0)][E(\mu_0) - E^{(1)}(\mu_0)] = 0$$

because

$$E(\lambda_0) = E^{(1)}(\lambda_0) \leq E(\lambda_0) \leq E^{(1)}(\mu_0) ,$$

$$E(\mu_0) - E^{(1)}(\mu_0) \leq 1 - E^{(1)}(\mu_0) .]$$

Without ascertaining the precise conditions under which this is true, we note the following: if an interval μ_1, μ_2 exists, in which $E(\lambda)$ increases continuously (i.e.,

$\mu_1 < \mu_2$, $E(\lambda)$ continuous in $\mu_1 \leq \lambda \leq \mu_2$, $E(\mu_1) \neq E(\mu_2)$) , then it is certainly not the case. Because, for $\lambda \leq \mu_1$, $E(\lambda) - E^{(1)}(\lambda) \leq E(\lambda) \leq E(\mu_1)$ while for $\mu_1 < \lambda \leq \mu_2$, $E(\lambda) - E^{(1)}(\lambda) = 0$ because of continuity, and for $\mu_2 < \lambda$, $E(\lambda) - E^{(1)}(\lambda) \leq 1 - E^{(1)}(\lambda) \leq 1 - E(\mu_2)$. Therefore $E(\lambda) - E^{(1)}(\lambda)$ is always orthogonal to $E(\mu_2) - E(\mu_1)$. Let $E(\mu_2) - E(\mu_1) = P_{\mathfrak{N}}$. Then all $\mathfrak{M}_\lambda$ are orthogonal to $\mathfrak{N}$. If a complete orthonormal set were to be chosen from this, then $\mathfrak{N}$ would contain only the zero, i.e., $E(\mu_2) - E(\mu_1) = 0$, contrary to assumption.

The discontinuities of $E(\lambda)$ are known as the discrete spectrum of A . They are the same λ for which $Af = \lambda f$, $f \neq 0$ is solvable. If we choose from each $\mathfrak{M}_\lambda \neq 0$ an f with $||f|| = 1$, then, because of the orthogonality of the $\mathfrak{M}_\lambda$, an orthonormal set is obtained. By THEOREM 3. in II.2., this is finite, or is a sequence. Therefore the λ of the discrete spectrum form at most a sequence.

All points in whose neighborhood $E(\lambda)$ is not constant form the spectrum of A . We have seen that if there are intervals into which the spectrum, but not the point spectrum of A , penetrates, i.e., intervals of continuity of $E(\lambda)$, in which it is not constant -- then the eigenvalue problem is certainly not solvable in the same sense in which it was formulated at the beginning of II.6. The precise conditions of this insolvability we do not investigate further, because insolvability may also arise under certain other circumstances, when the discrete spectrum does penetrate into all intervals in which points of the spectrum lie. The separation of the discrete spectrum from the rest of the spectrum is then considerably more laborious, and is beyond the scope of this work. (The reader will find these investigations in Hilbert's papers, which have been referred to previously.)

On the other hand, we want to show how in the case of the existence of a complete orthonormal set

$\phi_1, \phi_2, \ldots$ of solutions of $A\phi = \lambda\phi$ (with $\lambda = \lambda_1, \lambda_2, \ldots$ for corresponding $\phi = \phi_1, \phi_2, \ldots$ -- in the case of the pure discrete spectrum, as we shall say -- the $E(\lambda)$ are to be constructed. We have

$$E(\lambda) = \sum_{\lambda_\rho \leq \lambda} P_{[\phi_\rho]} \quad ^{80}$$

(The sum Σ can have 0 summands, then $E(\lambda) = 0$; or a positive finite number, then its meaning is clear; or infinitely many, in which case it converges according to the final considerations of II.4.)

Indeed $\overline{S_2}$. is evident, because, for $\lambda' \leq \lambda''$,

$$E(\lambda'') - E(\lambda') = \sum_{\lambda' < \lambda_\rho \leq \lambda''} P_{[\phi_\rho]}$$

is a projection, therefore $E(\lambda') \leq E(\lambda'')$ (THEOREM 14.). We prove $\overline{S_1}$. as follows: for each f

$$\sum_\rho \|P_{[\phi_\rho]} f\|^2 = \sum_\rho |(f, \phi_\rho)|^2 = \text{Note}^{81} = \|f\|^2$$

(THEOREM 7.), i.e., $\Sigma_\rho \|P_{[\phi_\rho]} f\|^2$ is convergent. Therefore for each $\epsilon > 0$ we can give a finite number of ρ so that the sum Σ_ρ taken over these alone is $> \|f\|^2 - \epsilon$, and therefore each Σ'_ρ from which they are missing is $< \epsilon$. Then also

---

[80]This is the precise restatement of the definition of $E(\lambda; \xi, \eta)$ given in II.7.

[81]As the construction carried out in the proof of THEOREM 10. shows, $P_{[\phi]} f = (f, \phi)\cdot\phi$ (if $\|\phi\| = 1$) , therefore $\|P_{[\phi]} f\| = |(f, \phi)| = |(\phi, f)|$.

$$\|\sum_{\rho}{}' P_{[\phi_\rho]} f\|^2 = \sum_{\rho}{}' \|P_{[\phi_\rho]} f\|^2 < \epsilon$$

From this it follows in particular that

$$\|\sum_{\lambda_\rho \leq \lambda} P_{[\phi_\rho]} f\|^2 < \epsilon \,, \quad \|\sum_{\lambda_\rho > \lambda} P_{[\phi_\rho]} f\|^2 < \epsilon \,,$$

$$\|\sum_{\lambda_0 < \lambda_\rho \leq \lambda} P_{[\phi_\rho]} f\|^2 < \epsilon$$

if λ is taken sufficiently small, sufficiently large, or near enough to λ_0 (and $\geq \lambda_0$) respectively. Then

$$E(\lambda) f = \sum_{\lambda_\rho \leq \lambda} P_{[\phi_\rho]} f \longrightarrow 0 \quad \text{for} \quad \lambda \longrightarrow -\infty \,,$$

$$f - E(\lambda) f = \sum_{\lambda_\rho > \lambda} P_{[\phi_\rho]} f \longrightarrow 0 \,[82] \quad \text{for} \quad \lambda \longrightarrow +\infty \,,$$

$$E(\lambda) f - E(\lambda_0) f = \sum_{\lambda_0 < \lambda_\rho \leq \lambda} P_{[\phi_\rho]} f \longrightarrow 0 \quad \text{for} \quad \lambda \longrightarrow \lambda_0 \,, \quad \lambda \geq \lambda_0$$

i.e., $\overline{S}_1$. is satisfied.

In order to convince ourselves of the validity of $\overline{S}_3$., we set $f = x_1\phi_1 + x_2\phi_2 + \cdots$. Then $Af = \lambda_1 x_1 \phi_1 + \lambda_2 x_2 \phi_2 + \cdots$. In order that Af be defined,

$$\sum_{\rho=1}^{\infty} \lambda_\rho^2 |x_\rho|^2$$

---

[82] We have (THEOREM 7.):

$$f = \sum_{\rho} (f, \phi_\rho) \cdot \phi_\rho = \sum_{\rho} P_{[\phi_\rho]} f$$

This also follows from the final considerations of II.4.

must then be finite. But

$$\int_{-\infty}^{\infty}\lambda^2 d\|E(\lambda)f\|^2 = \int_{-\infty}^{\infty}\lambda^2 d\left(\sum_{\lambda_\rho \leq \lambda}|x_\rho|^2\right) = \text{Note } ^{83}$$

$$= \sum_{\rho=1}^{\infty}\lambda_\rho^2|x_\rho|^2 ,$$

$$\int_{-\infty}^{\infty}\lambda d(E(\lambda)f,\ g) = \int_{-\infty}^{\infty}\lambda d\left(\sum_{\lambda_\rho \leq \lambda}x_\rho\overline{y}_\rho\right) = \text{Note } ^{83}$$

$$= \sum_{\rho=1}^{\infty}\lambda_\rho x_\rho \overline{y}_\rho = (Af,\ g)$$

---

[83] By Note 73,

$$\int_{-\infty}^{\infty}\lambda^2 d\left(\sum_{\lambda_\rho \leq \lambda}|x_\rho|^2\right) = \lim \sum_{\tau=1}^{k}\Lambda_\tau^2 \cdot \sum_{\Lambda_{\tau-1} < \lambda_\rho \leq \Lambda_\tau}|x_\rho|^2 .$$

If throughout, $\Lambda_\tau^2 - \Lambda_{\tau-1}^2 < \epsilon$ (i.e. if the $\Lambda_0, \ldots, \Lambda_k$ mesh is sufficiently fine), then this changes by

$$< \epsilon\sum_{\rho=1}^{\infty}|x_\rho|^2 = \epsilon\|f\|^2$$

if we replace it by

$$\sum_{\tau=1}^{k}\ \sum_{\Lambda_{\tau-1} < \lambda_\rho \leq \Lambda_\tau}\lambda_\rho^2|x_\rho|^2 = \sum_{\Lambda_0 < \lambda_\rho \leq \Lambda_k}\lambda_\rho^2|x_\rho|^2$$

and if Λ_0 is small enough, and Λ_k large enough, this is arbitrarily close to

$$\sum_{\tau=1}^{\infty}\lambda_\rho^2|x_\rho|^2$$

This sum is then the desired limit, i.e., the value of the integral.

Consequently, $\overline{S_3}$. is also satisfied.

Let us consider two cases of a pure continuous spectrum, i.e., such where there exists no discrete spectrum. Let $\Re_\infty$ be the space of all functions $f(q_1, \ldots, q_l)$ with finite

$$\int_{-\infty}^{\infty} \cdots \int_{-\infty}^{\infty} |f(q_1, \ldots, q_l)|^2 dq_1 \ldots dq_l$$

and A the operator $q_j \cdots$ the Hermitian character of which is evident.

We see: $Af = \lambda f$ implies that

$$(q_j - \lambda) f(q_1, \ldots, q_l) = 0$$

i.e., $f(q_1, \ldots, q_l) = 0$ everywhere, with the possible exception of the $l - 1$ dimensional plane $q_j = \lambda$. However, this plane is (according to the discussion in II.3. relative to condition **B.**) unimportant, because its Lebesgue measure (i.e., volume), is 0. Then $f \equiv 0$.[84] Consequently there exists no solution of $Af = \lambda f$ which $\neq 0$.

The next integral formula is proved in exactly the same way.

[84]At this point, the correct mathematical method followed by us deviates from the symbolic method of Dirac (cf. e.g., his book mentioned in Note 1). The latter method in essence is to consider the f with $(q - \lambda) f(q) \equiv 0$ as solutions (for simplicity, we set $l = j = 1$, $q_j = q$). But since each $(f, g) = \int f(q)\overline{g(q)}dq = 0$ and $f \neq 0$, $f(q)$ is infinite at the point $q = \lambda$ (the only one where it differs from zero!), and indeed, so strongly infinite that the $(f, g) \neq 0$. Since for $q \neq \lambda$, $f(q) = 0$, $\int f(q)\overline{g(q)}dq$ can depend only on $\overline{g(\lambda)}$, and indeed it is clear that the integral, because of its additive property, must be proportional to $\overline{g(\lambda)}$. Therefore, it must equal

But we also see (inexactly) where the solution is to be expected. $(q_j - \lambda)f(q_1,\ldots,q_l) = 0$ implies that only for $q_j = \lambda$ may $f \neq 0$. A linear combination of the solutions for several λ, say $\lambda = \lambda', \lambda'', \ldots, \lambda^{(s)}$, would then be an f which may be $\neq 0$ only for

$$q_j = \lambda', \lambda'', \ldots, \lambda^{(s)}$$

We can then consider an f as a linear combination of all solutions with $\lambda \leq \lambda_0$, if it is $\neq 0$ only for $\lambda \leq \lambda_0$. But in the case of the pure discrete spectrum, we had

---

$c\overline{g(\lambda)}$, and c must be different from zero. If we replace $f(q)$ by $f(q)/c$, we obtain $c = 1$. We then have a fictitious function $f(q)$ for which $\int f(q)\overline{g(q)}dq = \overline{g(\lambda)}$.

It suffices, of course, to consider the case $\lambda = 0$. We write $f(q) = \delta(q)$ which is defined by

$$\Delta \quad q\delta(q) \equiv 0, \quad \int \delta(q)f(q)dq = f(0)$$

For arbitrary λ, $\delta(q - \lambda)$ is the solution -- although a function δ with the property Δ. does not exist, there are function sequences which converge toward such a behavior (of course the limiting function does not exist). For example,

$$f_\epsilon(q) = \begin{cases} \frac{1}{2\epsilon}, & \text{for } |x| < \epsilon \\ 0, & \text{for } |x| \geq \epsilon \end{cases} \quad \text{for } \epsilon \longrightarrow +0$$

or

$$f_a(q) = \sqrt{\frac{a}{\pi}}\, e^{-ax^2} \quad \text{for } a \longrightarrow +\infty$$

(Cf. also I.3., in particular, Note 32).

$$E(\lambda_0) = \sum_{\lambda_\rho \leq \lambda_0} P_{[\phi_\rho]} = P_{\mathfrak{N}_{\lambda_0}}, \quad \mathfrak{N}_{\lambda_0} = \left[\phi_\rho (\lambda_\rho \leq \lambda_0)\right]$$

i.e., $\mathfrak{N}_\lambda$ consisted of the linear combinations of all ϕ_ρ with $\lambda_\rho \leq \lambda_0$ i.e., of all solutions of $Af = \lambda f$ with $\lambda \leq \lambda_0$. Consequently -- to be sure, inexactly and heuristically -- it is to be expected that now

$$E(\lambda_0) = P_{\mathfrak{N}_{\lambda_0}}$$

where $\mathfrak{N}_{\lambda_0}$ consists of those f which $\neq 0$ only for $q_j \leq \lambda_0$. $\mathfrak{N}_\infty - \mathfrak{N}_\infty$ then clearly consists of those f which always $= 0$ for $q_j > 0$ -- consequently,

$$E(\lambda_0) f(q_1, \ldots, q_l) = \begin{cases} f(q_1, \ldots, q_l), & \text{for } q_j \leq \lambda_0 \\ 0, & \text{for } q_j > \lambda_0 \end{cases}$$

In a rather inexact way then, we have found a family of projections $E(\lambda)$, of which it is to be supposed that they satisfy $\overline{S_1}$. - $\overline{S_3}$. for our A. In fact, $\overline{S_1}$., $\overline{S_2}$. are satisfied in a trivial fashion, and indeed for $\overline{S_1}$., the $\lambda \longrightarrow \lambda_0$ case holds, even without the condition $\lambda \geq \lambda_0$ -- i.e., $E(\lambda)$ is continuous in λ everywhere. In order to see that $\overline{S_3}$. also is satisfied, it suffices to prove the validity of the equations

$$\int_{-\infty}^{\infty}\lambda^2 d\|E(\lambda)f\|^2 = \int_{-\infty}^{\infty}\lambda^2 d\left(\int_{-\infty}^{\infty}\cdots\int_{-\infty}^{\lambda}\cdots\int_{-\infty}^{\infty}|f(q_1,\ldots,q_j,\ldots,q_l)|^2 dq_1\ldots dq_j\ldots dq_l\right)$$

$$= \int_{-\infty}^{\infty}\lambda^2\left(\int_{-\infty}^{\infty}\cdots\int_{-\infty}^{\infty}|f(q_1,\ldots,q_{j-1},\lambda,q_{j+1},\ldots,q_l)|^2 dq_1\ldots dq_{j-1}dq_{j+1}\ldots dq_l\right)d\lambda$$

$$= \int_{-\infty}^{\infty}\cdots\int_{-\infty}^{\infty}q_j^2|f(q_1,\ldots,q_{j-1},q_j,q_{j+1},\ldots,q_l)|^2 dq_1\ldots dq_{j-1}dq_j dq_{j+1}\ldots dq_l = \|Af\|^2 ,$$

$$\int_{-\infty}^{\infty}\lambda d(E(\lambda)f, g) = \int_{-\infty}^{\infty}\lambda d\left(\int_{-\infty}^{\infty}\cdots\int_{-\infty}^{\lambda}\cdots\int_{-\infty}^{\infty}f(q_1,\ldots,q_j,\ldots,q_l)\overline{g(q_1,\ldots,q_j,\ldots,q_l)}dq_1\ldots dq_j\ldots dq_l\right)$$

$$= \int_{-\infty}^{\infty}\lambda\left(\int_{-\infty}^{\infty}\cdots\int_{-\infty}^{\infty}f(q_1,\ldots,q_{j-1},\lambda,q_{j+1},\ldots,q_l)\overline{g(q_1,\ldots,q_{j-1},\lambda,q_{j+1},\ldots,q_l)}dq_1\ldots dq_{j-1}dq_{j+1}\ldots dq_l\right)d\lambda$$

$$= \int_{-\infty}^{\infty}\cdots\int_{-\infty}^{\infty}q_j f(q_1,\ldots,q_{j-1},q_j,q_{j+1},\ldots,q_l)\overline{g(q_1,\ldots,q_{j-1},q_j,q_{j+1},\ldots,q_l)}$$

$$\times\, dq_1\ldots dq_{j-1}dq_j dq_{j+1}\ldots dq_l = (Af, g)$$

We again recognize that the discrete spectrum or the old definition of the eigenvalue problem must fail, since $E(\lambda)$ increases continuously everywhere.

This example also indicates, more generally, a possible way to find the $E(\lambda)$ in the continuous spectrum: one may determine (incorrectly!) the solutions of $Af = \lambda f$ (since λ lies in the continuous spectrum, these f do not belong to the $\mathfrak{R}_\infty$!), and form their linear combinations for all $\lambda \leq \lambda_0$. These belong in part to $\mathfrak{R}_\infty$ again, and perhaps form a closed linear manifold $\mathfrak{R}_{\lambda_0}$. Then we may set $E(\lambda_0) = P_{\mathfrak{R}_{\lambda_0}}$, -- if we have handled it properly, then it is possible to verify $\overline{\mathbf{S}}_1$. - $\overline{\mathbf{S}}_3$. from this (for A and these $E(\lambda)$), and so transform the heuristic argument into an exact one.[85]

The second example which we wish to consider is the other important operator of wave mechanics, $\frac{h}{2\pi i}\frac{\partial}{\partial q_j}$. In order to avoid unnecessary complications, let $l = j = 1$

---

[85]The exact formulation of this idea (to be considered here only as an heuristic statement) is found in the papers of Hellinger [J. f. Math., 136 (1909)], and Weyl [Math. Ann. 68 (1910)].

(the treatment is the same for other values). We must then investigate the operator

$$A'f(g) = \frac{h}{2\pi i}\frac{\partial}{\partial q}f(q)$$

If the domain of q is $-\infty < q < +\infty$, then this is an Hermitian operator, as we saw in II.5. On the other hand, for a finite domain $a \leq q \leq b$, this is not the case:

$$\begin{aligned}
(A'f,\ g) - (f,\ A'g) \\
&= \int_a^b \frac{h}{2\pi i} f'(q)\overline{g(q)}dq - \int_a^b f(q)\,\frac{h}{2\pi i}\,\overline{g'(q)}dq \\
&= \frac{h}{2\pi i}\int_a^b \left\{ f'(q)\overline{g(q)} + f(q)\overline{g(q)'} \right\}dq \\
&= \frac{h}{2\pi i}\Big[f(q)\overline{g(q)} \Big]_a^b \\
&= \frac{h}{2\pi i}\Big[f(b)\overline{g(b)} - f(a)\overline{g(a)} \Big]
\end{aligned}$$

In order that this vanish, the domain of $\frac{h}{2\pi i}\frac{\partial}{\partial q}$ must be limited in such a way that for two $f,\ g$ picked arbitrarily from it, $f(a)\overline{g(a)} = f(b)\overline{g(b)}$. That is,

$$f(a):f(b) = \overline{g(b)}:\overline{g(a)}$$

If we vary f for a fixed g, then we see that $f(a):f(b)$ must be the same number θ throughout the entire domain (θ may even be 0 or ∞); and substitution for $f,\ g$ then gives $\theta = \frac{1}{\bar{\theta}}$ i.e., $|\theta| = 1$. That is, in order that $\frac{h}{2\pi i}\frac{\partial}{\partial q}$ be Hermitian, we must postulate a "boundary condition" of the form

$$f(a):f(b) = \theta$$

(θ any fixed number of absolute value 1).

First we take the interval $-\infty < q < +\infty$. The solutions of $A\phi = \lambda\phi$, i.e., $\frac{h}{2\pi i}\phi'(q) = \lambda\phi(g)$ are the functions

$$\phi(q) = ce^{\frac{2\pi i}{h}\lambda q}$$

However, these cannot be used for our purposes without further discussion, since

$$\int_{-\infty}^{\infty}|\phi(q)|^2 dq = \int_{-\infty}^{\infty}|c^2|dq = +\infty$$

(otherwise, $c = 0$, $\phi \equiv 0$). We may observe that in the first example we found the solution $\delta(q - \lambda)$, i.e., a fictitious, non-existent function (cf. Note 84). Now we find

$$e^{\frac{2\pi i}{h}\lambda q}$$

which is an entirely well-behaved function, but which does not belong to $\mathfrak{R}_\infty$ because of the unboundedness of the integral of the square of its absolute value. From our point of view, these two facts have the same meaning; because what does not belong to $\mathfrak{R}_\infty$ for us does not exist.[86]

As in the first case, we now form the linear combinations of the solutions belonging to the $\lambda \le \lambda_0$, i.e., the functions

$$f(q) = \int_{-\infty}^{\lambda_0} c(\lambda)e^{\frac{2\pi i}{h}\lambda q}d\lambda$$

It is to be hoped that among these, functions in $\mathfrak{R}_\infty$ will be present, furthermore, that these will form a closed linear manifold $\mathfrak{N}'_{\lambda_0}$, and finally, that the projections

[86] Of course, only success in the physical application can justify this point of view or its use in quantum mechanics.

$$E(\lambda_0) = P_{\mathfrak{N}'_{\lambda_0}}$$

will form the resolution of the identity belonging to A'. We obtain an example confirming the first surmise if we set

$$f(q) = \int_{\lambda_1}^{\lambda_0} e^{\frac{2\pi i}{h}\lambda q} d\lambda = \frac{e^{\frac{2\pi i}{h}\lambda_0 q} - e^{\frac{2\pi i}{h}\lambda_1 q}}{\frac{2\pi i}{h} q}$$

$$c(\lambda) = \begin{cases} 1, \text{ for } \lambda \geq \lambda_1 \\ 0, \text{ for } \lambda < \lambda_1 \end{cases} \Bigg| \; \lambda_1 < \lambda_0$$

because this $f(q)$ is everywhere regular for finite values, and for $q \to \pm\infty$ it behaves like $1/q$, so that

$$\int_{-\infty}^{\infty} |f(q)|^2 dq$$

is finite. But the other surmises also prove to be correct, and actually follow from the theory of the Fourier integral. Indeed this theory asserts the following:[87]

Let $f(x)$ be any function with finite

$$\int_{-\infty}^{\infty} |f(x)|^2 dx$$

Then a function

$$Lf(x) = F(y) = \frac{1}{\sqrt{2\pi}} \int_{-\infty}^{\infty} e^{ixy} f(x) dx$$

can be formed, and for this,

$$\int_{-\infty}^{\infty} |F(y)|^2 dy$$

---

[87] Plancherel: Circ. Math. di. Pal. 30 (1910): Titchmarsh, Lond. Math. Soc. Proc. 22 (1924).

is also finite, and is actually equal to

$$\int_{-\infty}^{\infty} |f(x)|^2 dx$$

Moreover, $LLf(x) = f(-x)$. (This is the so-called Laplace transform which plays an important role elsewhere in the theory of differential equations.)

If we replace x, y by

$$\sqrt{\frac{2\pi}{h}}\, q, \quad \sqrt{\frac{2\pi}{h}}\, p$$

then we obtain the transform

$$Mf(q) = F(p) = \frac{1}{\sqrt{h}} \int_{-\infty}^{\infty} e^{\frac{2\pi i}{h} pq} f(q) dq$$

which has the same properties. Consequently, it maps $\mathfrak{R}_\infty$ on itself [$Mf(q) = g(p)$ is solvable for each $g(p)$ of $\mathfrak{R}_\infty$: $f(q) = Mg(-p)$] , leaves $||f||$ invariant, and is linear. By II.5., this M is unitary. Therefore $M^2 f(q) = f(-q)$, and then $M^{-1}f(q) = M^* f(q) = Mf(-q)$, so that M commutes with M^2 , i.e., with the operation $f(q) \rightarrow f(-q)$.

What we then had in mind for $\mathfrak{R}'_{\lambda_0}$ was the following: $f(q)$ belongs to $\mathfrak{R}'_{\lambda_0}$, if $F(p) = M^{-1}(q)$ is equal to zero for all $p > \lambda_0$. (Here,

$$F(p) = \sqrt{h}\, c(p)$$

with the $c(\lambda)$ from above.) But these $F(p)$, as we know, form the closed linear manifold $\mathfrak{R}_{\lambda_0}$. Therefore the image $\mathfrak{R}'_{\lambda_0}$ of this $\mathfrak{R}_{\lambda_0}$ obtained by M is also a closed linear manifold. $E'(\lambda_0)$ is formed from $E(\lambda_0)$, just as $\mathfrak{R}'_{\lambda_0}$ from $\mathfrak{R}_{\lambda_0}$; by transformation of the entire $\mathfrak{R}_\infty$ with M . Therefore $E'(\lambda_0) = ME(\lambda_0)M^{-1}$. Then $E'(\lambda)$, as well as $E(\lambda)$, has the properties $\overline{S_1}$., $\overline{S_2}$. It still remains to prove $\overline{S_3}$., i.e., that the resolution of the identity $E(\lambda)$ belongs to A'.

In this, we limit ourselves to the demonstration of the following: If $f(q)$ is differentiable without special convergence difficulties, and

$$\int_{-\infty}^{\infty} \left| \frac{h}{2\pi i} f'(q) \right|^2 dq$$

is finite, then

$$\int_{-\infty}^{\infty} \lambda^2 d||E'(\lambda)f||^2$$

is finite, and

$$(A'f, g) = \int_{-\infty}^{\infty} \lambda d(E'(\lambda)f, g) \quad [88]$$

Indeed $(M^{-1}f(q) = F(p))$

$$A'f(q) = \frac{h}{2\pi i} f'(q) = \frac{h}{2\pi i} \frac{\partial}{\partial q} (MF(p))$$

$$= \frac{h}{2\pi i} \frac{\partial}{\partial q} \left(\frac{1}{\sqrt{h}} \int F(p) e^{\frac{2\pi i}{h} pq} dp \right)$$

$$= \frac{\sqrt{h}}{2\pi i} \int F(p) \frac{\partial}{\partial q} \left(e^{\frac{2\pi i}{h} pq} \right) dp = \frac{1}{\sqrt{h}} \int F(p) \cdot p \cdot e^{\frac{2\pi i}{h} pq} dp$$

$$= M(pF(p))$$

Therefore, for the f mentioned, $A' = MAM^{-1}$ (A is the

---

[88] That is, $E'(\lambda)$ does not belong to $A' = \frac{h}{2\pi i} \frac{\partial}{\partial q}$ itself, but to an operator whose domain includes that of A', and which coincides in this domain with A'. Cf. the devel ,p-ments regarding this in II.9.

operator $q\ldots$, or, since we use the variable p here, $p\ldots$) . Since the above propositions hold for A , $E(\lambda)$, they also are valid after the transformation of $\mathfrak{R}_\infty$ with M . Therefore they also hold for $A' = MAM^{-1}$ and the $E'(\lambda) = ME(\lambda)M^{-1}$.

The situation for $\frac{h}{2\pi i}\frac{\partial}{\partial q}$ in the interval $a \leq q \leq b$ ($a < b$, a, b finite) is essentially different. In this case, as we know, a boundary condition

$$f(a):f(b) = \theta \qquad (|\theta| = 1)$$

is also necessary to establish the Hermitian nature. Again, $\frac{h}{2\pi i}\frac{\partial}{\partial q} f(q) = \lambda f(g)$ is solved by

$$f(q) = ce^{\frac{2\pi i}{h}\lambda q}$$

but now

$$\int_a^b |f(q)|^2 dq = \int_a^b |c|^2 dq = (b - a)|c|^2$$

is finite, so that $f(q)$ always belongs to $\mathfrak{R}_\infty$. On the other hand, there is the boundary condition

$$f(a):f(b) = e^{\frac{2\pi i}{h}\lambda(a - b)} = \theta$$

to be satisfied; or, if we set $\theta = e^{-i\alpha}$ $(0 \leq \alpha < 2\pi)$,

$$\frac{2\pi i}{h}\lambda(a - b) = -i\alpha - 2k\pi i \quad (k = 0, \pm 1, \pm 2, \ldots) ,$$

$$\lambda = \frac{h}{b - a}\left(\frac{\alpha}{2\pi} + k\right)$$

Therefore, we have a discrete spectrum, and the normalized solutions are then determined by $(b - a)|c|^2 = 1$ whence $c = \frac{1}{\sqrt{b - a}}$:

$$\phi_k(q) = \frac{1}{\sqrt{b - a}} e^{\frac{2\pi i}{h} \lambda q} = \frac{1}{\sqrt{b - a}} e^{\frac{2\pi i}{a-b} (\frac{\alpha}{2\pi} + k)q} ,$$

$$k = 0, \pm 1, \pm 2, \ldots$$

This is then an orthonormal set, but is also complete. For, if $f(q)$ is orthogonal to all $\phi_k(q)$, then

$$e^{\frac{\alpha i}{b-a} q} f(q)$$

is orthogonal to all

$$e^{\frac{2\pi i}{b-a} kq}$$

therefore

$$e^{\frac{\alpha i}{2\pi} x} f(\frac{b - a}{2\pi} x)$$

is orthogonal to all e^{ikx} i.e., to $1, \cos x \sin x, \cos 2x, \sin 2x, \ldots$. Moreover, it is defined in the interval

$$a \leq \frac{b - a}{2\pi} x \leq b$$

whose length is 2π, so that it must vanish, according to well-known theorems.[89] Therefore $f(q) \equiv 0$.

Consequently, we have a pure discrete spectrum, a case which we treated in general at the beginning of this section. One should observe how the "boundary condition" -- i.e., θ or α -- affects the eigenvalues and eigenfunctions.

In conclusion, we can also consider the case of a one-way infinite interval (domain), say $0 \leq q < +\infty$.

---

[89] All Fourier coefficients disappear, therefore, the function itself vanishes (cf. for example, Courant-Hilbert, in the reference given in Note 30).

First of all, we must again prove the Hermitian nature of the operator. We have

$$(A'f, g) - (f, A'g) = \frac{h}{2\pi i} \int_0^\infty (f'(q)\overline{g(q)} + f(q)\overline{g'(q)})dq$$

$$= \frac{h}{2\pi i} \left[f(q)\overline{g(q)} \right]_0^\infty$$

We show that $f(q)\overline{g(q)}$ approaches zero as $q \longrightarrow +\infty$ just as we did in II.5. in the case of the (both ways) infinite interval (domain). Therefore the requirement has to be $f(0)\overline{g(0)} = 0$. If we set $f = g$, it can be seen that the "boundary condition" is $f(0) = 0$.

In this case, serious difficulties present themselves. The solutions of $A'\phi = \lambda\phi$ are the same as in the interval $-\infty < q < +\infty$, namely the

$$ce^{\frac{2\pi i}{h}\lambda q}$$

but they do not belong to $\mathfrak{R}_\infty$, and they do not satisfy the boundary condition. The latter is suspicious. Nevertheless, it is rather surprising that we must necessarily obtain the same $E(\lambda)$ by the method sketched earlier, as in the interval $-\infty < q < +\infty$ since the (improper, i.e., not belonging to $\mathfrak{R}_\infty$) solutions are the same. How is this to be reconciled with the fact that the operator is a different one? Also, they are not what we need. For if we again define M, M^{-1} in the Hilbert space F_Ω $f(q)$ $(0 \leq q < +\infty$, $\int_0^\infty |f(q)|^2 dq$ finite):

$$Mf(q) = F(p) = \frac{1}{\sqrt{h}} \int_0^\infty e^{\frac{2\pi i}{h}pq} f(q)dq ,$$

$$M^{-1}F(p) = f(q) = \frac{1}{\sqrt{h}} \int_{-\infty}^\infty e^{\frac{-2\pi i}{h}pq} F(p)dp \ (= MF(-p))$$

then M maps the Hilbert space $F_{\Omega'}$ of all $f(q)$,

$$0 \leq q < \infty \qquad \int_0^\infty |f(q)|^2 dq$$

on another Hilbert space: the $F_{\Omega''}$ of all $F(p)$,

$$-\infty < p < \infty , \qquad \int_{-\infty}^\infty |F(p)|^2 dp$$

While $||Mf(q)|| = ||f(q)||$ always holds (this follows from the theorems previously mentioned if we set $f(q) = 0$ for $-\infty < q < 0$) , $||M^{-1}F(p)|| = ||F(p)||$ is not the case in general -- because, by reason of the theorems previously discussed, if we define $f(q)$ for $q < 0$ also by

$$f(q) = \frac{1}{\sqrt{h}} \int_{-\infty}^\infty e^{\frac{-2\pi i}{h} pq} F(p) dp$$

then

$$||F||^2 = \int_{-\infty}^\infty |F(p)|^2 dp = \int_{-\infty}^\infty |f(q)|^2 dq ,$$

$$||M^{-1}F||^2 = ||f||^2 = \int_0^\infty |f(q)|^2 dq$$

-- therefore $||M^{-1}F|| < ||F||$, unless, by chance, $f(q)$ (defined as above) vanishes for all $q < 0$. Therefore $E'(\lambda) = ME(\lambda)M^{-1}$ is not a resolution of the identity[90] -- the method has failed.

---

[90]It is indeed true that $M^{-1}Mf(q) = f(q)$ (it suffices to define $f(q) = 0$ for $q < 0$, and to make use of earlier theorems), but it is not always true that $MM^{-1}F(p) = F(p)$ -- because in general $||M^{-1}F|| < ||F||$, therefore $||MM^{-1}F|| < ||F||$. Consequently $M^{-1}M = 1$, $MM^{-1} \neq 1$, i.e., M^{-1} is not the true reciprocal of M . (Also, no other one can exist because if there were one, then, since $M^{-1}M = 1$, it would have to be still equal to our M^{-1}) . As a consequence of this, for $E'(\lambda) = ME(\lambda)M^{-1}$ for ex-

We shall soon see (in Note 105) that this lies in the nature of the case, because no resolution of the identity belongs to this operator.

Before we conclude these introductory discussions, we shall give a few formal rules of calculation for operators, which are put in the symbolic form

$$A = \int_{-\infty}^{\infty} \lambda dE(\lambda)$$

by use of the resolution of the identity.

First, let F be a projection commutable with all $E(\lambda)$. Then for all $\lambda' < \lambda''$,

$$||E(\lambda'')Ff - E(\lambda')Ff||^2 = ||(E(\lambda'') - E(\lambda'))Ff||^2$$

$$= ||F(E(\lambda'') - E(\lambda'))f||^2 \leq ||(E(\lambda'') - E(\lambda'))f||^2$$

therefore, since $E(\lambda'')$, $E(\lambda')$, $E(\lambda'') - E(\lambda')$ as well as

$$E(\lambda'')F,\ E(\lambda')F,\ E(\lambda'')F - E(\lambda')F = (E(\lambda'') - E(\lambda'))F$$

are projections,

$$||E(\lambda'')Ff||^2 - ||E(\lambda')Ff||^2 \leq ||E(\lambda'')f||^2 - ||E(\lambda')f||^2$$

Therefore

$$\int_{-\infty}^{\infty} \lambda^2 d||E(\lambda)f||^2 \geq \int_{-\infty}^{\infty} \lambda^2 d||E(\lambda)Ff||^2$$

Then by $\overline{\mathbf{S}}_3$., AFf is defined if Af is. Furthermore,

---

ample, the conclusion

$$E'^2(\lambda) = E'(\lambda)$$

can still be made (since only $M^{-1}M$ is involved), but $E'(\lambda) \longrightarrow MM^{-1} \neq 1$ (for $\lambda \longrightarrow +\infty$) .

$$AF = \int_{-\infty}^{\infty} \lambda d(E(\lambda)F) = \int_{-\infty}^{\infty} \lambda d(FE(\lambda)) = FA$$ [91]

i.e., A, F also commute. In particular, we can take each $E(\lambda)$ for F (because of $\overline{\mathbf{S}}_2$ ·). Then

$$AE(\lambda) = \int_{-\infty}^{\infty} \lambda' d(E(\lambda')E(\lambda)) = \int_{-\infty}^{\infty} \lambda' d(E(\text{Min}\ (\lambda,\ \lambda')))$$

$$= \int_{-\infty}^{\lambda} + \int_{\lambda}^{\infty} = \int_{-\infty}^{\lambda} \lambda' d(\lambda') + \int_{\lambda}^{\infty} \lambda' dE(\lambda)$$

and since $\int_{\lambda}^{\infty} = 0$ (because the function following the differential is a constant),

$$AE(\lambda) = E(\lambda)A = \int_{-\infty}^{\lambda} \lambda' dE(\lambda')\ .$$

In addition, it follows from this that

$$A^2 = \int_{-\infty}^{\infty} \lambda d(E(\lambda)A) = \int_{-\infty}^{\infty} \lambda d \left(\int_{-\infty}^{\lambda} \lambda' dE(\lambda') \right) = \text{Note}$$ [92]

$$= \int_{-\infty}^{\infty} \lambda^2 dE(\lambda)$$

---

[91] Actually, this must be proved non-symbolically with the help of the rigorous equation

$$(Af,\ g) = \int \lambda d(E(\lambda)f,\ g)$$

Then the calculation runs as follows:

$$(AFf,\ g) = \int \lambda d(E(\lambda)Ff,\ g) = \int \lambda d(FE(\lambda)f,\ g)$$

$$= \int \lambda d(E(\lambda)f,\ Fg) = (Af,\ Fg) = (FAf,\ g)$$

From this it follows that $AF \equiv FA$.

[92] This follows from the equation

In general, we have

$$A^n = \int_{-\infty}^{\infty} \lambda^n dE(\lambda)$$

because we can reason inductively from $n - 1$ to n :

$$A^n = A^{n-1}A = \int_{-\infty}^{\infty} \lambda^{n-1} d(E(\lambda)A) = \int_{-\infty}^{\infty} \lambda^{n-1} d\left(\int_{-\infty}^{\lambda} \lambda' dE(\lambda')\right)$$

$$= \text{Note } 92 = \int_{-\infty}^{\infty} \lambda^{n-1} . \lambda dE(\lambda) = \int_{-\infty}^{\infty} \lambda^n dE(\lambda)$$

Then, if $p(x) = a_0 + a_1 x + \dots + a_n x^n$ is any polynomial, we have

$$p(A) = \int_{-\infty}^{\infty} p(\lambda) dE(\gamma)$$

(By $p(A)$ we mean, of course, $p(A) = a_0 1 + a_1 A + \dots + a_n A^n$.

$\int_{-\infty}^{\infty} dE(\lambda) = 1$ follows from $\overline{S}_1$.)

Furthermore, we have the following: if $r(\lambda)$, $s(\lambda)$ are any two functions, and if we define two operators B, C (symbolically) by

---

$$\int f(\lambda) d\left(\int^{\lambda} g(\lambda') dh(\lambda')\right) = \int f(\lambda) g(\lambda) dh(\lambda)$$

which is generally valid for the Stieltjes integral. This equation is clear without further discussion, by reason of the reciprocal relation between d and $\int^{\lambda}$. A rigorous proof has been given by the author: Annals of Mathematics, 32 (1931).

$$B = \int_{-\infty}^{\infty} r(\lambda)dE(\lambda)\ ,\quad C = \int_{-\infty}^{\infty} s(\lambda)dE(\lambda)$$ [93]

then it follows that

$$BC = \int_{-\infty}^{\infty} r(\lambda)s(\lambda)dE(\lambda)$$

For the proof we proceed exactly as in the special case $B = C = A$:

$$BE(\lambda) = \int_{-\infty}^{\infty} r(\lambda')d(E(\lambda')E(\lambda)) = \int_{-\infty}^{\infty} r(\lambda')d(E(\text{Min}\ (\lambda,\ \lambda')))$$

$$= \int_{-\infty}^{\lambda} + \int_{\lambda}^{\infty} = \int_{-\infty}^{\lambda} r(\lambda')dE(\lambda') + \int_{\lambda}^{\infty} r(\lambda')dE(\lambda)$$

$$= \int_{-\infty}^{\lambda} r(\lambda')dE(\lambda')\ ,$$

$$CB = \int_{-\infty}^{\infty} s(\lambda)d(BE(\lambda)) = \int_{-\infty}^{\infty} s(\lambda)d\ \left(\int_{-\infty}^{\lambda} r(\lambda')dE(\lambda')\right)$$

$$= \int_{-\infty}^{\infty} s(\lambda)\cdot r(\lambda)dE(\lambda) = \int_{-\infty}^{\infty} s(\lambda)r(\lambda)dE(\lambda)$$

The following relations may easily be verified:

$$B^* = \int_{-\infty}^{\infty} \overline{r(\lambda)}dE(\lambda)\ ,\quad aB = \int_{-\infty}^{\infty} ar(\lambda)dE(\lambda)\ ,$$

$$B \pm C = \int_{-\infty}^{\infty} (r(\lambda) \pm s(\lambda))dE(\lambda)$$

There then exists no formal obstacle to writing

---

[93]That is,

$$(Bf,\ g) = \int_{-\infty}^{\infty} r(\lambda)d(E(\lambda)f,\ g)\ ,\quad (Cf,\ g) = \int_{-\infty}^{\infty} s(\lambda)d(E(\lambda)f,\ g).$$

$B = r(A)$ for such functions $r(\lambda)$.[94] Particularly noteworthy are the (discontinuous!) functions

$$e_\lambda(\lambda') = \begin{cases} 1 & \text{for } \lambda' \leq \lambda \\ 0 & \text{for } \lambda' > \lambda \end{cases}$$

For these, we have (by $\overline{S_1}$.)

$$e_\lambda(A) = \int_{-\infty}^{\infty} e_\lambda(\lambda')dE(\lambda) = \int_{-\infty}^{\lambda} dE(\lambda') = E(\lambda)$$

(At the beginning of this section we discussed the operator $A = q_j \cdots$. Its $E(\lambda)$ was the multiplication with 1 or 0 respectively for $q \leq \lambda$ or $> \lambda$ respectively, i.e., multiplication by $e_\lambda(q)$. Consequently, $e_\lambda(q_j \cdots) = e_\lambda(q_j) \cdots$. This example is well suited to furnish an intuitive picture of the above concepts.)

9. DIGRESSION ON THE EXISTENCE AND UNIQUENESS OF THE SOLUTIONS OF THE EIGENVALUE PROBLEM

The last section gave only a qualitative survey, emphasizing special cases, of the problem as to which resolutions of the identity $E(\lambda)$ correspond to a given Hermitian operator A. A systematic investigation of the question still remains to be effected. To do this in mathematical completeness goes beyond the scope of this volume. We limit ourselves to the proof of a few points, and the statement of the rest -- particularly since a precise knowledge of these circumstances is not absolutely

---

[94] The precise foundation of this function concept was given by the author in Annals of Math. 32 (1931). F. Riesz was the first one to define general operator functions by limiting processes applied to polynomials.

necessary for understanding quantum mechanics.[95]

In THEOREM 18., it was shown that the continuity in linear operators is expressed by

$$(\mathbf{Co.})\quad \|Af\| \leq C.\|f\|$$

(C arbitrary, but fixed).

By THEOREM 18., there are several equivalent forms for the condition **Co.**:

$$(\mathbf{Co_1.})\quad |(Af, g)| \leq C.\|f\|\,\|g\|$$

$$(\mathbf{Co_2.})\quad |(Af, f)| \leq C.\|f\|^2$$

(the latter only for Hermitian A).

The condition $\mathbf{Co_1}$., equivalent to continuity, is the Hilbert concept of boundedness. Hilbert has formulated and solved the eigenvalue problem for bounded (i.e., continuous) Hermitian operators (cf. Note 70). But before we discuss this case, we must introduce an additional concept.

A Hermitian operator A is said to be closed if it has the following property: Assume that if $f_1, f_2, \ldots$ is a point sequence, if all Af_n are defined, and $f_n \to f_1$, $Af_n \to f^*$. Then Af is also defined, and is equal to f^*. It should be observed that continuity could be defined in a way which is closely related to the above definition, namely the following: If all Af_n, Af are defined, and if $f_n \to f$, then $Af_n \to Af$. The difference between the two definitions is that for closure, the existence of a limit f^* for the Af_n is required, and its equality with Af is asserted only under this assump-

[95]The theory of unbounded Hermitian operators, to which reference will be made in the following (in addition to the Hilbert theory of bounded operators), was developed by the author (see reference in Note 78). M. Stone (Proc. Nat. Ac. 1929 and 1930) arrived at similar results independently.

tion. In continuity on the other hand, even the existence of f^* is asserted.

A few examples: Let $\mathfrak{R}_\infty$ again be the space of all $f(q)$ with finite

$$\int_{-\infty}^{\infty} |f(q)|^2 dq$$

$(-\infty < q < \infty)$; A the operator $q\cdots$, defined for all $f(q)$ with finite

$$\int_{-\infty}^{\infty} |f(q)|^2 dq \quad \text{and} \quad \int_{-\infty}^{\infty} q^2 |f(q)|^2 dq$$

A' the operator $\frac{h}{2\pi i}\frac{\partial}{\partial q}$ defined for all functions differentiable everywhere, with finite

$$\int_{-\infty}^{\infty} |f(q)|^2 dq \quad \text{and} \quad \int_{-\infty}^{\infty} |\frac{h}{2\pi i} f'(q)|^2$$

as we know, both of these are Hermitian. A is closed; because, let $f_n \to f$, $Af_n \to f^*$, i.e.,

$$\int_{-\infty}^{\infty} |f_n(q) - f(q)|^2 dq \to 0 , \quad \int_{-\infty}^{\infty} |qf_n(q) - f^*(q)|^2 dq \to 0$$

By reason of the discussion in II.3. on the proof **D.**, there exists a subsequence $f_{n_1}, f_{n_2}, \ldots$ of the $f_1, f_2, \ldots$ which converges to a limit everywhere with the exception of a q set of measure 0 : $f_{n_\nu}(q) \to g(q)$. Therefore

$$\int_{-\infty}^{\infty} |g(q) - f(q)|^2 dq = 0 , \quad \int_{-\infty}^{\infty} |qg(q) - f^*(q)|^2 dq = 0$$

i.e., except for a set of measure 0 , $g(q) = f(q)$ and also $qg(q) = f^*(q)$, therefore $qf(q) = f^*(q)$ -- i.e., $f^*(q)$ and $qf(q)$ are not essentially different. But since $f^*(q)$ belongs to $\mathfrak{R}_\infty$ by assumption, $qf(q)$ does also. Consequently $Af(q)$ is defined and is equal to $qf(q) = f^*(q)$. On the other hand, A' is not closed; set

$$f_n(q) = e^{-\sqrt{q^2 + \frac{1}{n}}}, \quad f(q) = e^{-|q|}$$

Clearly, all Af_n are defined, but not Af (f has no derivative at $q = 0$). Nevertheless, $f_n \longrightarrow f$, $Af_n \longrightarrow f^*$, as can easily be calculated if we set $f^*(q) = -\operatorname{sgn}(q)e^{-|q|}$ ($\operatorname{sgn}(q) = -1, 0, +1$ for $q <, =, > 0$).

We now show: in contrast to continuity, closure is a property which can always be achieved with little difficulty for Hermitian operators. This is done by the process of extension -- i.e., we leave the operator unchanged at all points of $\mathfrak{R}_\infty$ where it is defined, but in addition define it for some of those points where it was not previously defined.

Indeed, let A be an arbitrary Hermitian operator. We define an operator $\tilde{A}$ as follows: $\tilde{A}f$ is defined, if a sequence $f_1, f_2, \ldots$ with defined Af_n exists, in such a way that f is the limit of the f_n, and the Af_n also possess a limit f^*. Then $\tilde{A}f = f^*$. This definition is admissible only if it is unique, i.e., if it follows from $f_n \longrightarrow f$, $g_n \longrightarrow f$, $Af_n \longrightarrow f^*$, $Ag_n \longrightarrow g^*$ that $f^* = g^*$. In fact, if Ag is defined,

$$(f^*, g) = \lim(Af_n, g) = \lim(f_n, Ag) = (f, Ag),$$

$$(g^*, g) = \lim(Ag_n, g) = \lim(g_n, Ag) = (f, Ag)$$

therefore $(f^*, g) = (g^*, g)$. But these g are everywhere dense, consequently $f^* = g^*$. Therefore we have defined $\tilde{A}$ correctly. This $\tilde{A}$ is an extension of A, i.e., whenever Af is defined, then $\tilde{A}f$ is also defined, and is $= Af$. It suffices to set all $f_n = f$ and $f^* = Af$. From the fact that A is linear and Hermitian, the same follows for $\tilde{A}$ (by continuity). Finally, $\tilde{A}$ is closed: Indeed, let all $\tilde{A}f_n$ be defined, $f_n \longrightarrow f$, $\tilde{A}f_n \longrightarrow f^*$. Then there are sequences $f_{n,1}, f_{n,2}, \ldots$ with defined

$Af_{n,m}$, $f_{n,m} \to f_n$, $Af_{n,m} \to f_n^*$ and $\tilde{A}f_n = f_n^*$. For each n there is an N_n such that for $m \geq N_n$, $||f_{n,m} - f_n|| \leq \frac{1}{n}$, $||Af_{n,m} - f_n^*|| \leq \frac{1}{n}$. Therefore $f_{n,N_n} - f_n \to 0$, $Af_{n,N_n} - \tilde{A}f_n \to 0$, and then $f_{n,N_n} - f \to 0$, $Af_{n,N_n} - f^* \to 0$. From this it follows by definition that $\tilde{A}f = f^*$.

(It should be noted that a discontinuous operator can never be made continuous by extension.)

If an operator B extends an operator A, i.e., if as often as Af is defined, Bf is also defined and is $= Af$, then we write $B \succ A$ or $A \prec B$. We then have just proved that $A \prec \tilde{A}$, and that $\tilde{A}$ is Hermitian and closed. It is evident without further discussion that for each closed B with $A \prec B$, $\tilde{A} \prec B$ must also hold. Consequently, $\tilde{A}$ is the smallest closed extension of A. (Therefore $\tilde{\tilde{A}} = \tilde{A}$.)

The close relation between A and $\tilde{A}$ makes it plain that A can be replaced by $\tilde{A}$ in all considerations, since $\tilde{A}$ extends the domain of A in a logical way or, looking at it from the opposite point of view, A restricts the domain of $\tilde{A}$ in an unnecessary manner. Let this replacement of A by $\tilde{A}$ take place. Then, as a consequence, we may assume that all Hermitian operators with which we have to deal are closed.

Let us again consider a continuous Hermitian operator A. In this case, closure is equivalent to the closure of the domain. Now the condition $||Af|| \leq C \cdot ||f||$, characteristic for continuity, clearly holds for $\tilde{A}$ also. Therefore $\tilde{A}$ is also continuous -- and since the domain of $\tilde{A}$ is then closed, but on the other hand is everywhere dense, it is equal to $\mathfrak{R}_\infty$. That is, $\tilde{A}$ is defined everywhere, and consequently each closed and continuous operator is also. The converse holds too: if a closed operator is defined everywhere, then it is continuous (this is the theorem of Toeplitz,[96] into the proof of which we do not enter here).

Hilbert's result runs as follows: to each continuous operator there corresponds one and only one resolution of the identity (see the reference in Note 70). Since a continuous operator is always defined,

$$\int \lambda^2 d\|E(\lambda)f\|^2$$

must always be finite; since, in addition, this equals $\|Af\|^2$, and consequently, by **Co.**, $\leq C^2\|f\|^2$, we have:

$$0 \geq \|Af\|^2 - C^2\|f\|^2 = \int_{-\infty}^{\infty} \lambda^2 d\|E(\lambda)f\|^2 - C^2 \int_{-\infty}^{\infty} d\|E(\lambda)f\|^2$$

$$= \int_{-\infty}^{\infty} (\lambda^2 - C^2) d\|E(\lambda)f\|^2$$

Now let $f = E(-C-\epsilon)g$. Then $E(\lambda)f = E[\mathrm{Min}(\lambda, -C-\epsilon)]g$ and therefore for $\lambda \geq -C-\epsilon$ it is constant, so that we need only consider $\int_{-\infty}^{-C-\epsilon}$. In this case, $E(\lambda)f = E(\lambda)g$ and

$$\lambda^2 - C^2 \geq (C+\epsilon)^2 - C^2 > 2C\epsilon$$

so that

$$0 \geq 2C\epsilon \int_{-\infty}^{-C-\epsilon} d\|E(\lambda)g\|^2 = 2C\epsilon\|E(-C-\epsilon)g\|^2,$$

$$\|E(-C-\epsilon)g\|^2 \leq 0, \quad E(-C-\epsilon)g = 0$$

In the same way, it may be shown for $f = g - E(C+\epsilon)g$ that

$$g - E(C+\epsilon)g = 0$$

---

[96] Math. Ann. 69 (1911).

Consequently, for all $\epsilon > 0$, $E(-C-\epsilon) = 0$, $E(C+\epsilon) = 1$ i.e., $E(\lambda) = 0$ - for $\lambda < -C$ and $= 1$ for $\lambda > C$. (Because of $\overline{S_2}$., the latter also holds for $\lambda = C$.) That is, $E(\lambda)$ is variable only in the range $-C \leq \lambda \leq C$.

Conversely, this has the continuity of A as a consequence:

$$||Af||^2 = \int_{-\infty}^{\infty} \lambda^2 d||E(\lambda)f||^2 = \int_{-C}^{C} \lambda^2 d||E(\lambda)f||^2$$

$$\leq C^2 \int_{-C}^{C} d||Ef||^2 = C^2 \int_{-\infty}^{\infty} d||E(\lambda)f||^2 = C^2||f||^2 ,$$

$$||Af|| \leq C||f||$$

We see therefore that the continuous A are entirely exhausted by resolutions of the identity variable only in a finite interval of λ. But what is the situation with the other, discontinuous Hermitian operators? There are still available all resolutions of the identity which are variable for arbitrarily large λ. Now, do these exhaust the Hermitian operators mentioned?

The circumstances under which these operators cannot be defined everywhere must first be assessed correctly.

It is perfectly possible that a Hermitian operator may not be defined at points in Hilbert space at which this would be actually feasible. For example, our operator $A' = \frac{h}{2\pi i}\frac{\partial}{\partial q}$ was undefined for $f(q) = e^{-|q|}$, and we could also have limited $\frac{h}{2\pi i}\frac{\partial}{\partial q}$ to analytic functions (in $-\infty < q < +\infty$, q real),[97] etc. The domain was protected

[97]Even the $f(q)$ which are analytic in $-\infty < q < +\infty$ (with finite $\int_{-\infty}^{\infty}|f(q)|^2 dq, \int_{-\infty}^{\infty}|f'(q)|^2 dq, \ldots$) are everywhere dense in $\mathfrak{R}_\infty$. Indeed, by II.3., **D.**, the linear combinations of the

from entirely arbitrary contractions by the fact that we required it to be everywhere dense. Furthermore, we can restrict ourselves to closed operators. Still, even this is not sufficiently effective. Indeed, let us take, for example, the operator $A' = \frac{h}{2\pi i}\frac{\partial}{\partial q}$ in the interval $0 \leq q \leq 1$. Then let $f(q)$ be assumed differentiable everywhere, $\int_0^1 |f(q)|^2 dq$, $\int_0^1 |f'(q)|^2 dq$ finite. In order that A' be Hermitian, a boundary condition $f(0):f(1) = e^{-i\alpha}$ $(0 \leq \alpha < 2\pi)$ must be imposed; let the set of these $f(q)$ be $\mathfrak{A}_\alpha$, and A' itself, thus restricted, becoming A'_α. Furthermore, let us consider the boundary condition $f(1) = f(0) = 0$. We call the $f(q)$-set $\mathfrak{A}^0$ and the A', restricted accordingly, A'^0. All $\tilde{A}'_\alpha$ are extensions of $\tilde{A}'^0$ (which is then Hermitian, its domain is everywhere dense),[98] and therefore the closed

---

$$f_{a,b}(q) = \begin{cases} 1, & \text{for } a < q < b \\ 0, & \text{elsewhere} \end{cases}$$

are everywhere dense. Therefore, it suffices to approximate these arbitrarily well by the above $f(q)$. In fact, for example,

$$f_{a,b}^{(\epsilon)}(q) = \frac{1}{2} - \frac{1}{2}\tanh\frac{(x-a)(x-b)}{\epsilon} = \frac{1}{e^{2\frac{(x-a)(x-b)}{\epsilon}} + 1}$$

is of the desired type, and converges to $f_{a,b}(q)$ for $\epsilon \to +0$.

[98]It is again sufficient to approximate the $f_{a,b}(q)$, $0 \leq a < b \leq 1$, with functions from $\mathfrak{A}^0$. For example, the

$$f_{a,b}^{(\epsilon)}(q) = \frac{1}{2} - \frac{1}{2}\tanh\left(\frac{1}{\epsilon}\frac{(x-a-\epsilon)(x-b+\epsilon)}{x(1-x)}\right)$$

with $\epsilon \to +0$ can be used as such.

$\tilde{A}'_\alpha$ are also extensions of $\tilde{A}'^{\circ}$. All are different from one another and from $\tilde{A}'^{\circ}$. Indeed, the clearly unitary operation $f(q) \longrightarrow e^{i\beta q}f(q)$ transforms A' into $A' + \frac{h\beta}{2\pi} 1$, and $\mathfrak{A}_\alpha$ into $\mathfrak{A}_{\alpha+\beta}$, $\mathfrak{A}^{\circ}$ into $\mathfrak{A}^{\circ}$; therefore A'_α into $A'_{\alpha-\beta} + \frac{h\beta}{2\pi} 1$, A'° into $A'^{\circ} + \frac{h\beta}{2\pi} 1$; therefore $\tilde{A}'_\alpha$ into $\tilde{A}'_{\alpha+\beta} + \frac{h\beta}{2\pi} 1$, $\tilde{A}'^{\circ}$ into $\tilde{A}'^{\circ} + \frac{h\beta}{2\pi} 1$. Hence it would follow from $\tilde{A}'_\alpha = \tilde{A}'^{\circ}$ that $\tilde{A}'_{\alpha-\beta} = \tilde{A}'^{\circ}$, i.e., all $\tilde{A}'_\gamma$ would then be equal to one another. Consequently it suffices to show that $\tilde{A}'_\alpha \neq \tilde{A}'_\gamma$ for $\alpha \neq \gamma$, and this is certainly the case if A'_α, A'_γ possess no common Hermitian extension, i.e., if A' is not Hermitian in the union of $\mathfrak{A}_\alpha$, $\mathfrak{A}_\gamma$. Since $e^{i\alpha q}$ belongs to $\mathfrak{A}_\alpha$, $e^{i\gamma q}$ to $\mathfrak{A}_\gamma$, and

$$(A'e^{i\alpha q}, e^{i\gamma q}) - (e^{i\alpha q}, A'e^{i\gamma q})$$

$$= i\alpha \int_0^1 e^{i(\alpha-\gamma)q}dq - i\gamma \int_0^1 e^{i(\alpha-\gamma)q}dq$$

$$= \int_0^1 e^{i(\alpha-\gamma)q}i(\alpha - \gamma)dq = e^{i(\alpha-\gamma)} - 1 \neq 0$$

this is indeed the case. Consequently the closed Hermitian operator $\tilde{A}'^{\circ}$ is defined in too restricted a region, because there exist proper (i.e., different from $\tilde{A}'^{\circ}$) closed Hermitian extensions of it: the $\tilde{A}'_\alpha$ -- and therefore the extension process is infinitely many valued, since each $\tilde{A}'_\alpha$ can be used and each produces another solution of the eigenvalue problem. (It is always a pure discrete spectrum, but this is dependent on α: the set $h(\frac{\alpha}{2\pi} + k)$, $k = 0, \pm 1, \pm 2, \ldots$.) On the other hand, with the operator $\tilde{A}'^{\circ}$ itself, we in general expect no reasonable solutions of the eigenvalue problem. In fact we shall show in the course of this section that a Hermitian operator which belongs to a resolution of the identity (i.e., has a solvable eigenvalue problem), possesses no proper extensions. An operator which possesses no proper extensions -- which is already defined at all points where it could be defined in a reasonable fashion, -- i.e., without violation of its Hermitian nature -- we call a maximal operator. Then, by the above, a resolution of the identity can belong only to maximal operators.

On the other hand, the following theorem holds: each Hermitian operator can be extended to a maximal Hermitian operator. (In fact, a non-maximal but closed operator may always be so extended in an infinite number of different ways. That is, the only unique step of the extension process is the closure $A \longrightarrow \tilde{A}$. See Note 95.) There the most generally valid solution of the problem which we can expect is this: to each maximal Hermitian operator belongs one and only one resolution of the identity (each closed continuous operator is defined everywhere in $\mathfrak{R}_\infty$ and is therefore maximal).

Therefore it is necessary to answer these questions: Does a resolution of the identity belong to a maximal Hermitian operator? Can it ever happen, that several belong to the same operator?

We begin by stating the answers: to a given maximal Hermitian operators, there belongs none or exactly one resolution of the identity, and the former does occur -- i.e., the eigenvalue problem is certainly unique, but under certain conditions it is involvable. Nevertheless, the latter case is to be regarded in a certain sense as an exception. The method which leads to this result will now be sketched in broad outlines.

If we consider a rational function $f(\lambda)$ of a matrix A (of infinite dimensions, and capable of transformation to the diagonal form by a unitary transformation), then the eigenvectors are preserved, and the eigenvalues $\lambda_1, \ldots, \lambda_n$ go over into $f(\lambda_1), \ldots, f(\lambda_n)$.[99] If now the

[99]Since the function $f(\lambda)$ can be approximated by polynomials, it suffices to consider polynomials, and therefore their components, simple powers, $f(\lambda) = \lambda^s$ $(s = 0, 1, 2, \ldots)$.

$f(\lambda)$ maps the real axis (in the complex plane) on the circumference of the unit circle, then the matrices with exclusively real eigenvalues go over into those with eigenvalues of absolute value 1 -- i.e., the Hermitian go over into the unitary.[100] For example, $f(\lambda) = \frac{\lambda - i}{\lambda + i}$ has this property -- the corresponding transformation

$$U = \frac{A - i1}{A + i1}, \quad A = -i\frac{U + 1}{U - 1}$$

is known as the Cayley transformation. We shall now try the effect of this transformation for the Hermitian operators of $\mathfrak{R}_\infty$, i.e., we will define an operator U as follows: Uf is defined if and only if $f = (A + i1)\phi = A\phi + i\phi$, and then $Uf = (A - i1)\phi = A\phi - i\phi$. We hope that this definition will yield a single-valued Uf for all f, and that U will be unitary. The proof in the $\mathfrak{R}_n$ is naturally not relevant now, since it presumes the transformability to the diagonal form, i.e., the solvability of the eigenvalue problem, in fact with a pure discrete spectrum. But if the statements about U prove to be

---

Since a unitary transformation does not matter here, we may assume A to be a diagonal matrix; since the diagonal elements are the eigenvalues, they are $\lambda_1, \lambda_2, \ldots, \lambda_n$. We must then show only that A^s is also diagonal, and that it has the diagonal elements $\lambda_1^s, \lambda_2^s, \ldots, \lambda_n^s$ -- but this is obvious.

[100]That these properties are characteristic for the Hermitian or unitary character respectively, we again need only to verify them for the diagonal matrices. For the diagonal matrix A with elements $\lambda_1, \ldots, \lambda_n$, the diagonal matrix A^* with the elements $\overline{\lambda}_1, \ldots, \overline{\lambda}_n$ is the transposed conjugate; therefore $A = A^*$ implies that $\lambda_1 = \overline{\lambda}_1 \ldots, \lambda_n = \overline{\lambda}_n$, i.e., that $\lambda_1, \ldots, \lambda_n$ are real; and $AA^* = A^*A = 1$ implies that $\lambda_1\overline{\lambda}_1 = 1, \ldots, \lambda_n\overline{\lambda}_n = 1$, i.e., $|\lambda_1| = \ldots = |\lambda_n| = 1$.

correct, then we can solve the eigenvalue problem of A in the following way:

For U, the eigenvalue problem is solvable in the following form: there is a unique family of projections $E(\sigma)$ $(0 \leq \sigma \leq 1)$ which satisfies the following conditions:

$\bar{\bar{S}}_1$. $E(0) = 0$, $E(1) = 1$, and for $\sigma \to \sigma_0$, $\sigma \geq \sigma_0$, $E(\sigma)f \to E(\sigma_0)f$.

$\bar{\bar{S}}_2$. From $\sigma' \leq \sigma''$ it follows that $E(\sigma') \leq E(\sigma'')$.

$\bar{\bar{S}}_3$. It is always true that

$$(Uf, g) = \int_0^1 e^{2\pi i\sigma} d(E(\sigma)f, g)$$

(Uf is defined everywhere, and the integral on the right is always absolutely convergent.)[101]

---

[101] For proof of this fact, see the work of the author referred to in Note 78, also A. Wintner: Math. Z. <u>30</u> (1929). The absolute convergence of all integrals

$$\int_0^1 f(\sigma) d(E(\sigma)f, g)$$

with bounded $f(\sigma)$ is shown as follows. It is sufficient to observe $\text{Re}\,(E(\sigma)f, g)$, since substitution of if, g for f, g changes this into $\text{Im}\,(E(\sigma)f, g)$. Because

$$\text{Re}\,(E(\sigma)f, g) = \left(E(\sigma)\frac{f+g}{2}, \frac{f+g}{2}\right) - \left(E(\sigma)\frac{f-g}{2}, \frac{f-g}{2}\right)$$

only the $(E(\sigma)f, g)$ need be investigated. In

$$\int_0^1 f(\sigma) d(E(\sigma)f, g)$$

the integrand is bounded and the σ function behind the differential sign is monotonic; therefore the proposition is demonstrated.

This is proved in the framework and with the method of the Hilbert theory. This is made possible by the fact that the unitary operator U is always continuous (cf. references in Notes 70, 101). The analogy to the formulation $\overline{S_1}$. - $\overline{S_3}$. for Hermitian operators comes to mind. The only differences are: instead of the real integrands λ, $> -\infty$, $< +\infty$, the complex integrand $e^{2\pi i\sigma}$ here is taken around the circumference of the unit circle (even in $\mathfrak{R}_n$ the relation Hermitian-unitary possessed a far-reaching analogy to that of the real axis-unit circle contour; cf. Note 100), and the description of the operator-domain in $\overline{S_3}$. is superfluous, because unitary operators are defined everywhere.

Because of $\overline{\overline{S_1}}$., $E(\sigma)f \longrightarrow E(0)f = 0$ for $\sigma \geq 0$ (since $\sigma \geq 0$ by its nature), while for $\sigma \longrightarrow 1$ (since $\sigma \leq 1$) there need not be $E(\sigma)f \longrightarrow E(1)f = f$. If this is actually not the case, then the $E(\sigma)$ is discontinuous at $\sigma = 1$. But since a projection E' exists, such that for $\sigma \longrightarrow 1$, $\sigma < 1$, $E(\sigma)f \longrightarrow E'f$ (cf. THEOREM 17. in II.4., as well as Note 79), this means that $E' \neq E(1) = 1$ i.e., that $E'f = 0$ also possesses solutions $f \neq 0$. Because of $E(\sigma) \leq E'$ it follows from $E'f = 0$ that $E(\sigma)f = 0$ for all $\sigma < 1$. Conversely, by the definition of E', the former is also a consequence of the latter. If all $E(\sigma)f = 0$ $(\sigma < 1)$, then we see, just as at the beginning of II.8., that $(Uf, g) = (f, g)$ for all g, therefore $Uf = f$. Conversely, if $Uf = f$, then

$$\int_0^1 e^{2\pi i\sigma} d(E(\sigma)f, f) = (Uf, f) = (f, f),$$

$$\int_0^1 e^{2\pi i\sigma} d(E(\sigma)f, f) = (f, f), \quad \int_0^1 (1 - \cos(2\pi\sigma)) d(E(\sigma)f, f) = 0$$

$$\int_0^1 (1 - \cos(2\pi\sigma)) d(\|E(\sigma)f\|^2) = 0$$

From this we get, exactly as at the beginning of II.8.,

$E(\sigma)f = 0$ for all $\sigma < 1$ (and > 0, but this also holds for $\sigma = 0$). Consequently, the discontinuity of $E(\sigma)$ for $\sigma = 1$ means that $Uf = f$ is solvable with $f \neq 0$.

With our Cayley transforms U, we now have $\phi = Af + if$, $U\phi = Af - if$, and from $U\phi = \phi$ it then follows that $f = 0$, $\phi = 0$. Here $E(\sigma)f \longrightarrow f$ must also hold for $\sigma \longrightarrow 1$. Consequently, by the mapping

$$\lambda = -i\,\frac{e^{2\pi i\sigma} + 1}{e^{2\pi i\sigma} - 1} = -\cot \pi\sigma, \quad \sigma = -\frac{1}{\pi}\cot^{-1}\lambda$$

(which maps the intervals $0 < \sigma < 1$ and $-\infty < \lambda < +\infty$ one-to-one and monotonically on each other), we can produce a resolution of the identity $F(\lambda)$ from $E(\sigma)$ in the sense of $\overline{\mathbf{S}}_1$., $\overline{\mathbf{S}}_2$.:

$$\text{(C.)} \quad F(\lambda) = E\left(-\frac{1}{\pi}\cot^{-1}\lambda\right), \quad E = F(-\cot \pi\sigma)$$

We now want to show that $F(\lambda)$ satisfies $\overline{\mathbf{S}}_3$. for A if and only if $E(\sigma)$ satisfies $\overline{\overline{\mathbf{S}}}_3$. for U. In this way, the questions of uniqueness and existence for the solutions of the eigenvalue problem of the (possibly discontinuous) Hermitian operator A are reduced to the corresponding questions for the unitary operator U. These, however, as has been described, are answered in the most favorable way.

Therefore, let A be Hermitian and U be its Cayley transform. To begin with, we discuss the case that U is unitary. Then its $E(\sigma)$ must exist with $\overline{\overline{\mathbf{S}}}_1$., $\overline{\overline{\mathbf{S}}}_2$., as well as with $\overline{\overline{\mathbf{S}}}_3$. We form the $F(\lambda)$ according to **C.**, and then $\overline{\mathbf{S}}_1$., $\overline{\mathbf{S}}_2$. are fulfilled. If Af is defined, then

$$Af + if = \phi, \quad Af - if = U\phi$$

and therefore

$$f = \frac{\phi - U\phi}{2i}, \quad Af = \frac{\phi + U\phi}{2}$$

We calculate, in part symbolically:[102]

$$= \frac{1}{2i}(\phi - U\phi) = \frac{1}{2i}\left(\phi - \int_0^1 e^{2\pi i\sigma} dE(\sigma)\phi\right) = \int_0^1 \frac{1-e^{2\pi i\sigma}}{2i} dE(\sigma)\phi ,$$

$$E(\sigma)f = \int_0^1 \frac{1-e^{-2\pi i\sigma'}}{2i} d(E(\sigma)E(\sigma')\phi)$$

$$= \int_0^1 \frac{1-e^{2\pi i\sigma'}}{2i} d(E(\mathrm{Min}(\sigma, \sigma'))\phi) = \int_0^\sigma \frac{1-e^{2\pi i\sigma'}}{2i} dE(\sigma')\phi ,$$

$$\|E(\sigma)f\|^2 = (E(\sigma)f, f) = \int_0^\sigma \frac{1-e^{2\pi i\sigma'}}{2i} d(E(\sigma')\phi, f)$$

$$= \int_0^\sigma \frac{1-e^{2\pi i\sigma'}}{2i} \overline{d(E(\sigma')f, \phi)}$$

$$= \int_0^\sigma \frac{1-e^{2\pi i\sigma'}}{2i} d\left(\int_0^{\sigma'} \frac{1-e^{-2\pi i\sigma''}}{-2i} \overline{d(E(\sigma'')\phi, \phi)}\right)$$

$$= \int_0^\sigma \frac{1-e^{2\pi i\sigma'}}{2i} \cdot \frac{1-e^{-2\pi i\sigma'}}{-2i} \cdot \overline{d(E(\sigma')\phi, \phi)}$$

$$= \int_0^\sigma \frac{(1-e^{2\pi i\sigma'})(1-e^{-2\pi i\sigma'})}{4} d(\|E(\sigma')\phi\|^2)$$

$$= \int_0^\sigma \sin^2(\pi\sigma') d(\|E(\sigma')\phi\|^2)$$

therefore the integral given in $\overline{S_3}$. is

---

[102] We apply the Stieltjes integral to the elements of $\mathfrak{R}_\infty$ instead of to numbers. All our relations are so to be understood that they hold if we choose a fixed g from $\mathfrak{R}_\infty$ and substitute for each element of $\mathfrak{R}_\infty$ present in it its inner product with g . This holds for all g . In contrast to the operator-Stieltjes integrals in II.7., this is a half-symbolic process; instead of the one g of $\mathfrak{R}_\infty$, there two f, g were to be chosen arbitrarily from $\mathfrak{R}_\infty$, and instead of $(\ldots, g)$, $(\ldots f, g)$ was to be formed (the dots indicate the operator).

$$\int_{-\infty}^{\infty}\lambda^2 d||F(\lambda)f||^2 = \int_0^1 \cot^2(\pi\sigma)d||E(\sigma)f||^2$$

$$= \int_0^1 \cot^2(\pi\sigma)d\left(\int_0^\sigma \sin^2(\pi\sigma')d||E(\sigma')\phi||^2\right)$$

$$= \int_0^1 \cot^2(\pi\sigma)\sin^2(\pi\sigma')d||E(\sigma')\phi||^2$$

$$= \int_0^1 \cos^2(\pi\sigma)d||E(\sigma)\phi||^2$$

But since this is dominated absolutely by

$$\int_0^1 d||E(\sigma)\phi||^2 = ||\phi||^2$$

the result is finite. Furthermore,

$$Af = \tfrac{1}{2}(\phi + U\phi) = \tfrac{1}{2}(\phi + \int_0^1 e^{2\pi i\sigma}dE(\sigma)\phi) = \int_0^1 \frac{1+e^{2\pi i\sigma}}{2}dE(\sigma)\phi$$

$$= \int_0^1 - i\,\frac{e^{2\pi i\sigma}+1}{e^{2\pi i\sigma}-1}\cdot\frac{1-e^{2\pi i\sigma}}{2i}dE(\sigma)\phi$$

$$= \int_0^1 - \cot(\pi\sigma)d\left(\int_0^\sigma \frac{1-e^{2\pi i\sigma'}}{2i}dE(\sigma')\phi\right)$$

$$= \int_0^1 - \cot(\pi\sigma)dE(\sigma)f = \int_{-\infty}^{\infty}\lambda dF(\lambda)f$$

i.e., the final relation of $\overline{S}_3$. also holds. Consequently, A is in any case an extension of that operator which, by $\overline{S}_3$., belongs to F(λ), but since this is maximal (as we shall show,) A must be equal to it.[103]

---

[103]There is an implied assumption here that there actually exists such an operator for each given resolution of the identity F(λ). That is, for finite

We now discuss the converse. Let $F(\lambda)$ belong to A by $\bar{S}_1. - \bar{S}_3.$ What about U ? We first define $E(\sigma)$ by **C**. It therefore satisfies $\bar{\bar{S}}_1.$, $\bar{\bar{S}}_2.$ Let ϕ be arbitrary. We write (again symbolically)

$$f = \int_{-\infty}^{\infty} \frac{1}{\lambda+i}\, dF(\lambda)\phi = \int_0^1 \frac{1}{-\cot(\pi\sigma)+i}\, dE(\sigma)\phi$$

$$= \int_0^1 \frac{1-e^{2\pi i\sigma}}{2i}\, dE(\sigma)\phi$$

(since $\frac{1}{\lambda+i}$ or $\frac{1-e^{2\pi i\sigma}}{2i}$ is bounded, all integrals converge). Then

$$F(\lambda)f = E(\sigma)f = \int_0^1 \frac{1-e^{2\pi i\sigma'}}{2i}\, d(E(\sigma)E(\sigma')\phi)$$

$$= \int_0^1 \frac{1-e^{2\pi i\sigma'}}{2i}\, d(E(\mathrm{Min}(\sigma, \sigma'))\phi) = \int_0^{\sigma} \frac{1-e^{2\pi i\sigma'}}{2i}\, dE(\sigma')\phi ,$$

$$Af = \int_{-\infty}^{\infty} \lambda dF(\lambda)f = \int_0^1 -\cot(\pi\sigma)dE(\sigma)f$$

$$= \int_0^1 -i\,\frac{e^{2\pi i\sigma}+1}{e^{2\pi i\sigma}-1}\, d\left(\int_0^{\sigma} \frac{1-e^{2\pi i\sigma'}}{2i}\, dE(\sigma')\phi\right)$$

$$= \int_0^1 -i\,\frac{e^{2\pi i\sigma}+1}{e^{2\pi i\sigma}-1}\,\frac{1-e^{2\pi i\sigma}}{2i}\, dE(\sigma)\phi = \int_0^1 \frac{1+e^{2\pi i\sigma}}{2}\, dE(\sigma)\phi$$

$$\int_{-\infty}^{\infty} \lambda^2 d||F(\lambda)f||^2$$

it is assumed that an f^* can be found such that for all g ,

$$(f^*, g) = \int_{-\infty}^{\infty} \lambda d(F(\lambda)f, g)$$

and also that the f with this property is everywhere dense. (The Hermitian character of the operator thus defined follows from $\bar{S}_3$.: We exchange f, g in the final equation and take the complex conjugate.) Both of these propositions are proved in the reference given in Note 78.

therefore

$$Af + if = \int_0^1 dE(\sigma)\phi = \phi \ , \ Af - if = \int_0^1 e^{2\pi i\sigma} dE(\sigma)\phi$$

Hence $U\phi$ is defined and is equal to $\int_0^1 e^{2\pi i\sigma} dE(\sigma)\phi$. Since ϕ was arbitrary, U is then defined everywhere. When we form the inner product with any ψ and take the complex conjugate, we see that $U^*\psi = \int_0^1 e^{-2\pi i\sigma} dE(\sigma)\psi$. The final calculation of II.8. then shows that $UU^* = U^*U = 1$, i.e., U is unitary and belongs to $E(\sigma)$.

The solvability of the eigenvalue problem of A is then equivalent to the unitary nature of its Cayley transform U , and its uniqueness is established. The only questions remaining are: Can we always form U , and if so, is it unitary? To decide these questions, we begin again with a closed Hermitian operator A .

U was defined as follows: If $\phi = Af + if$, and only then, $U\phi$ is defined and is equal to $Af - if$. But first it must be shown that this definition is admissible in general, i.e. that several f cannot exist for one ϕ . That is, that it follows from $Af + if = Ag + ig$ that $f = g$, or, because of the linearity of A , that $f = 0$ follows from $Af + if = 0$.

We have

$$||Af \pm if||^2 = (Af \pm if, Af \pm if) =$$

$$= (Af, Af) \pm (if, Af) \pm (Af, if) + (if, if) =$$

$$= ||Af||^2 \pm i(Af, f) \mp i(Af, f) + ||f||^2 = ||Af||^2 + ||f||^2$$

Hence $Af + if = 0$ has $||f||^2 \leq ||Af + if||^2 = 0$, $f = 0$ as a consequence, and therefore our mode of definition is justified. Second, $||Af - if|| = ||Af + if||$, i.e. $||U\phi|| = ||\phi||$. Therefore U is continuous in so far as it is defined. Furthermore, let $\mathfrak{E}$ be the domain of U (therefore the set of all $Af + if$) and $\mathfrak{F}$ the range of U (the set of all $U\phi$, therefore the set of all $Af - if$). Since A and U are linear, $\mathfrak{E}$ and $\mathfrak{F}$ are linear manifolds, but they are also closed. Indeed, let ϕ be a limit point of $\mathfrak{E}$ or $\mathfrak{F}$ respectively. Then there is a sequence $\phi_1, \phi_2, \ldots$ from $\mathfrak{E}$ or $\mathfrak{F}$ respectively, with $\phi_m \longrightarrow \phi$. Hence $\phi_n = Af_n \pm if_n$. Since the ϕ_n converge, they satisfy the Cauchy convergence criterion (cf. **D.** in II.1.), and because

$$||f_m - f_n||^2 \leq ||A(f_m - f_n) \pm i(f_m - f_n)||^2 = ||\phi_m - \phi_n||^2$$

the f_n certainly satisfy this condition, and since

$$||Af_m - Af_n||^2 = ||A(f_m - f_n)||^2 \leq$$

$$\leq ||A(f_m - f_n) \pm i(f_m - f_n)||^2 = ||\phi_m - \phi_n||^2$$

the Af_n do also. Therefore the $f_1, f_2, \ldots$ and the $Af_1, Af_2, \ldots$ (by **D.** in II.1.) converge: $f_m \longrightarrow f$, $Af_m \longrightarrow f^*$. Since A is closed, Af is defined and is equal to f^*. Consequently we have

$$\phi_n = Af_n \pm if_n \longrightarrow f^* \pm if = Af \pm if \ , \ \phi_n \longrightarrow \phi$$

Therefore $\phi = Af \pm if$, i.e., ϕ also belongs to $\mathfrak{E}$ or $\mathfrak{F}$, respectively.

Therefore U is defined in the closed linear manifold $\mathfrak{E}$, and maps this on the closed linear manifold $\mathfrak{F}$. U is linear, and because

$$||Uf - Ug|| = ||U(f - g)|| = ||f - g||$$

it leaves all distances invariant. We then say that it is isometric. Consequently, $Uf \neq Ug$ follows from the fact that $f \neq g$, i.e., the mapping is one-to-one. It is also true that $(f, g) = (Uf, Ug)$, which we prove just as the analogous relation was proved in II.5. with unitary operators. Therefore U also leaves all inner products invariant. But U is clearly unitary if and only if $\mathfrak{E} = \mathfrak{F} = \mathfrak{R}_\infty$.

If A, B are two closed Hermitian operators, U, V their Cayley transforms, and the above sets are $\mathfrak{E}$, $\mathfrak{F}$ and $\mathfrak{G}$, $\mathfrak{H}$ respectively, then we see immediately that if B is a proper extension of A, then V is also a proper extension of U. Therefore $\mathfrak{E}$ is a proper subset of $\mathfrak{G}$ and $\mathfrak{F}$ a proper subset of $\mathfrak{H}$. Consequently $\mathfrak{E} \neq \mathfrak{R}_\infty$, $\mathfrak{F} \neq \mathfrak{R}_\infty$. Then U is not unitary, and the eigenvalue problem of A is unsolvable. Thus we have proved the theorem repeatedly cited before: If the eigenvalue problem of A is solvable, then there are no proper extensions of A, i.e., A is maximal.

Let us now turn back again to the closed Hermitian operator A and to its $\mathfrak{E}$, $\mathfrak{F}$, U. If Af is defined, then for $Af + if = \phi$, $U\phi$ is defined and $Af - if = U\phi$, therefore $f = \frac{1}{2i}(\phi - U\phi)$, $Af = \frac{1}{2}(\phi + U\phi)$, i.e., if we set $\psi = \frac{1}{2i}\phi$, $f = \psi - U\psi$, $Af = i(\psi + U\psi)$. Conversely, for $f = \psi - U\psi$, Af is certainly defined; because, since $U\psi$ is defined, $\psi = Af' + if'$ (Af' defined!), $U\psi = Af' - if'$ therefore $f = \psi - U\psi = 2if'$. The domain of A is then the set of all $\psi - U\psi$, and for $f = \psi - U\psi$, $Af = i(\psi + U\psi)$. Consequently, A is also uniquely determined by U (as well as $\mathfrak{E}$, $\mathfrak{F}$). Simultaneously we see that the $\psi - U\psi$ must be everywhere dense (as domain of A).

Conversely, we start out now from two closed linear manifolds E, F, and a linear, isometric mapping U of E on F. Is there a Hermitian operator A whose Cayley transform is this U? Since it is necessary that the $\psi - U\psi$ lie everywhere dense, this will also be assumed. The A in question is then uniquely determined by the foregoing, except that the question still presents itself as to whether this definition is possible, whether this A is really Hermitian, and whether U actually is its Cayley transform. The first is quite certainly correct if f determines the ϕ (whenever this in general exists) in $f = \phi - U\phi$ uniquely, i.e., if $\phi = \psi$ follows from $\phi - U\phi = \psi - U\psi$, or $\phi = 0$ from $\phi - U\phi = 0$. In fact, let $\phi - U\phi = 0$. Then it follows from $g = \psi - U\psi$ that

$$(\phi, g) = (\phi, \psi) - (\phi, U\psi) = (U\phi, U\psi) - (\phi, U\psi) = (U\phi - \phi, U\psi) = 0$$

and since these g are everywhere dense, $\phi = 0$.

Second, we must prove $(Af, g) = (f, Ag)$, i.e., that (Af, g) goes over into its complex conjugate upon the exchange of f, g. Let $f = \phi - U\phi$, $g = \psi - U\psi$, then $Af = i(\phi + U\phi)$, and

$$\begin{aligned}(Af, g) &= (i(\phi + U\phi), \psi - U\psi) \\ &= i(\phi, \psi) + i(U\phi, \psi) - i(\phi, U\psi) - i(U\phi, U\psi) \\ &= i[(U\phi, \psi) - \overline{(U\psi, \phi)}] = i(U\phi, \psi) + \overline{i(U\psi, \phi)}\end{aligned}$$

and this accomplishes what is desired in the exchange of f, g, i.e., of ϕ, ψ. The answer to the third question is seen in the following way. Let the Cayley transform of A be called V. Its domain is the set of all

$$Af + if = i(\phi + U\phi) + i(\phi - U\phi) = 2i\phi$$

i.e., the domain of U, and in this domain,

$$V(2i\phi) = V(Af + if) = Af - if = i(\phi + U\phi) - i(\phi - U\phi) = 2iU\phi$$

i.e., $V\phi = U\phi$. Therefore $V = U$.

The (closed) Hermitian operators therefore correspond to our linear isometric U, with everywhere dense $\phi - U\phi$, in a one-to-one correspondence -- if we order to each A its Cayley transform U .[104] We can now characterize all Hermitian extensions B of A , since all isometric extensions V of U can be found without difficulty (the $\phi - V\phi$ are automatically everywhere dense, since the $\phi - U\phi$, which are a subset of the former, are everywhere dense. In order that A be maximal, U must be also, and conversely. If U is not maximal, then $\mathfrak{E} \neq \mathfrak{R}_\infty$, $\mathfrak{F} \neq \mathfrak{R}_\infty$. These inequalities in turn imply that U is not maximal; indeed then $\mathfrak{R}_\infty - \mathfrak{E} \neq 0$, $\mathfrak{R}_\infty - \mathfrak{F} \neq 0$. We can therefore select a ϕ_0 from $\mathfrak{R}_\infty - \mathfrak{E}$ and a ψ_0 from $\mathfrak{R}_\infty - \mathfrak{F}$ with $\phi_0 \neq 0$, $\psi_0 \neq 0$, and if we replace these by

$$\frac{\phi_0}{||\phi_0||}, \frac{\psi_0}{||\psi_0||}$$

we even have $||\phi_0|| = ||\psi_0|| = 1$. We now define an operator V in $[\mathfrak{E}, \phi_0]$ such that for $f = \phi + a\phi_0$ (ϕ from $\mathfrak{E}$, a a number) $Vf = U\phi + a\psi_0$ -- V is clearly linear; since ϕ is orthogonal to ϕ_0 and $U\phi$ to ψ_0 , $||f||^2 = ||\phi||^2 + |a|^2$, $||Vf||^2 = ||U\phi||^2 + |a|^2$, therefore $||Vf|| = ||f||$ and V is isometric. Finally, V is a proper extension of U . Consequently it is characteristic for the maximal nature of A that $\mathfrak{E} = \mathfrak{R}_\infty$ or $\mathfrak{F} = \mathfrak{R}_\infty$.

---

[104] In order that the eigenvalue problem of A may always be solvable, the unitary character of U , i.e., $\mathfrak{E} = \mathfrak{F} = \mathfrak{R}_n$ or $\mathfrak{R}_\infty$ must follow from this. This is not the case in $\mathfrak{R}_\infty$ as we deduced from the existence of non-maximal A . In $\mathfrak{R}_n$ on the other hand, this must be the case. This can also be seen directly: since each linear manifold of $\mathfrak{R}_n$ is closed, that of the $\phi - U\phi$ is also; since it is everywhere dense, it is $= \mathfrak{R}_n$. $\mathfrak{E}$, the set of the ϕ , has no fewer dimensions that its linear image, the set of the $\phi - U\phi$, i.e., the maximum dimension number n . This latter must also hold for $\mathfrak{F}$ as a linear, one-to-one image of $\mathfrak{E}$. But for finite n it follows from this that $\mathfrak{E} = \mathfrak{F} = \mathfrak{R}_n$.

If, on the other hand, A is not maximal, then the closed linear manifolds $\mathfrak{N}_\infty - \mathfrak{E}$, $\mathfrak{N}_\infty - \mathfrak{F}$ are both $\neq 0$. Let the orthonormal sets spanning them be $\phi_1, \ldots, \phi_p$ and $\psi_1, \ldots, \psi_q$ respectively (cf. THEOREM 9., II.2.; $p = 1, 2, \ldots, \infty$; $q = 1, 2, \ldots, \infty$; for p or $q = \infty$, the ϕ or ψ series respectively does not terminate). Let $r = \mathrm{Min}(p, q)$, then we define a V in $[\mathfrak{E}, \phi_1, \ldots, \phi_r]$ as follows: for

$$f = \phi + \sum_{\nu=1}^{r} a_\nu \phi_\nu$$

(ϕ from $\mathfrak{E}$, $a_1, \ldots, a_r$ numbers),

$$Vf = U\phi + \sum_{\nu=1}^{r} a_\nu \psi_\nu$$

Furthermore, it can easily be seen that V is linear and isometric, as well as a proper extension of U. Its domain is $[\mathfrak{E}, \phi_1, \ldots, \phi_r]$, therefore for $r = p$, it is equal to $[\mathfrak{E}, \mathfrak{N}_\infty - \mathfrak{E}] = \mathfrak{N}_\infty$; its range is $[\mathfrak{F}, \psi_1, \ldots, \psi_r]$, therefore for $r = q$, it is equal to $[\mathfrak{F}, \mathfrak{N}_\infty - \mathfrak{F}] = \mathfrak{N}_\infty$. One of the two is then certainly equal to $\mathfrak{N}_\infty$. Let V be the Cayley transform of the Hermitian operator B. According to the discussion, B is the extension of A and maximal. We may observe that the ϕ and ψ can be chosen in an infinite number of ways (for example, we can replace ψ_1 by any $\theta\psi_1$, $|\theta| = 1$). Hence V and B can also be so chosen.

In this way we have examined the eigenvalue problem to its conclusion, and with the following result: if it is solvable, then it has only one solution, but for non-maximal operators, it is certainly not solvable. Non-maximal operators can always be extended in an infinite number of ways to maximal ones (we are discussing Hermitian operators throughout). But the maximality condition is not exactly the same as the solvability condition of the eigenvalue problem. The former is equivalent to $\mathfrak{E} = \mathfrak{N}_\infty$ or

$\mathfrak{F} = \mathfrak{R}_\infty$, the latter to $\mathfrak{E} = \mathfrak{R}_\infty$ and $\mathfrak{F} = \mathfrak{R}_\infty$.

We do not wish to investigate in any greater detail those operators for which the former but not the latter is the case. These are operators for which the eigenvalue problem is insolvable, and since no proper extensions exist (because of the maximality) this state of affairs is a final one. These operators are characterized by $\mathfrak{E} = \mathfrak{R}_\infty$, $\mathfrak{F} \neq \mathfrak{R}_\infty$ or $\mathfrak{E} \neq \mathfrak{R}_\infty$, $\mathfrak{F} = \mathfrak{R}_\infty$. Such operators in fact exist, and they all can be generated from two simple normal forms, so that they can be regarded as exceptional cases when compared with the maximal operators with a solvable eigenvalue problem. The reader will find more on this subject in the paper of the author mentioned in Note 95. In any case, such operators must be eliminated for the present from quantum mechanical considerations. The reason for this is that the resolution of the identity belonging to a Hermitian operator enters (as we shall see later) so essentially into all quantum mechanical concepts that we cannot dispense with its existence, i.e., with the solvability of the eigenvalue problem.[105] We shall accordingly admit only such Hermitian operators in general whose

---

[105]Nevertheless, as the author has pointed out (reference in Note 78), the following operator is maximal, but not hypermaximal: let $\mathfrak{R}_\infty$ be the space of all $f(q)$ defined in $0 \leq q < +\infty$ with finite

$$\int_0^\infty |f(q)|^2 dq$$

and let R be the operator $i\frac{d}{dq}$ which is defined for all continuously differentiable $f(q)$ with finite

$$\int_{-\infty}^\infty |f'(q)|^2 dq$$

and $f(0) = 0$, and which is closed. It is then equal to $-\frac{2\pi}{h}A'$ if we take the A' in II.8. for the interval

eigenvalue problem is solvable. Since this property is a sharpening of the maximality, these will be called hypermaximal operators.[106]

In conclusion, mention should be made of two classes of (closed) Hermitian operators which are certainly hypermaximal, too. First, the continuous operators: these are defined everywhere, and are therefore maximal, and since their eigenvalue problem is solvable according to Hilbert (see reference in Note 70), they are even hypermaximal. Second, the real operators, in any realization of $\mathfrak{R}_\infty$, if they are maximal. The only difference between $\mathfrak{E}$, $\mathfrak{F}$ in their definition was the sign of i which, if everything else is real, can make no difference. Therefore $\mathfrak{F} = \mathfrak{R}_\infty$ follows from $\mathfrak{E} = \mathfrak{R}_\infty$ and conversely, i.e., the hypermaximality from the maximality. Without assumption of the maximal property, we can in any event say that $\mathfrak{R}_\infty - \mathfrak{E}$ and $\mathfrak{R}_\infty - \mathfrak{F}$ have equally many dimensions. Therefore (in the terminology employed above in the investigation of the extension relations) $p = q$, therefore $r = p = q$ and

$$[\mathfrak{E}, \phi_1, \ldots, \phi_r] = [\mathfrak{E}, \mathfrak{R}_\infty - \mathfrak{E}] = \mathfrak{R}_\infty ,$$

$$[\mathfrak{F}, \psi_1, \ldots, \psi_r] = [\mathfrak{F}, \mathfrak{R}_\infty - \mathfrak{F}] = \mathfrak{R}_\infty ,$$

i.e., the extensions obtained at that time were hypermaximal. In any case, real operators possess hypermaximal extensions. In the reference given in Note 95 it is shown that the same holds true for all definite operators.

$0, \infty$. This R is now maximal but not hypermaximal. This can be verified by effective calculation of $\mathfrak{E}$, $\mathfrak{F}$. This is noteworthy because $A' = \frac{h}{2\pi} R$ can be interpreted physically as the momentum operator in the half space bounded on one side by the plane $q = 0$.

[106] This concept originated with Erhard Schmidt. Cf. the reference in Note 78.

10. COMMUTATIVE OPERATORS

Two operators R, S commute, by reason of the definition given in II.4., if $RS = SR$; and also, if the two are not defined everywhere, the domains on either side must coincide. To begin, we restrict ourselves to Hermitian operators, and in order to avoid difficulties with regard to the domains, we also limit ourselves to those operators which are defined everywhere -- hence to continuous operators. Along with R, S , we also consider the resolutions of the identity belonging to them: $E(\lambda)$, $F(\lambda)$.

The commutativity of R, S means that $(RSf, g) = (SRf, g)$ for all f, g , i.e., $(Sf, Rg) = (Rf, Sg)$. Furthermore, the commutativity of R^n, S $(n = 0,1,2,\ldots)$ follows from that of R, S , and hence the commutativity of $p(R), S$ $[p(x) = a_0 + a_1x + \ldots + a_nx^n]$ also follows.

Now symbolically,

$$R = \int_{-C}^{C} \lambda dE(\lambda), \quad s(R) = \int_{-C}^{C} s(\lambda)dE(\lambda)$$

(C is the constant introduced in II.9. for the continuous operator R , which was called A at that time; $s(x)$ is any function, cf. II.8. in particular, Note 94). For polynomials $s(x)$, we have $(s(R)f, Sg) = (Sf, s(R)g)$, therefore

$$*. \qquad \int_{-C}^{C} s(\lambda)d(E(\lambda)f, Sg) = \int_{-C}^{C} s(\lambda)d(Sf, E(\lambda)g)$$

Since we can approximate every continuous function $s(x)$ arbitrarily well by polynomials (uniformly in $-C \leq x \leq C$) , *. also holds for continuous $s(x)$. Now let

$$s(x) = \begin{cases} \lambda_0 - x, & \text{for } x \leq \lambda_0 \\ 0, & \text{for } x \geq \lambda_0 \end{cases}$$

then *. gives

$$\int_{-C}^{\lambda_0} (\lambda_0 - \lambda) d(E(\lambda)f,\ Sg) = \int_{-C}^{\lambda_0} (\lambda_0 - \lambda) d(Sf,\ E(\lambda)g)$$

If we replace λ_0 by $\lambda_0 + \epsilon$ $(\epsilon > 0)$ here, then subtraction and division by ϵ gives:

$$\int_{-C}^{\lambda_0} d(E(\lambda)f,\ Sg) + \int_{\lambda_0}^{\lambda_0+\epsilon} \frac{\lambda-\lambda_0}{\epsilon} d(E(\lambda)f,\ Sg)$$

$$= \int_{-C}^{\lambda_0} d(Sf,\ E(\lambda)g) + \int_{\lambda_0}^{\lambda_0+\epsilon} \frac{\lambda-\lambda_0}{\epsilon} d(Sf,\ E(\lambda)g)$$

and as $\epsilon \longrightarrow 0$ (recall $\overline{\mathbf{S}}_1$ .1):

$$\int_{-C}^{\lambda_0} d(E(\lambda)f,\ Sg) = \int_{-C}^{\lambda_0} d(Sf,\ E(\lambda)g)\ ,$$

$$(E(\lambda_0)f,\ Sg) = (Sf,\ E(\lambda_0)g)$$

Consequently, all $E(\lambda_0)$, $-C \leq \lambda_0 \leq C$ commute with S , but this is all the more true for the remaining $E(\lambda_0)$, since for $\lambda_0 < -C$ and $> C$ respectively, $E(\lambda_0) = 0$ and 1 respectively.

Therefore, if R commutes with S , then all $E(\lambda_0)$ do likewise. Conversely, if all $E(\lambda)$ commute with S , then *. holds for each function $s(x)$. Consequently, all $s(R)$ commute with S . From this we may conclude: first, that R commutes with S if and only if all $E(\lambda)$ do; and second, that in this case, all functions of R [the $s(R)$] commute with S also.

But an $E(\lambda)$ commutes with S if and only if this holds for $E(\lambda)$ and all $F(\mu)$ (we apply our theorem to $S, E(\lambda)$ -- instead of R, S --) . Therefore this is also characteristic for the commuting of R, S : all $E(\lambda)$ should commute with all $F(\mu)$. Furthermore, the commutativity of R, S , has, from the above, the commutativity of $r(R)$, S as a consequence. If we replace R, S by S, $r(R)$, we then obtain the commutativity of $r(R)$, $s(S)$

If the Hermitian operators R, S are not restricted by the condition of continuity, then the situation is more complicated, since the domains of RS and SR may now offer considerable involvements. For example, $R.0$ is always defined ($0f = 0$, $R.0f = R(0f) = R.0 = 0$), while on the other hand, $0.R$ is defined only if R is defined (cf. the comments on this in II.5.). Therefore for R not defined everywhere, $R.0 \neq 0.R$ because of the difference in the domains. That is, taking it literally, R, 0 do not commute. Such a state of affairs is unsatisfactory for our later purposes: 0 should commute not only with all continuous Hermitian operators, but also with all Hermitian operators.[107] We therefore want to define commutativity for discontinuous R, S in a different way. We limit ourselves to the hypermaximal R, S, which, by II.9., are the only operators of interest to us. We then define that R, S shall be called commutative in the new sense if all $E(\lambda)$ commute with all $F(\mu)$ (where these are again the respective resolutions of the identity) in the old sense. For continuous R, S, the new definition is identical with the old, as we have seen, while in the case of discontinuity of R or S (or both) it differs, the two definitions may differ. An example of the latter case is R, 0: in the old sense they did not commute, but they do in the new,

---

[107] Since (cf. II.5.) $R.1$, $1.R$ are defined if and only if R is defined, the same holds for $R.a1$, $a1.R$ ($a \neq 0$). Then these two products are equal, i.e., R and $a1$ commute. Consequently the commutativity of R and $a1$ holds with a single exception: $a = 0$, R not defined everywhere. This is unfortunate and accounts for the change in the commutativity definition.

since for 0, each $F(\mu)$ equals 0 or 1,[108] therefore each $F(\mu)$ commutes with the $E(\lambda)$.

We have proved above that if R, S are two commutative (continuous) Hermitian operators, then each function $r(R)$ of R commutes with each functions $s(S)$ of S. Since the hypothesis is always satisfied for $R = S$, two functions $r(R), s(R)$ of the same operator always commute [this also follows from the multiplication formula at the end of II.8.: $r(R)s(R) = t(R)$ with $r(x)s(x) = t(x)$]. If $r(x), s(x)$ are real, by the way, then $r(R)$, $s(R)$ are Hermitian [by II.8.: if $r(x)$ is real, $(r(R))^* = \overline{r}(R) = r(R)$].

The converse of the former is also valid. If A, B are two commuting Hermitian operators, then there exists a Hermitian operator R, of which both are functions, i.e., $A = r(R), B = s(R)$. Indeed, even more is true: if an arbitrary (finite or infinite) set of commuting Hermitian operators is given, $A, B, C, \ldots,$ then there exists a Hermitian operator R, of which all $A, B, C, \ldots$ are functions. We give no proof of this theorem here, and can only refer to the literature on the subject.[109] For

---

[108] This resolution of the identity belongs to a.1:

$$F(\mu) = \begin{cases} 1, & \text{for } \mu \geq a \\ 0, & \text{for } \mu < a \end{cases}$$

This can easily be verified.

[109] For two Hermitian operators A, B, which belong to a special class (the so-called totally continuous class, cf. reference in Note 70). Toeplitz proved (cf. reference in Note 33) a theorem from which the above follows. (Namely, the existence of a complete orthonormal set from the common eigenfunctions of A, B.) The general theorem for

our purposes, this theorem is of importance only for a finite number of A, B, C, ... with pure discrete spectra. It shall be proved for this case in the following; regarding the general case we can give only a few orienting remarks.

Therefore, let A, B, C, ... be a finite number of Hermitian operators with pure discrete spectra. If λ is any number, then let the closed linear manifold spanned by all solutions of $Af = \lambda f$ be called $\mathfrak{L}_\lambda$, and its projection E_λ. Then λ is a discrete eigenvalue of A if and only if solutions $f \neq 0$ exist, hence for $\mathfrak{L}_\lambda \neq (0)$, i.e., $E_\lambda \neq 0$. Correspondingly we form $\mathfrak{M}_\lambda$, F_λ for B, $\mathfrak{N}_\lambda$, G_λ for C, ... From $Af = \lambda f$ it follows that $ABf = BAf = B(\lambda f) = (\lambda Bf)$, i.e., along with f, Bf also belongs to $\mathfrak{L}_\lambda$. Since $E_\lambda f$ always belongs to $\mathfrak{L}_\lambda$, $BE_\lambda f$ does also, therefore $E_\lambda BE_\lambda f = BE_\lambda f$. This holds identically; $E_\lambda BE_\lambda = BE_\lambda$. Application of *. results in $E_\lambda BE_\lambda = E_\lambda B$, therefore $E_\lambda B = BE_\lambda$. In the same way as we just now deduced the commutativity of B, E_λ from that of A, B, that of E_λ, F_μ follows from that of B, E_λ. Since A, B are in no way distinguished from the other A, B, C, ..., we can then say that all E_λ, F_μ, G_ν, ... commute with each other. Consequently $K(\lambda\mu\nu\ldots) = E_\lambda F_\mu G_\nu \ldots$ is also a projection. Let its closed linear manifold be called $\mathfrak{K}(\lambda\mu\nu\ldots)$. By THEOREM 14, (II.4.), $\mathfrak{K}(\lambda\mu\nu\ldots)$ is the intersection of $\mathfrak{L}_\lambda$, $\mathfrak{M}_\mu$, $\mathfrak{N}_\nu$, ..., i.e., the totality of common solutions of

$$Af = \lambda f, \quad Bf = \mu F, \quad Cf = \nu f, \quad \ldots$$

Let $\lambda, \mu, \nu, \ldots$ and $\lambda', \mu', \nu', \ldots$ be two different sets of numbers, i.e., $\lambda \neq \lambda'$ or $\mu \neq \mu'$ or $\nu \neq \nu'$, ... If f belongs to $\mathfrak{K}(\lambda\mu\nu\ldots)$, f' to $\mathfrak{K}(\lambda'\mu'\nu'\ldots)$, then f, f' are orthogonal. For $\lambda \neq \lambda'$

---

arbitrary A, B, or A, B, C, ... has been proved by the author (see Note 94).

because $Af = \lambda f$, $Af' = \lambda' f'$, for $\mu \neq \mu'$ because $Bf = \mu f$, $Bf' = \mu' f'$, ... Consequently the entire $\mathfrak{K}(\lambda\mu\nu\ldots)$ is orthogonal to the entire $\mathfrak{K}(\lambda'\mu'\nu'\ldots)$.

Since A has a pure discrete spectrum, the $\mathfrak{L}_\lambda$ spans the entire $\mathfrak{R}_\infty$ (as closed linear manifold). An $f \neq 0$ cannot therefore be orthogonal to all $\mathfrak{L}_\lambda$, i.e., for at least one $\mathfrak{L}_\lambda$ its projection in $\mathfrak{L}_\lambda$ must be $\neq 0$, i.e., $E_\lambda f \neq 0$. In the same way, a μ must exist with $F_\mu f \neq 0$, moreover, a ν with $G_\nu f \neq 0$, ... Consequently, for each $f \neq 0$ we can find a λ with $E_\lambda f \neq 0$, hence a μ with $F_\mu(E_\lambda f) \neq 0$, then a ν with $G_\nu(F_\mu(E_\lambda f)) \neq 0$, ... Finally, $\ldots G_\nu F_\mu E_\lambda f \neq 0$, $E_\lambda F_\mu G_\nu \ldots f \neq 0$, $K(\lambda\mu\nu\ldots)f \neq 0$ i.e., f is not orthogonal to $\mathfrak{K}(\lambda\mu\nu\ldots)$. Therefore an f orthogonal to all $\mathfrak{K}(\lambda\mu\nu\ldots)$ is $= 0$. Consequently, the $\mathfrak{K}(\lambda\mu\nu\ldots)$ together span all $\mathfrak{R}_\infty$ as closed linear manifold.

Now let $\phi^{(1)}_{(\lambda\mu\nu\ldots)}$, $\phi^{(2)}_{(\lambda\mu\nu\ldots)}$, ... be an orthonormal set which spans the linear manifold $\mathfrak{K}(\lambda\mu\nu\ldots)$. (This sequence may or may not terminate, depending on whether or not $\mathfrak{K}(\lambda\mu\nu\ldots)$ has a finite or infinite number of dimensions. On the other hand, if $\mathfrak{K}(\lambda\mu\nu\ldots) = 0$ then it consists of 0 terms.) Each $\phi^{(n)}_{(\lambda\mu\nu\ldots)}$ belongs to a $\mathfrak{K}(\lambda\mu\nu\ldots)$ and is therefore an eigenfunction of all A, B, C, ... Two different are always mutually orthogonal; if they have the same $\lambda,\mu,\nu,\ldots$-set, by reason of their definition, if they have different $\lambda,\mu,\nu,\ldots$-sets, then because they belong to different $\mathfrak{K}(\lambda\mu\nu\ldots)$. All $\phi^{(n)}_{(\lambda\mu\nu\ldots)}$ span the same linear manifold as all $\mathfrak{K}(\lambda\mu\nu\ldots)$: $\mathfrak{R}_\infty$. Consequently the $\phi^{(n)}_{(\lambda\mu\nu\ldots)}$ form a complete orthonormal set.

We have then produced a complete orthonormal set from the common eigenfunctions of A, B, C, ... We now call these $\psi_1, \psi_2, \ldots$, and we write the corresponding eigenvalue equations

$$A\psi_m = \lambda_m\psi_m, \quad B\psi_m = \mu_m\psi_m, \quad C\psi_m = \nu_m\psi_m, \quad \ldots$$

We now take any set of pairwise different numbers $\kappa_1, \kappa_2, \kappa_3, \ldots$ and form a Hermitian operator R with the pure discrete spectrum, $\kappa_1, \kappa_2, \ldots$ and with the corresponding eigenfunctions $\psi_1, \psi_2, \ldots$. That is,

$$R\left(\sum_{m=1}^{\infty} x_m\psi_m\right) = \sum_{m=1}^{\infty} x_m\kappa_m\psi_m$$ [110]

Now let $F(\kappa)$ be a function defined in $-\infty < \kappa < +\infty$ for which $F(\kappa_m) = \lambda_m$ $(m = 1,2,\ldots)$ [at all other points κ, $F(\kappa)$ may be arbitrary]. Also, let $G(\kappa)$ be a function with $G(\kappa_m) = \mu_m$, $H(\kappa)$ one with $H(\kappa_m) = \nu_m, \ldots$. We wish to show that

$$A = F(R),\ B = G(R),\ C = H(R),\ \ldots$$

We must then show that if R has a pure discrete spectrum $\kappa_1, \kappa_2, \ldots$ with the eigenfunctions $\psi_1, \psi_2, \ldots$, then $F(R)$ has a pure discrete spectrum $F(\kappa_1), F(\kappa_2), \ldots$

---

[110] The $\kappa_1, \kappa_2, \kappa_3, \ldots$ are to be chosen bounded (for example, $\kappa_m = 1/m$) in order that R be continuous. In fact,

$$\|R\psi_m\| = \|\kappa_m\psi_m\| = |\kappa_m| \leq C\cdot\|\psi_m\| = C,\quad |\kappa_m| < C$$

follows immediately from the continuity of R, i.e., from $\|Rf\| \leq C\cdot\|f\|$. Conversely, from $|\kappa_m| \leq C$ $(m = 1,2,\ldots)$ it follows that

$$\|Rf\|^2 = \left\|R\left(\sum_{m=1}^{\infty} x_m\psi_m\right)\right\|^2 = \left\|\sum_{m=1}^{\infty} x_m\kappa_m\psi_m\right\|^2 = \sum_{m=1}^{\infty} |x_m|^2|\kappa_m|^2,$$

$$\|f\|^2 = \left\|\sum_{m=1}^{\infty} x_m\psi_m\right\|^2 = \sum_{m=1}^{\infty} |x_m|^2$$

Therefore, $\|Rf\|^2 \leq C^2\cdot\|f\|^2$, $\|Rf\| \leq C\cdot\|f\|$, i.e., R is continuous.

with the same eigenfunctions $\psi_1, \psi_2, \cdots$. But since these also form a complete orthonormal set, it suffices to show that $F(R)\psi_m = F(\kappa_m)\cdot\psi_m$.

Let (by II.8.) $E(\lambda) = \sum_{\kappa_m \leq \lambda} P_{[\psi_m]}$ be the resolution of the identity belonging to R. Then symbolically,

$$R = \int \lambda dE(\lambda)$$

and by definition

$$F(R) = \int F(\lambda) dE(\lambda)$$

Furthermore,

$$E(\lambda)\psi_m = \begin{cases} \psi_m, & \text{for } \kappa_m \leq \lambda \\ 0, & \text{for } \kappa_m > \lambda \end{cases}$$

From this it follows that

$$(F(R)\psi_m, g) = \int F(\lambda) d(E(\lambda)\psi_m, g) = F(\kappa_m)\cdot(\psi_m, g)$$

for all g, therefore it is actually true that $F(R)\psi_m = F(\kappa_m)\cdot\psi_m$.

With this, the problem is settled for the case of the pure discrete spectra, as we asserted previously. In the case of the continuous spectra, we must be content with the reference of Note 109, and we shall emphasize here only an especially characteristic case.

Let $\Re_\infty$ be the space of all $f(q_1, q_2)$ with finite $\iint |f(q_1, q_2)|^2 dq_1 dq_2$, and let the square $0 \leq q_1, q_2 \leq 1$ be their domain of variation. We form the operators $A = q_1\cdots$, $B_1 = q_2\cdots$. They are Hermitian for this q_1, q_2-region, also continuous (but not for $-\infty < q_1, q_2 < +\infty$!) and they commute. Therefore both must be functions of an R. Consequently, this R commutes with A, B, from which it follows (although we will not prove this here) that R has the form $s(q_1, q_2)\cdots$ [$s(q_1, q_2)$ a bounded function]. Consequently R^n ($n = 0,1,2,\ldots$) is equal to $(s(q_1, q_2))^n\cdots$, and $F(R)$ is equal to $F(s_1(q_1, q_2))\cdots$, if $F(\kappa)$ is a polynomial. But this formula can be extended to all $F(\kappa)$, which we again will not discuss in detail. It also follows from $F(R) = A$, $G(R) = B$ that

$$F(s(q_1, q_2)) = q_1, \; G = (s(q_1, q_2)) = q_2 \quad {}^{111}$$

That is, the mappings $s(q_1, q_2) = \kappa$ and $F(\kappa) = q_1$, $G(\kappa) = q_2$ (which are reciprocal to each other) must map the square surface $0 \leq q_1, q_2 \leq 1$ on the linear number set of the κ uniquely -- something which conflicts with the ordinary geometric intuition.

But on the basis of our proof previously mentioned, we know that this must nevertheless be possible -- and indeed a mapping of the desired type is accomplished by means of the so-called Peano curve.[112] A more rigorous treatment of the proof given in Note 109 actually shows that this leads in the present case to the Peano curve or to constructs that are closely related to it.

11. THE TRACE

Several important invariants of operators shall be defined here.

For a matrix $\{a_{\mu\nu}\}$ of the $\mathfrak{R}_n$, the trace

---

[111] Exceptions may occur in a q_1, q_2-set of Lebesgue measure 0.

[112] Cf., for example, the reference in Note 45.

$\sum_{\mu=1}^{n} a_{\mu\mu}$ is one such invariant. It is unitary-invariant, that is, it does not change if we transform $\{a_{\mu\nu}\}$ into another (cartesian) coordinate system.[113] But if we replace the matrix $\{a_{\mu\nu}\}$ by the corresponding operator

$$A\{x_1, \ldots, x_n\} = \{y_1, \ldots, y_n\}, \quad y_\mu = \sum_{\nu=1}^{n} a_{\mu\nu} x_\nu$$

then the $a_{\mu\nu}$ are expressed as follows, with the help of A:

---

[113] $\{a_{\mu\nu}\}$ replaces the transformation (i.e., the operator)

$$\eta_\mu = \sum_{\nu=1}^{n} a_{\mu\nu} \xi_\nu$$

($\mu = 1, \ldots, n$, cf. the developments in II.7.). If we transform by

$$\xi_\mu = \sum_{\nu=1}^{n} x_{\nu\mu} \mathfrak{x}_\nu, \quad \eta_\mu = \sum_{\nu=1}^{n} x_{\nu\mu} \mathfrak{y}_\nu \quad (\mu = 1, \ldots, n)$$

then we obtain

$$\mathfrak{y}_\mu = \sum_{\nu=1}^{n} \mathfrak{a}_{\mu\nu} \mathfrak{x}_\nu \quad (\mu = 1, \ldots, n)$$

with

$$\mathfrak{a}_{\mu\nu} = \sum_{\rho,\sigma=1}^{n} a_{\rho\sigma} \overline{x}_{\mu\rho} x_{\nu\sigma} \quad (\mu, \nu = 1, \ldots, n)$$

$\{\mathfrak{a}_{\mu\nu}\}$ is the transformed matrix. Clearly,

$$\sum_{\mu=1}^{n} \mathfrak{a}_{\mu\mu} = \sum_{\mu,\rho,\sigma=1}^{n} a_{\rho\sigma} \overline{x}_{\mu\rho} x_{\mu\sigma} = \sum_{\rho,\sigma=1}^{n} a_{\rho\sigma} \left(\sum_{\mu=1}^{n} \overline{x}_{\mu\rho} x_{\mu\sigma} \right) = \sum_{\rho=1}^{n} a_{\rho\rho}$$

i.e. the trace is invariant.

$$\phi_1 = \{1,0,\ldots,0\}, \phi_2 = \{0,1,\ldots,0\}, \ldots, \phi_n = \{0,0,\ldots,1\}$$

form a complete orthonormal set, and obviously $a_{\mu\nu} = (A\phi_\nu, \phi_\mu)$ (cf. II.5., in particular, Note 60). The trace is therefore $\sum_{\mu=1}^{n} (A\phi_\mu, \phi_\mu)$ and its unitary invariance means that its value is the same for each complete orthonormal set.

We can immediately consider the analogy of this concept in the $\mathfrak{R}_\infty$. Let A be a linear operator. We then take any complete orthonormal set $\phi_1, \phi_2, \ldots$ for which all $A\phi_\mu$ are defined (this is certainly possible, if the domain of A is everywhere dense -- it suffices to orthogonalize a dense sequence $f_1, f_2, \ldots$ in it, by II.2., THEOREM 8.), and set trace (A) [or $\mathrm{Tr}\,A$] $= \sum_{\mu=1}^{\infty} (A\phi_\mu, \phi_\mu)$. We must then show that this actually depends only on A (and not on the ϕ_μ !).

For this purpose, we first introduce two complete orthonormal sets $\phi_1, \phi_2, \ldots$ and $\psi_1, \psi_2, \ldots$, and set

$$\mathrm{Tr}(A;\ \phi,\ \psi) = \sum_{\mu,\nu=1}^{\infty} (A\phi_\mu, \psi_\nu)(\psi_\nu, \phi_\mu) .$$

From II.2., THEOREM 7.γ. it follows that this is equal to $\sum_{\mu=1}^{\infty} (A\phi_\mu, \phi_\mu)$ so that this depends only apparently on the ψ_ν. Furthermore,

$$\sum_{\mu,\nu=1}^{\infty} (A\phi_\mu, \psi_\nu)(\psi_\nu, \phi_\mu) = \sum_{\mu,\nu=1}^{\infty} (\phi_\mu, A^*\psi_\nu)(\psi_\nu, \phi_\mu)$$

$$= \overline{\sum_{\mu,\nu=1}^{\infty} (A^*\psi_\nu, \phi_\mu)(\phi_\mu, \psi_\nu)}$$

i.e., $\mathrm{Tr}\,(A;\ \phi,\ \psi) = \overline{\mathrm{Tr}\,(A^*;\ \psi,\ \phi)}$. Since the right side, according to the above, depends only apparently on the ϕ_μ, the same holds for the left: their dependence on ϕ_μ and ψ_ν is therefore only apparent. In reality, therefore, the trace depends only on A. Consequently, we may designate $\operatorname{Tr}(A; \phi, \psi)$ with $\operatorname{Tr} A$. Since this is equal to $\sum_{\mu=1}^{\infty}(A\phi_\mu, \phi_\mu)$, the desired invariance proof has been achieved. But from the last equation it also follows that $\operatorname{Tr} A = \overline{\operatorname{Tr} A^*}$.

The relations

$$\operatorname{Tr}(aA) = a \operatorname{Tr} A, \quad \operatorname{Tr}(A \pm B) = \operatorname{Tr} A \pm \operatorname{Tr} B$$

are obvious. Furthermore

$$\operatorname{Tr}(AB) = \operatorname{Tr}(BA)$$

holds, even for non-commuting A, B. This may be shown as follows:

$$\operatorname{Tr}(AB) = \sum_{\mu=1}^{\infty}(AB\phi_\mu, \phi_\mu) = \sum_{\mu=1}^{\infty}(B\phi_\mu, A^*\phi_\mu)$$

$$= \sum_{\mu,\nu=1}^{\infty}(B\phi_\mu, \psi_\nu)(\psi_\nu, A^*\phi_\mu) = \sum_{\mu,\nu=1}^{\infty}(B\phi_\mu, \psi_\nu)(A\psi_\nu, \phi_\mu)$$

in which $\phi_1, \phi_2, \ldots$ and $\psi_1, \psi_2, \ldots$ can be two arbitrary complete orthonormal sets, and the symmetry of this expression in A, B (in the case of simultaneous interchange of the ϕ, ψ) is evident. Consequently, for the Hermitian operators A, B,

$$\operatorname{Tr}(AB) = \overline{\operatorname{Tr}[(AB)^*]} = \overline{\operatorname{Tr}(B^*A^*)}$$

$$= \overline{\operatorname{Tr}(BA)} \quad = \overline{\operatorname{Tr}(AB)}$$

Therefore Tr AB is real. (Tr A is of course real, too).

If $\mathfrak{M}$ is a closed linear manifold, and E its projection, then Tr E is determined as follows: let

$\psi_1, \ldots, \psi_k$ be an orthonormal set which spans the closed linear manifold $\mathfrak{M}$ and $\chi_1, \ldots, \chi_l$ one which spans $\mathfrak{R}_\infty - \mathfrak{M}$ (of course one of k or l or both must be infinite) -- then $\psi_1, \ldots, \psi_k, \chi_1, \ldots, \chi_l$ together span $\mathfrak{R}_\infty$, i.e., they form a complete orthonormal set (THEOREM 7.α. in II.2.). Therefore

$$\operatorname{Tr} E = \sum_{\mu=1}^{k} (E\psi_\mu, \psi_\mu) + \sum_{\mu=1}^{l} (E\chi_\mu, \chi_\mu) = \sum_{\mu=1}^{k} (\psi_\mu, \psi_\mu) + \sum_{\mu=1}^{l} (0, \chi_\mu)$$

$$= \sum_{\mu=1}^{k} 1 = k$$

i.e., Tr E is the dimension number of $\mathfrak{M}$.

If A is definite, then all $(A\phi_\mu, \phi_\mu) \geq 0$, therefore $\operatorname{Tr} A \geq 0$. If in this case, $\operatorname{Tr} A = 0$, then all the $(A\phi_\mu, \phi_\mu)$ must vanish, therefore $A\phi_\mu = 0$ (THEOREM 19. in II.5.). If $\|\phi\| = 1$, then we can find a complete orthonormal set $\phi_1, \phi_2, \ldots$ with $\phi_1 = \phi$. (Indeed, let $f_1, f_2, \ldots$ be everywhere dense. We then orthogonalize $\phi, f_1, f_2, \ldots$ -- cf. the proof of THEOREM 7. in II.2. -- by which means we obtain a complete orthonormal set beginning with ϕ). Therefore $A\phi = 0$. If now f is arbitrary, then $Af = 0$ holds for $f = 0$, and for $f \neq 0$, it follows from the above with

$$\phi = \frac{1}{\|f\|} f$$

-- therefore $A = 0$. That is, if A is definite, then $\operatorname{Tr} A > 0$.

In the brevity and simplicity of our observations relative to the trace, our treatment has not been mathematically rigorous. For example, we have considered the series

$$\sum_{\mu, \nu=1}^{\infty} (A\phi_\mu, \psi_\nu)(\psi_\nu, \phi_\mu)$$

and

$$\sum_{\mu=1}^{\infty} (A\phi_\mu, \phi_\mu)$$

without examination of their convergence, and we have transformed one into the other. In short, everything has been done which one should not do when working in correct mathematical fashion. As a matter of fact, this kind of negligence is present elsewhere in theoretical physics, and the present treatment actually will produce no disastrous consequences in our quantum mechanical applications. Nevertheless it must be understood that we have been careless.

It is therefore the more important to point out that in the fundamental statistical assertions of quantum mechanics, the trace is employed only for operators of the form AB, where A, B are both definite -- and that this concept can be established with complete rigor. In the remainder of this section we shall therefore assemble those facts concerning the trace which are capable of proof with absolute mathematical rigor.

We first consider the trace of A^*A [A arbitrary, A^*A is Hermitian by II.4., and, since $(A^*Af, f) = (Af, Af) \geq 0$, it is definite]. Then

$$\mathrm{Tr}\,(A^*A) = \sum_{\mu=1}^{\infty} (A^*A\phi_\mu, \phi_\mu) = \sum_{\mu=1}^{\infty} (A\phi_\mu, A\phi_\mu) = \sum_{\mu=1}^{\infty} ||A\phi_k||^2$$

Since this series has only terms ≥ 0, it is convergent or diverges to $+\infty$, and therefore it is in any case defined. We now want to show, independently of the previous discussion, that its sum is independent of the choice of the $\phi_1, \phi_2, \ldots$. In this case, only series with terms ≥ 0 will appear, therefore all will be defined, and all resummations are permissible.

Let $\phi_1, \phi_2, \ldots$ and $\psi_1, \psi_2, \ldots$ be two complete orthonormal sets. We define

$$\Sigma\,(A;\ \phi_\mu,\ \psi_\nu) = \sum_{\mu,\nu=1}^{\infty} |(A\phi_\mu,\ \psi_\nu)|^2\ .$$

By THEOREM 7.γ. in II.2., this is equal to

$$\sum_{\mu=1}^{\infty} ||A\phi_\mu||^2$$

i.e., $\Sigma\,(A;\ \phi_\mu,\ \psi_\nu)$ depends only apparently on the ψ_ν. Furthermore (the $A\phi_\mu$ and the $A^*\psi_\nu$ must be defined),

$$\Sigma\,(A;\ \phi_\mu,\ \psi_\nu) = \sum_{\mu,\nu=1}^{\infty} |(A\phi_\mu,\ \psi_\nu)|^2 = \sum_{\mu,\nu=1}^{\infty} |(\phi_\mu,\ A^*\psi_\nu)|^2$$

$$= \sum_{\mu,\nu=1}^{\infty} |(A^*\psi_\nu,\ \phi_\mu)|^2 = \Sigma\,(A^*;\ \psi_\nu,\ \phi_\mu)$$

Therefore the dependence on the ϕ_μ is also only apparent, because this is the case on the right side of the formula. Consequently, $\Sigma\,(A;\ \phi_\mu,\ \psi_\nu)$ depends in general only on A. We then call it $\Sigma\,(A)$. By the above proofs,

$$\Sigma\,(A) = \sum_{\mu=1}^{\infty} ||A\phi_\mu||^2 = \sum_{\mu,\nu=1}^{\infty} |(A\phi_\mu,\ \psi_\nu)|^2$$

and $\Sigma\,(A) = \Sigma\,(A^*)$. Therefore $\mathrm{Tr}\,(A^*A)$ is correctly re-defined as $\Sigma\,(A)$.

We now prove independently several properties of $\Sigma(A)$ which also follow from the general properties of the $\mathrm{Tr}\,A$ previously derived.

From the definition it follows in general that $\Sigma(A) \geq 0$; and that for $\Sigma(A) = 0$ all $A\phi_\mu$ must be $= 0$, from which it follows, just as before, that $A = 0$. That is, for $A \neq 0$, $\Sigma(A) > 0$.

Obviously, $\Sigma(aA) = |a|^2 \Sigma(A)$. If $A^*B = 0$, then

$$||(A + B)\phi_\mu||^2 - ||A\phi_\mu||^2 - ||B\phi_\mu||^2 = (A\phi_\mu, B\phi_\mu) + (B\phi_\mu, A\phi_\mu)$$

$$= 2 \operatorname{Re} (A\phi_\mu, B\phi_\mu) = 2 \operatorname{Re} (\phi_\mu, A^*B\phi_\mu) = 0$$

therefore, after summation $\sum_{\mu=1}^{\infty}$:

$$\Sigma(A + B) = \Sigma(A) + \Sigma(B) .$$

This relation does not change if we interchange A, B . Therefore it is also valid for $B^*A = 0$. Furthermore, we can replace A, B in it by A^*, B^* . Then $AB^* = 0$ or $BA^* = 0$ is likewise sufficient. For Hermitian A (or B) , we can therefore write $AB = 0$ or $BA = 0$.

If E is the projection of the closed linear manifold $\mathfrak{M}$, then for the $\psi_1, \dots, \psi_k, \chi_1, \dots, \chi_l$ considered in the determination of $\operatorname{Tr} E$,

$$\Sigma(E) = \sum_{\mu=1}^{k} ||E\psi_\mu||^2 + \sum_{\mu=1}^{l} ||E\chi_\mu||^2 = \sum_{\mu=1}^{k} \psi_\mu||^2 + \sum_{\mu=1}^{l} ||0||^2$$

$$= \sum_{\mu=1}^{k} 1 = k .$$

That is, $\Sigma(E)$ is also the dimension number of $\mathfrak{M}$ (because $E^*E = EE = E$, this is just what would be expected).

For two definite (Hermitian) operators A, B , our $\operatorname{Tr} AB$ is now reducible to Σ . That is, there are two operators A', B' of the same category with $A'^2 = A$, $B'^2 = B$ [114] -- we call them $\sqrt{A}, \sqrt{B}$. We have the formal

---

[114] The precise proposition runs as follows: If A is hyper-

relations

$$\mathrm{Tr}\ (AB) = \mathrm{Tr}\ (\sqrt{A}\,\sqrt{A}\,\sqrt{B}.\sqrt{B}) = \mathrm{Tr}\ (\sqrt{B}.\sqrt{A}\,\sqrt{A}\,\sqrt{B})$$

$$= \mathrm{Tr}\ ((\sqrt{A}\,\sqrt{B})^*(\sqrt{A}\,\sqrt{B})) = \Sigma\ (\sqrt{A}\,\sqrt{B})$$

---

maximal and definite, then there exists one and only one operator A' of the same category with $A'^2 = A$. We prove the existence.

Let $A = \int_{-\infty}^{\infty} \lambda dE(\lambda)$ be the eigenvalue representation of A. Since A is definite, then $E(\lambda)$ is constant for $\lambda < 0$ (therefore, $= 0$, by $\overline{S_1}$.). For otherwise, $E(\lambda_2) - E(\lambda_1) \neq 0$ for suitable $\lambda_1 < \lambda_2 < 0$. Therefore an $f \neq 0$ can be chosen with

$$(E(\lambda_2) - E(\lambda_1))f = f$$

But it follows from this, as we have deduced several times previously, that

$$E(\lambda)f = \begin{cases} f\ , & \text{for } \lambda \geq \lambda_2 \\ 0\ , & \text{for } \lambda \leq \lambda_1 \end{cases},$$

therefore

$$(Af,\ f) = \int_{-\infty}^{\infty} \lambda d(E(\lambda)f,\ f)$$

$$= \int_{\lambda_1}^{\lambda_2} \lambda d(E(\lambda)f,\ f) \leq \int_{\lambda_1}^{\lambda_2} \lambda_2 d(E(\lambda)f,\ f)$$

$$= \lambda_2((E(\lambda_2) - E(\lambda_1))f,\ f) = \lambda_2(f,\ f) < 0$$

Consequently,

This $\Sigma(\sqrt{A}\sqrt{B})$, by reason of its own definition and without consideration of the relation with the trace, has all the properties which one expects of $\operatorname{Tr} AB$ -- namely:

$$\Sigma(\sqrt{A}\sqrt{B}) = \Sigma(\sqrt{B}\sqrt{A})$$

$$\Sigma(\sqrt{A}\sqrt{B+C}) = \Sigma(\sqrt{A}\sqrt{B}) + \Sigma(\sqrt{A}\sqrt{C})$$

$$\Sigma(\sqrt{A+B}\sqrt{C}) = \Sigma(\sqrt{A}\sqrt{C}) + \Sigma(\sqrt{B}\sqrt{C})$$

The first follows from the fact that $\Sigma(XY)$ is symmetric in X, Y:

$$\Sigma(XY) = \sum_{\mu,\nu=1}^{\infty} |(XY\phi_\mu, \psi_\nu)|^2 = \sum_{\mu,\nu=1}^{\infty} |(Y\phi_\mu, X\psi_\nu)|^2$$

The second follows from the third by reason of the first property. Therefore this third property is the only one which need be proved -- that is, that $\Sigma(\sqrt{A}\sqrt{B})$ is additive in A. But this can be seen if we write $\Sigma(\sqrt{A}\sqrt{B})$ as:

$$\Sigma(\sqrt{A}\sqrt{B}) = \sum_{\mu=1}^{\infty} \|\sqrt{A}\sqrt{B}\phi_\mu\|^2 = \sum_{\mu=1}^{\infty} (\sqrt{A}\sqrt{B}\phi_\mu, \sqrt{A}\sqrt{B}\phi_\mu)$$
$$= \sum_{\mu=1}^{\infty} (\sqrt{A}.\sqrt{A}\sqrt{B}\phi_\mu, \sqrt{B}\phi_\mu) = \sum_{\mu=1}^{\infty} (A\sqrt{B}\phi_\mu, \sqrt{B}\phi_\mu)$$

$$A = \int_{-\infty}^{\infty} \lambda dE(\lambda) = \int_{0}^{\infty} \lambda dE(\lambda) = \int_{0}^{\infty} \mu^2 dE(\mu^2)$$

and $A' = \int_{1}^{\infty} \mu dE(\mu^2)$ yields the desired result.

We may observe that we have deduced from the property of definiteness that $E(\lambda) = 0$ for $\lambda < 0$, and since definiteness clearly follows from this, the fact that the entire spectrum is ≥ 0 is characteristic for definiteness.

In this way, we have established a rigorous foundation of the concept of the trace to the extent which was desired.

In addition, the last formula permits the following conclusion: if A, B are definite, then $AB = 0$ is a consequence of $\mathrm{Tr}\, AB = 0$. For the latter implies that $\sum(\sqrt{A}\,\sqrt{B}) = 0$ and therefore that $\sqrt{A}\,\sqrt{B} = 0$ (cf. the discussion on p. 184, and also the considerations given above with regard to $\sum$). Therefore

$$AB = \sqrt{A}.\sqrt{A}\,\sqrt{B}.\sqrt{B} = 0$$

For a definite Hermitian operator A, the calculation with the trace is correct even in its original form. Indeed, let $\phi_1, \phi_2, \ldots$ be a complete orthonormal set. Then

$$\sum_{\mu=1}^{\infty} (A\phi_\mu, \phi_\mu)$$

(the sum which should define the trace) is a sum with all terms ≥ 0, and is therefore convergent or divergent to $+\infty$. Two cases are then possible: either this sum is infinite for each choice of the $\phi_1, \phi_2, \ldots$, and therefore the trace is actually defined independently of $\phi_1, \phi_2, \ldots$ and is $= +\infty$; or the sum is finite for at least one choice of $\phi_1, \phi_2, \ldots$, say $\overline{\phi}_1, \overline{\phi}_2, \ldots$. Then, since

$$\left(\sum_{\mu=1}^{\infty} (A\overline{\phi}_\mu, \overline{\phi}_\mu)\right)^2 = \sum_{\mu,\nu=1}^{\infty} (A\overline{\phi}_\mu, \overline{\phi}_\mu)(A\overline{\phi}_\nu, \overline{\phi}_\nu)$$

$$\geq \sum_{\mu,\nu=1}^{\infty} |(A\overline{\phi}_\mu, \overline{\phi}_\nu)|^2 = \sum(A)$$

$\sum(A)$ is also finite, and is equal to some C^2. If $\psi_1, \psi_2, \ldots$ is any complete orthonormal set, then

$$\sum(A) = \sum_{\mu=1}^{\infty} ||A\psi_\mu||^2 = C^2 \ , \ ||A\psi_1||^2 \leq C^2 \ , \ ||A\psi_1|| \leq C$$

Since each ψ with $||\psi|| = 1$ can be chosen as the ψ_1 of such a system, it follows from $||\psi|| = 1$ that $||A\psi|| \leq C$. In general then, $||Af|| \leq C.||f||$: for $f = 0$, this is obvious, while for $f \neq 0$, it suffices to set $\psi = \frac{1}{||f||} . f$. Consequently A satisfies the condition **Co.** from II.9. It is therefore a continuous operator. But actually even more is true.

Because of the finite nature of $\Sigma(A)$, A belongs to the class of the so-called totally continuous operators. Hilbert showed that the eigenvalue problem for such an operator is solvable in the original form, i.e., that a complete orthonormal set $\psi_1, \psi_2, \ldots$ with $A\psi_\mu = \lambda_\mu \psi_\mu$ exists (and for $\mu \longrightarrow \infty$, $\lambda_\mu \longrightarrow 0$).[115] Because the

---

[115]Cf. the reference in Note 64. A direct proof runs as follows. Let $\lambda_0 < \lambda_1 < \ldots < \lambda_n$, all $\geq \epsilon$ or $\leq -\epsilon$, $E(\lambda_0) \neq E(\lambda_1) \neq \ldots \neq E(\lambda_n)$. Then $E(\lambda_\nu) - E(\lambda_{\nu-1}) \neq 0$, therefore $\phi_\nu \neq 0$ can be chosen with $(E(\lambda_\nu) - E(\lambda_{\nu-1}))\phi_\nu = \phi_\nu$. It follows from this that

$$E(\lambda)\phi = \begin{cases} \phi_\nu \ , \text{ for } \lambda \geq \lambda_\nu \\ 0 \ , \text{ for } \lambda \leq \lambda_{\nu-1} \end{cases}$$

-- we can even make $||\phi_\nu|| = 1$. It follows from the above that $(\phi_\mu, \phi_\nu) = 0$ for $\mu \neq \nu$. The $\phi_1, \ldots, \phi_n$ consequently form an orthonormal set, and we can extend it to a complete one: $\phi_1, \ldots, \phi_n, \phi_{n+1}, \ldots$.

We have $(\nu = 1, \ldots, n)$

$$||A\phi_\nu||^2 = \int_{-\infty}^{\infty} \lambda^2 d||E(\lambda)\phi_\nu||^2 = \int_{\lambda_{\nu-1}}^{\lambda_\nu} \lambda^2 d||E(\lambda)\phi_\nu||^2$$

operator is definite, $\lambda_\mu = (A\psi_\mu, \psi_\mu) \geq 0$, and furthermore,

$$\sum_{\mu=1}^{\infty} \lambda_\mu{}^2 = \sum_{\mu=1}^{\infty} ||A\psi_\mu||^2 = \sum(A) = C^2$$

If $\phi_1, \phi_2, \ldots$ is any complete orthonormal set, then

$$\sum_{\mu=1}^{\infty} (A\phi_\mu, \phi_\mu) = \sum_{\mu=1}^{\infty} \left(\sum_{\nu=1}^{\infty} (A\phi_\mu, \psi_\nu)(\psi_\nu, \phi_\mu) \right)$$

$$= \sum_{\mu=1}^{\infty} \left(\sum_{\nu=1}^{\infty} (\phi_\mu, A\psi_\nu)(\psi_\nu, \phi_\mu) \right)$$

$$= \sum_{\mu=1}^{\infty} \left(\sum_{\nu=1}^{\infty} \lambda_\nu (\phi_\mu, \psi_\nu)(\psi_\nu, \phi_\mu) \right) = \sum_{\mu=1}^{\infty} \left(\sum_{\nu=1}^{\infty} \lambda_\mu |(\phi_\mu, \psi_\nu)|^2 \right.$$

$$\geq \int_{\lambda_{\nu-1}}^{\lambda_\nu} \epsilon^2 d||E(\lambda)\phi_\nu||^2 = \epsilon^2 (||E(\lambda_\nu)\phi_\nu||^2 - ||E(\lambda_{\nu-1})\phi_\nu||^2)$$

$$= \epsilon^2 ||\phi_\nu||^2 = \epsilon^2 ,$$

and therefore

$$\sum_{\mu=1}^{\infty} ||A\phi_\mu||^2 \begin{cases} \geq \sum_{\mu=1}^{n} ||A\phi_\mu||^2 \geq n\epsilon^2 \\ = \sum(A) = C^2 , \end{cases}$$

i.e., $n \leq \frac{C^2}{\epsilon^2}$. Then for $|\lambda| \geq \epsilon$, $E(\lambda)$ can in general assume only $\leq 2 \cdot \frac{C^2}{\epsilon^2}$ many different values. It can therefore change only in a finite number of places, the remaining space being made up of constancy intervals. That is, for $|\lambda| \geq \epsilon$ only a discrete spectrum exists. Since this

Since all terms are ≥ 0, we may rearrange the summations:

$$\sum_{\mu=1}^{\infty}(A\phi_{\mu}, \phi_{\mu}) = \sum_{\mu,\nu=1}^{\infty}\lambda_{\nu}|(\phi_{\mu}, \psi_{\nu})|^2 = \sum_{\nu=1}^{\infty}\lambda_{\nu}\left(\sum_{\mu=1}^{\infty}|(\phi_{\mu}, \psi_{\nu})|^2\right)$$

$$= \sum_{\nu=1}^{\infty}\lambda_{\nu}||\psi_{\nu}||^2 = \sum_{\nu=1}^{\infty}\lambda_{\nu}$$

In this case, therefore, $\sum_{\mu=1}^{\infty}(A\phi_{\mu}, \phi_{\mu})$ is again independent of $\phi_1, \phi_2, \cdots$, and is actually equal to the sum of the eigenvalues. Since the sum is finite for $\overline{\phi}_1, \overline{\phi}_2, \cdots$, it is therefore always finite. That is, Tr A is again unique, but this time it is also finite.

The calculation with the trace is therefore justified in both cases.

We now give several estimates relative to Tr A and $\Sigma(A)$. For all A with finite $\Sigma(A)$, $||Af|| \leq \sqrt{\Sigma(A)}\,||f||$; for all definite (Hermitian) A with finite Tr A, $||Af|| \leq \text{Tr}(A)\cdot||f||$. Now let A again be definite, Tr A = 1. For an appropriate ϕ with $||\phi|| = 1$, $||A\phi||^2 \geq 1 - \epsilon$ or $(A\phi, \phi) \geq 1 - \epsilon$. It suffices to consider the second case, since the first follows from the second because $(A\phi, \phi) \leq ||A\phi||\cdot||\phi|| = ||A\phi||$ [put $(1 - \epsilon)^2 \geq 1 - 2\epsilon$ in place of $1 - \epsilon$, therefore 2ϵ in place of ϵ].

Let ψ be orthogonal to ϕ, $||\psi|| = 1$. Then we can find a complete orthonormal set $\chi_1, \chi_2, \cdots$ with $\chi_1 = \phi$, $\chi_2 = \psi$. Therefore:

$$\sum_{\mu=1}^{\infty}||A\chi_{\mu}||^2 \begin{cases} = \Sigma(A) \leq [\text{Tr } A]^2 = 1\,, \\ \geq ||A\phi||^2 + ||A\psi||^2 \geq 1 - 2\epsilon + ||A\psi||^2\,, \end{cases}$$

---

holds for all $\epsilon > 0$, only a pure discrete spectrum is present in general.

$$||A\psi||^2 \leq 2\epsilon \ , \ ||A\psi|| \leq \sqrt{2\epsilon} \ .$$

For an arbitrary f orthogonal to ϕ, it follows that $||Af|| \leq \sqrt{2\epsilon}\ ||f||$. (For $f = 0$ this is obvious. Otherwise, $\psi = \frac{1}{||f||} \cdot f$). If we now remember that $(Af, g) = (f, Ag)$ then we obtain $|(Af, g)| \leq \sqrt{2\epsilon}\ ||f|| \cdot ||g||$ if either f or g is orthogonal to ϕ.

Now let f, g be arbitrary. Then

$$f = \alpha\phi + f', \ g = \beta\phi + g'$$

where f', g' are orthogonal to ϕ, and $\alpha = (f, \phi)$, $\beta = (g, \phi)$. Then

$$(Af, g) = \alpha\overline{\beta}(A\phi, \phi) + \alpha(A\phi, g') + \overline{\beta}(Af', \phi) + (Af', g')$$

therefore, if we set $(A\phi, \phi) = c$

$$|(Af, g) - c\alpha\overline{\beta}| \leq |\alpha| \cdot |(A\phi, g')| + |\beta| \cdot |(Af', \phi)| + |(Af', g')|$$

and according to the above estimates,

$$\begin{aligned} |(Af, g) - c\alpha\overline{\beta}| &\leq \sqrt{2\epsilon} \cdot (|\alpha| \cdot ||g'|| + |\beta| \cdot ||f'|| + ||f'|| \cdot ||g'||) \\ &\leq \sqrt{2\epsilon} \cdot (|\alpha| + ||f'||)(|\beta| + ||g||) \\ &\leq 2\sqrt{2\epsilon} \cdot \sqrt{|\alpha|^2 + ||f'||^2}\ \sqrt{|\beta|^2 + ||g'||^2} \\ &= 2\sqrt{2\epsilon} \cdot ||f|| \cdot ||g|| \end{aligned}$$

On the other hand,

$$(Af, g) - c\alpha\overline{\beta} = (Af, g) - c(f, \phi)(\phi, g) = ((A - cP_{[\phi]})f, g)$$

In general then, $|((A - cP_{[\phi]})f, g)| \leq 2\sqrt{2\epsilon} \cdot ||f|| \cdot ||g||$. Therefore, as we know from II.9.,

$$||(A - cP_{[\phi]})f|| \leq 2\sqrt{2\epsilon}.||f||.$$

For $f = \phi$ this implies that

$$||A\phi - c\phi|| \leq 2\sqrt{2\epsilon},$$

$$c = ||c\phi|| \begin{cases} \leq ||A\phi - c\phi|| + ||A\phi|| \leq 2\sqrt{2\epsilon} + 1, \\ \geq -||A\phi - c\phi|| + ||A\phi|| \geq -2\sqrt{2\epsilon} + (1 - \epsilon), \end{cases}$$

$$1 - (\epsilon + 2\sqrt{2\epsilon}) \leq c \leq 1 + 2\sqrt{2\epsilon}$$

($c = (A\phi, \phi)$ is real and ≥ 0). Consequently,

$$||(A - P_{[\phi]})f|| \leq ||(A - cP_{[\phi]})f|| + ||(c - 1)P_{[\phi]}f||$$

$$\leq 2\sqrt{2\epsilon}.||f|| + (\epsilon + 2\sqrt{2\epsilon})||P_{[\phi]}f||$$

$$\leq (\epsilon + 4\sqrt{2\epsilon}).||f||.$$

For $\epsilon \longrightarrow 0$ therefore, A converges uniformly to $P_{[\phi]}$.

In conclusion, let us consider $\mathrm{Tr}\, A$ and $\mathrm{Tr}\, B$ in the realizations F_Z and F_Ω of $\mathfrak{R}_\infty$ (cf. I.4. and II.3.), since physical applications occur for these cases.

In F_Z (set of all $\{x_1, x_2, \ldots\}$ with finite $\sum_{\mu=1}^{\infty} |x_\mu|^2$) A may be described by a matrix $\{a_{\mu\nu}\}$:

$$A\{x_1, x_2, \ldots\} = \{y_1, y_2, \ldots\}, \quad y_\mu = \sum_{\nu=1}^{\infty} a_{\mu\nu} x_\nu .$$

$\{1,0,0,\ldots\}, \{0,1,0,\ldots\}, \ldots$ form a complete orthonormal set $\phi_1, \phi_2, \ldots$ and $A\phi_\mu = \{a_{1\mu}, a_{2\mu}, \ldots\} = \sum_{\rho=1}^{\infty} a_{\rho\mu}\phi_\rho$. Therefore $(A\phi_\mu, \phi_\mu) = a_{\mu\mu}$, $||A\phi_\mu||^2 = \sum_{\rho=1}^{\infty} |a_{\rho\mu}|^2$. From this it follows immediately that

$$\text{Tr } A = \sum_{\mu=1}^{\infty} a_{\mu\mu} \ , \ \sum(A) = \sum_{\mu,\nu=1}^{\infty} |a_{\mu\nu}|^2 \ .$$

In F_Ω (set of all $f(P)$ defined in Ω with finite $\int_\Omega |f(P)|^2 dv$), let us consider only the integral operators

$$Af(P) = \int_\Omega a(P, P')f(P')dv'$$

($a(P, P')$ a two variable function defined in Ω, the "kernel" of A; cf. I.4.). Let $\phi_1(P), \phi_2(P), \ldots$ be any complete orthonormal set; then

$$\text{Tr } A = \sum_{\mu=1}^{\infty} (A\phi_\mu(P), \phi_\mu(P)) = \sum_{\mu=1}^{\infty} \int_\Omega \left[\int_\Omega a(P, P')\phi_\mu(P')dv' \right] \overline{\phi_\mu(P)}dv$$

and because in general [THEOREM 7.β.) in II.2., applied to $\overline{g(P)}$]

$$\sum_{\mu=1}^{\infty} \left(\int_\Omega \overline{g(P')} \ \overline{\phi_\mu(P')}dv' \right) \phi_\mu(P) = \overline{g(P)} \ ,$$

$$\sum_{\mu=1}^{\infty} \left(\int_\Omega g(P')\phi_\mu(P')dv' \right) \overline{\phi_\mu(P)} = g(P)$$

holds,

$$\text{Tr } A = \int_\Omega a(P, P)dv$$

Furthermore,

$$\sum(A) = \sum_{\mu=1}^{\infty} \int_\Omega \left| \int_\Omega a(P, P')\phi_\mu(P')dv' \right|^2 dv \ ,$$

therefore, because [THEOREM 7.γ.), II.2.]

$$\sum_{\mu=1}^{\infty} \left| \int_{\Omega} g(P')\phi_{\mu}(P')dv' \right|^2 = \sum_{\mu=1}^{\infty} \left| \int_{\Omega} \overline{g(P')}\ \overline{\phi_{\mu}(P')}dv' \right|^2$$

$$= \int_{\Omega} |\overline{g(P')}|^2 dv' = \int_{\Omega} |\overline{g(P')}|^2 dv' .$$

It is also true that

$$\sum(A) = \int_{\Omega} \int_{\Omega} |a(P, P')|^2 dvdv' .$$

We see that $\text{Tr}\ A$, $\sum(A)$ accomplish what was sought for by the use of mathematically doubtful artifices: in the transition from F_Z to F_{Ω}, $\sum_{\mu=1}^{\infty} \ldots$ is replaced by $\int_{\Omega} \ldots dv$.

With this we have concluded our mathematical treatment of Hermitian operators. The reader who is interested in the mathematics will find more on this subject in the literature relating to these topics.[116]

---

[116] In addition to the original papers mentioned in the course of the discussions above, the foremost reference is the encyclopedia article of Hellinger and Toeplitz referred to in Note 33.

CHAPTER III

THE QUANTUM STATISTICS

1. THE STATISTICAL ASSERTIONS OF QUANTUM MECHANICS

Let us now return to the analysis of the quantum mechanical theories, which was interrupted by the mathematical considerations of Chapter II. At that time, we discussed only how quantum mechanics makes possible the determination of all possible values of one particular physical quantity -- energy. These values are the eigenvalues of the energy operator (i.e., the numbers of its spectrum). On the other hand, no mention was made regarding what might be said about the values of other quantities, as well as regarding the causal or statistical relations among the values of several quantities. The statements of the theory relative to this problem should now be considered. We shall take as a basis the wave mechanical method of description since the equivalence of the two theories has already been ascertained.

In this method of description, it is evident that everything which can be said about the state of a system must be derived from its wave function $\phi(q_1, \ldots, q_k)$. (We suppose that the system has k degrees of freedom and employ $q_1, \ldots, q_k$ as the coordinates of its configuration.) Actually, this does not restrict us to the stationary states of the system (quantum orbits in which ϕ is an eigenfunction of H : $H\phi = \lambda\phi$, cf. I.3.), but also admits all other states of the system -- i.e., wave functions ϕ (which vary according to the Schrödinger time dependent differential equation $H\phi = -\frac{h}{2\pi i}\frac{\partial}{\partial t}\phi$, cf. I.2.). What pronouncements can now be made regarding a system which is in the state ϕ ?

First of all, we observe that ϕ was normalized (I.3.) by

$$\int_{-\infty}^{\infty}\cdots\int_{-\infty}^{\infty}|\phi(q_1,\ldots,q_k)|^2dq_1\ldots dq_k = 1$$

i.e., (in our present terminology) as a point of the Hilbert space $\mathfrak{R}_\infty$ of all $f(q_1,\ldots,q_k)$ with finite

$$\int_{-\infty}^{\infty}\cdots\int_{-\infty}^{\infty}|f(q_1,\ldots,q_k)|^2dq_1\ldots dq_k$$

(in an F_Ω !) , it is normalized by $\|\phi\| = 1$. In other words, it must lie on the surface of the unit sphere in Hilbert space.[117] We already know that a constant (i.e., independent of $q_1,\ldots,q_k$) factor in ϕ is physically meaningless. (This means that the substitution of $a\phi$ for ϕ , a a complex number. Because of the normalization $\|\phi\| = 1$, there must be $|a| = 1$.) Furthermore, it should be pointed out regarding this that while ϕ is dependent on the time t , as well as on the coordinates $q_1,\ldots,q_k$ of the configuration space of our system, nevertheless the Hilbert space involves only the $q_1,\ldots,q_k$ (because the normalization is related to these alone).

---

[117] By geometric analogy, the sphere with the center ϕ_0 and radius r is (in $\mathfrak{R}_\infty$) the set of points f with $\|f - \phi_0\| \leq r$, its interior the set of $\|f - \phi_0\| < r$ and its outer surface the set of $\|f - \phi_0\| = r$. For the unit sphere, $\phi_0 = 0$, $r = 1$.

Hence the dependence on t is not to be considered in forming the Hilbert space. Instead of this, it is rather to be regarded as a parameter. Consequently, ϕ , as a point of the $\mathfrak{R}_\infty$, depends on t , but is on the other hand independent of the $q_1, \ldots, q_k$: Indeed, as a point of the $\mathfrak{R}_\infty$, it represents the entire functional dependence. Because of this, we shall occasionally indicate the parameter t in ϕ (when ϕ is viewed as a point in $\mathfrak{R}_\infty$) by writing ϕ_t .

Let us now consider the state $\phi = \phi(q_1, \ldots, q_k)$. The statistical assertions which can then be made are as follows: the system is in the point $q_1, \ldots, q_k$ of the configuration space with the probability density $(q_1, \ldots, q_k)^2$ -- i.e., the probability that it is in the volume V of the configuration space is

$$\int_V \cdots \int |\phi(q_1, \ldots, q_k)|^2 dv$$

(This is one of the first and simplest examples by means of which the statistical character of quantum mechanics was recognized.[118] In addition, the relationship between this statement and Schrödinger's charge distribution assumption [cf. I.2.] is manifest.) Furthermore, if the energy of the system has the operator H , and if this operator has the eigenvalues $\lambda_1, \lambda_2, \ldots$ and the eigenfunctions $\phi_1, \phi_2, \ldots$, then the probability of the energy values λ_n in the state ϕ is equal to

$$\left| \int \cdots \int \phi(q_1, \ldots, q_k) \overline{\phi_n(q_1, \ldots, q_k)} dq_1 \ldots dq_k \right|^2$$

---

[118] The first statistical statements on the behavior of a system in the state ϕ originated with M. Born, and were treated in more detail by Dirac and by Jordan. See the references in Note 8 and Note 2.

(cf. the papers mentioned in Note 118). We now want to join these two statements, and put them in a unified form.

Let V be the k-dimensional cube

$$q_1' < q_1 \leq q_1'', \ldots, q_k' < q_k \leq q_k''$$

We denote the intervals $\{q_1', q_1''\}, \ldots, \{q_k', q_k''\}$ by $I_1, \ldots, I_k$ respectively. $q_1, \ldots, q_k$ have the operators $q_1 \cdots, \ldots, q_k \cdots$ respectively. The resolutions of the identity belonging to these are defined as follows (cf. II.8.): we call that resolution belonging to q_j $(j = 1, \ldots, k)$ $E_j(\lambda)$. For this,

$$E_j(\lambda) f(q_1, \ldots, q_k) = \begin{cases} f(q_1, \ldots, q_k) \ , & \text{for} \quad q_j \leq \lambda \\ 0 \ , & \text{for} \quad q_j > \lambda \end{cases}$$

We introduce the following general notation: If $F(\lambda)$ is a resolution of the identity, and I is an interval $\{\lambda', \lambda''\}$ then $F(I) = F(\lambda'') - F(\lambda')$ (a projection for $\lambda' \leq \lambda''$, $F(\lambda') \leq F(\lambda'')$) . The probability therefore that the system lies in the above V , i.e., that q_1 lies in $I_1, \ldots,$ q_k in I_k , amounts to

$$\int_{q_1'}^{q_1''} \cdots \int_{q_k'}^{q_k''} |\phi(q_1, \ldots q_k)|^2 dq_1 \ldots dq_k$$

$$= \int \cdots \int |E_1(I_1) \ldots E_k(I_k) \phi(q_1, \ldots, q_k)|^2 dq_1 \ldots dq_k$$

(because $E_1(I_1) \ldots E_k(I_k) \phi(q_1, \ldots, q_k) = \phi(q_1, \ldots, q_k)$ for q_1 from $I_1, \ldots,$ q_k from I_k , otherwise 0) , i.e.,

$$= \| E_1(I_1) \ldots E_k(I_k) \phi \|^2$$

In the second case, let us consider the probability that the energy lies in the interval $I = \{\lambda', \lambda''\}$.

The resolution of the identity, $E(\lambda)$, is defined (cf. II.8.) by

$$E(\lambda) = \sum_{\lambda_n \leq \lambda} P_{[\phi_n]}$$

Therefore

$$E(I) = E(\lambda'') - E(\lambda') = \sum_{\lambda' < \lambda_n \leq \lambda''} P_{[\phi_n]}$$

But since only the $\lambda_1, \lambda_2, \ldots$ appear as values of the energy, this latter probability is the sum of the probabilities of all λ_n with $\lambda' < \lambda_n \leq \lambda''$, therefore,

$$\sum_{\lambda' < \lambda_n \leq \lambda''} \left| \int \cdots \int \phi(q_1, \ldots, q_k) \overline{\phi_n(q_1, \ldots, q_k)} dq_1 \ldots dq_k \right|^2$$

$$= \sum_{\lambda' < \lambda_n \leq \lambda''} |(\phi, \phi_n)|^2 = \sum_{\lambda' < \lambda_n \leq \lambda''} (P_{[\phi_n]}\phi, \phi)$$

$$= \left(\left\{ \sum_{\lambda' < \lambda_n \leq \lambda''} P_{[\phi_n]} \right\} \phi, \phi \right) = (E(I)\phi, \phi) = \|E(I)\phi\|^2$$

In both cases we have then obtained a result which may be formulated as follows:

(**P.**) The probability that in the state ϕ the quantities with the operators $R_1, \ldots, R_l$ [119] take on

---

[119] We shall speak more expressly in IV.1. about this correspondence, which allows each physical quantity to correspond to a Hermitian operator. For the present, we know only (by reason of I.2.) that the operators $q_1, \ldots, q_k$ correspond to the coordinates, the operators

$$\frac{h}{2\pi i} \frac{\partial}{\partial q_1}, \ldots, \frac{h}{2\pi i} \frac{\partial}{\partial q_k}$$

to the momenta, and the "energy-operator" H to the energy.

values from the respective intervals $I_1, \ldots, I_l$ is

$$||E_1(I_1)\ldots E_l(I_l)\phi||^2$$

where $E_1(\lambda), \ldots, E_l(\lambda)$ are the resolutions of the identity belonging to $R_1, \ldots, R_l$ respectively.

The first case corresponds to $l = k$, $R_1 = q_1 \cdots, \ldots, R_k = q_k \cdots$, the second to $l = 1$, $R_1 = H$. We shall now assume this statement **P**. to be generally valid. It actually contains all the statistical assertions of quantum mechanics which have been made thus far.

However, a limitation of its validity is necessary. Since the order of the $R_1, \ldots, R_l$ is entirely arbitrary in the problem, it must also be arbitrary in the result. That is, all the $E_1(I_1), \ldots, E_l(I_l)$ or equivalently all the $E_1(\lambda_1), \ldots, E_l(\lambda_l)$ must commute. By II.10., this means that the $R_1, \ldots, R_l$ commute with each other. This condition is satisfied for $q_1 \cdots, \ldots, q_k \cdots$, while for $l = 1$, $R_1 = H$ it is vacuously satisfied.

Consequently, we postulate . for all commuting $R_1, \ldots, R_l$. Then the $E_1(I_1), \ldots, E_l(I_l)$ commute, and therefore $E_1(I_1)\ldots E_l(I_l)$ is a projection (THEOREM 14. in II.4.), and the probability in question becomes

$$P = ||E_1(I_1)\ldots E_l(I_l)\phi||^2 = (E_1(I_1)\ldots E_l(I_l)\phi, \phi)$$

(THEOREM 12. in II.4.).

Before we go any further, we must verify a few properties of **P**., which must hold in any reasonable statistical theory.

1. The order of the propositions is irrelevant.

2. Vacuous proportions can be inserted at will. Indeed, these are the ones where the interval I_j is equal to $-\infty, +\infty$, and they give rise only to a factor

$$E_j(I_j) = E_j(+\infty) - E_j(-\infty) = 1 - 0 = 1$$

3. The addition theorem of probability holds. That is, if we resolve an interval I_j into two intervals I_j', I_j'', then the old probability is the sum of the two new probabilities. For, let I_j, I_j', I_j'' be respectively $\{\lambda', \lambda''\}$, $\{\lambda', \lambda\}$, $\{\lambda, \lambda''\}$, then

$$E(\lambda'') - E(\lambda') = (E(\lambda) - E(\lambda')) + (E(\lambda'') - E(\lambda))$$

i.e., $E(I_j) = E(I_j') + E(I_j'')$ which, by reason of the second of the forms of **P**. given above (which is linear in $E_1(I_1)\ldots E_j(I_j)\ldots E_l(I_l)$ give the additivity of the probabilities.

4. For absurd propositions (one I_j empty), $\mathbf{P}. = 0$ -- because then the corresponding $E_j(I_j) = 0$. For trivially true propositions (all I_j equal to $\{-\infty, +\infty\}$), $\mathbf{P} = 1$ -- because then all $E_j(I_j) = 1$, $\mathbf{P} = ||\phi||^2 = 1$. We always have $0 \leq \mathbf{P} \leq 1$, because of THEOREM 13. in II.4.

Finally, we observe that **P**. contains the assertion that a quantity R_j can take on as values only its eigenvalues, i.e., the numbers of its spectrum. For if the interval $I_j = \{\lambda', \lambda''\}$ lies outside of the spectrum, then $E_j(\lambda)$ is constant in it, and therefore

$$E_j(I_j) = E_j(\lambda'') - E_j(\lambda') = 0$$

from which it follows that $\mathbf{P} = 0$.

We shall now set $l = 1$ and denote R_1 by R. Let $\mathfrak{R}$ be the physical quantity to which R corresponds (see Note 119). Let $F(\lambda)$ be any function. The expectation value of $F(\mathfrak{R})$ is then to be calculated.

For this purpose, we divide the interval $\{-\infty, +\infty\}$ into a sequence of subintervals $\{\lambda_n, \lambda_{n+1}\}$, $n = 0, \pm 1, \pm 2, \ldots$. The probability that $\mathfrak{R}$ lies in $\{\lambda_n, \lambda_{n+1}\}$ is

$$(\{E(\lambda_{n+1}) - E(\lambda_n)\}\phi, \phi) = (E(\lambda_{n+1})\phi, \phi) - (E(\lambda_n)\phi, \phi)$$

and the expectation value of $F(\mathfrak{R})$ is consequently

$$\sum_{-\infty}^{+\infty} F(\lambda_n')\{(E(\lambda_{n+1})\phi, \phi) - (E(\lambda_n)\phi, \phi)\}$$

if λ_n' is an appropriate intermediate value from $\{\lambda_n, \lambda_{n+1}\}$. But if we choose the subdivisions $\ldots, \lambda_{-2}, \lambda_{-1}, \lambda_0, \lambda_1, \lambda_2, \ldots$ closer and closer together, this sum converges to the Stieltjes integral

$$\int_{-\infty}^{\infty} F(\lambda)d(E(\lambda)\phi, \phi)$$

Therefore the expectation value in question is also equal to this quantity. By reason of the general definition of the operator functions in II.8., however, this integral is equal to $(F(R)\phi, \phi)$. Consequently, we have:

(**E₁**.) Let $\mathfrak{R}$ be any physical quantity, R its operator (cf. Note 119), and $F(\lambda)$ an arbitrary function. Then, for the expectation value of $F(\mathfrak{R})$ in the state ϕ, we have

$$\text{Exp}(F(\mathfrak{R}); \phi) = (F(R)\phi, \phi)$$

In particular, if we set $F(\lambda) = \lambda$, then:

(**E₂**.) Let $\mathfrak{R}$, R be as above. Then for the expectation value of $\mathfrak{R}$ in the state ϕ, we have

$$\text{Exp}(\mathfrak{R}; \phi) = (R\phi, \phi)$$

In the following, the relations among **P**., **E₁**., **E₂**. will be investigated.

We shall now deduce **E₁**. from **P**. and **E₂**. from **E₁**.

If we denote the operator of $F(\mathfrak{R})$ by S, then a comparison of **E₁**., **E₂**. gives

$$(S\phi, \phi) = (F(R)\phi, \phi)$$

for all states ϕ , i.e., for all ϕ with $||\phi|| = 1$. Consequently, in general,

$$(Sf, f) = (F(R)f, f)$$

(obvious for $f = 0$, while otherwise, $\phi = \frac{1}{||f||} \cdot f$) , and therefore

$$(Sf, g) = (F(R)f, g)$$

(if we replace f by $\frac{f+g}{2}$ and $\frac{f-g}{2}$ and subtract, this gives the equality of the real parts; if, g instead of f, g gives the equality of the imaginary parts). Therefore $S = F(R)$. We formulate this important result as follows:

(**F.**) If the quantity $\mathfrak{R}$ has the operator R , then the quantity $F(\mathfrak{R})$ must have the operator $F(R)$.

Because of **F.**, now $\mathbf{E_1}$. clearly follows from $\mathbf{E_2}$.

Consequently, (under assumption of **F.**), $\mathbf{E_1}$., $\mathbf{E_2}$. are equivalent assertions, and we shall now show that they are also equivalent to **P**. Since they follow from **P**., we need only show that **P**. follows from $\mathbf{E_1}$. or $\mathbf{E_2}$.

Let $R_1, \ldots, R_l$ be commuting operators belonging to the respective quantities $\mathfrak{R}_1, \ldots, \mathfrak{R}_n$. By II.10., they are functions of a Hermitian operator R :

$$R_1 = F_1(R), \ldots, R_l = F_l(R)$$

We assume that R also belongs to a quantity $\mathfrak{R}$. (We therefore make the assumption that a [hypermaximal] Hermitian operator R belongs to each quantity $\mathfrak{R}$ and conversely. Cf. Note 119 and IV.2.) Then by **F**.,

$$\mathfrak{R}_1 = F_1(\mathfrak{R}), \ldots, \mathfrak{R}_l = F_l(\mathfrak{R})$$

Now let $I_1, \ldots, I_l$ be the intervals involved in **P**., and

$$G_j(\lambda) = \begin{cases} 1 & \text{for } \lambda \text{ in } I_j \\ 0 & \text{otherwise} \end{cases} \qquad (j = 1, \ldots, l)$$

We set

$$H(\lambda) = G_1(F_1(\lambda)) \ldots G_l(F_l(\lambda))$$

and form the quantity

$$\mathfrak{S} = H(\mathfrak{R})$$

If $\mathfrak{R}_j$ lies in I_j, i.e., $F_j(\mathfrak{R})$ in I_j, then $G_j(F_j(\mathfrak{R}))$ is equal to 1; otherwise, it is equal to 0. $\mathfrak{S} = H(\mathfrak{R})$ is then equal to 1 if all $\mathfrak{R}_j$ lie in its I_j $(j = 1, \ldots, l)$, otherwise it is equal to 0. The expectation value of $\mathfrak{S}$ is therefore equal to the probability P that $\mathfrak{R}_1$ lies in $I_1, \ldots, \mathfrak{R}_l$ in I_l. Hence,

$$P = \mathrm{Exp}(\mathfrak{S}, \phi) = (H(R)\phi, \phi)$$

$$= (G_1(F_1(R)) \ldots G_l(F_l(R))\phi, \phi) = (G_1(R_1) \ldots G_l(R_l)\phi, \phi)$$

Let the resolution of the identity belonging to R_j again be called $E_j(\lambda)$, and let I_j be the interval $\{\lambda_j', \lambda_j''\}$. Then, by reason of the discussion at the end of II.8., and with the notation used there,

$$G_j(\lambda) = e_{\lambda_j''}(\lambda) - e_{\lambda_j'}(\lambda),$$

$$G_j(R_j) = e_{\lambda_j''}(R_j) - e_{\lambda_j'}(R_j) = E_j(\lambda_j'') - E_j(\lambda_j') = E_j(I_j)$$

and therefore

$$P = (E_1(I_1) \ldots E_l(I_l)\phi, \phi)$$

But this is precisely **P**.

Because of the simplicity of their form, **E**$_2$., **F**. are especially suited to be considered as the foundations on which the entire theory is built. We saw that the most general probability assertion possible, **P**., follows from these. But the statement **P**. has two striking features:

1. **P**. is statistical, and not causal, i.e., it does not tell us what values $\mathfrak{R}_1, \ldots, \mathfrak{R}_l$ have in the state ϕ, but only with what probability they take on all possible values.

2. The problem of **P**. cannot be answered for arbitrary quantities $\mathfrak{R}_1, \ldots, \mathfrak{R}_l$, but only for those whose operators $R_1, \ldots, R_l$ commute with one another.

Our next problem is to discuss the significance of these two facts.

2. THE STATISTICAL INTERPRETATION

Classical mechanics is a causal discipline, i.e., if we know exactly the state of a system in it -- for which, with k degrees of freedom, $2k$ numbers are necessary: the k space coordinates $q_1, \ldots, q_k$ and their k time derivatives $\frac{\partial q_1}{\partial t}, \ldots, \frac{\partial q_k}{\partial t}$, or, in place of these the k momenta $p_1, \ldots, p_k$ -- then we can give the value of each physical quantity (energy, torque, etc.) uniquely and with numerical exactness. Nevertheless, there also exists a statistical method of treatment of classical mechanics. But this is, as it were, a luxury or extra addition. That is, if we do not know all $2k$ variables $(q_1, \ldots, q_k, p_1, \ldots, p_k)$, but only several of them (and possibly some of these may not be known exactly), then, by averaging over the unknown variables in some fashion, we can at least make statistical assertions on all physical quantities. The same holds for the preceding or subsequent states of the system: if we know $q_1, \ldots, q_k, p_1, \ldots, p_k$ at the time $t = t_0$, then, by means of the classical equations of motion, we can calculate (causally) the state for every other time; but if we know only some of the variables, we must average over the rest, and we can then make only statistical statements for other values of the time.[120]

The statistical statements which we found in quantum mechanics have a different character. Here, for k degrees of freedom, the state is described by the wave function $\phi(q_1, \ldots, q_k)$ -- i.e., by a point ϕ of the $\mathfrak{R}_\infty$ suitably realized ($||\phi|| = 1$, and a numerical factor of absolute value 1 is unimportant). Although we believe that after having specified (ϕ) we know the state completely, nevertheless, only statistical statements can be made on the values of the physical quantities involved.

On the other hand this statistical character is limited to statements on the values of physical quantities, while the preceding and subsequent states ϕ_t can be calculated causally from $\phi_{t_0} = \phi$. The time dependent Schrödinger equation (cf. I.2.) makes this possible: because

---

[120]The kinetic theory of gases furnishes a good illustration of these relations.

A mole (32g) of oxygen contains $6 \cdot 10^{23}$ oxygen molecules, and, if we observe that each oxygen molecule is composed of 2 oxygen atoms (whose inner structure we shall neglect, so that they are to be treated as mass points with 3 degrees of freedom), one such mole is a mechanical system of $2 \cdot 3 \cdot 6 \cdot 10^{23} = 36 \cdot 10^{23} = k$ degrees of freedom. Its behavior can therefore be determined causally by the knowledge of $2k$ variables, but the gas theory uses only two: pressure and temperature, which are certain complicated functions of these $2k$ independent variables.

Consequently only statistical (probability) observations can be made. That these are in many cases nearly causal, i.e., the probabilities are near 0 or near 1, does not alter the fundamental nature of the situation.

$$\phi_{t_o} = \phi \; , \; \frac{h}{2\pi i}\frac{\partial}{\partial t}\phi_t = -H\phi_t$$

determine the entire path of ϕ_t. The solution of this differential equation is also possible in explicit form

$$\phi_t = e^{\frac{-2\pi i}{h}(t-t_o)H}\phi$$

($e^{\frac{-2\pi i}{h}(t-t_o)H}$ is unitary).[121] (In this formula, H was assumed to be independent of the time, but even with a time dependent H, ϕ_t is uniquely determined, since the differential equation is of first degree. In this case,

---

[121]If $F_t(\lambda)$ is a time dependent function, $\frac{\partial}{\partial t}F_t(\lambda) = G_t(\lambda)$, and H is a Hermitian operator, then $\frac{\partial}{\partial t}F_t(H) = G(H)$, because $\frac{\partial}{\partial t}$ is obtained by subtraction, division, and transition to the limit. For

$$F_t(\lambda) = e^{\frac{-2\pi i}{h}(t-t_o).\lambda}$$

this gives

$$\frac{\partial}{\partial t}\left(e^{\frac{-2\pi i}{h}(t-t_o).H}\right) = \frac{-2\pi i}{h}H.\,e^{\frac{-2\pi i}{h}(t-t_o).H}$$

which yields the desired differential equation, when it is applied to ϕ.

Because $|F_t(\lambda)| = 1$, $F_t(\lambda).\overline{F_t(\lambda)} = 1$, we have $F_t(H).\{F_t(H)\}^* = 1$, i.e., our

$$F_t(H) = e^{\frac{-2\pi i}{h}(t-t_o).H}$$

is unitary. Since it is obviously 1 for $t = t_o$, $\phi_{t_o} = \phi$ is also satisfied.

however, there are no simple solution formulas.)

If we want to explain the non-causal character of the connection between ϕ and the values of the physical quantities following the pattern of classical mechanics, then this interpretation is clearly the proper one: In reality, ϕ does not determine the state exactly. In order to know this state absolutely, additional numerical data are necessary. That is, the system has other characteristics or coordinates in addition to ϕ. If we were to know all of these, then we could give the values of all physical quantities exactly and with certainty. On the other hand, with the use of ϕ alone, just as in classical mechanics when only some of the $q_1, \ldots, q_k$, $p_1, \ldots, p_k$ are known, only statistical pronouncements are possible. Of course, this concept is only hypothetical. It is an attempt whose value depends on whether or not it actually succeeds in finding the additional coordinates contributing to ϕ, and in building, with their help, a causal theory which is in agreement with experiment, and which gives the statistical assertions of quantum mechanics when only ϕ is given (and an averaging is performed over the other coordinates).

It is customary to call these hypothetical additional coordinates "hidden parameters" or "hidden coordinates," since they must play a hidden role, in addition to the ϕ which alone have been uncovered by the investigation thus far. The explanation by means of the hidden parameters has (in classical physics) reduced many apparently statistical relations to the causal foundations of mechanics. An example of this is the kinetic theory of gases (cf. Note 120).

Whether or not an explanation of this type, by means of hidden parameters, is possible for quantum mechanics, is a much discussed question. The view that it will sometime be answered in the affirmative has at present prominent representatives. If it were correct, it would brand the present form of the theory as provisional, since then the description of the states would be essentially incomplete.

We shall show later (IV.2.) that an introduction of hidden parameters is certainly not possible without a basic change in the present theory. For the present, let us re-emphasize only these two things: The ϕ has an entirely different appearance and role from the $q_1, \ldots, q_k$, $p_1, \ldots, p_k$ complex in classical mechanics and the time dependence of ϕ is causal and not statistical: ϕ_{t_0} determines all ϕ_t uniquely, as we saw above.

Until a more precise analysis of the statements of quantum mechanics will enable us to prove objectively the possibility of the introduction of hidden parameters (which is carried out in the place quoted above), we shall abandon this possible explanation. We therefore adopt the opposite point of view. That is, we admit as a fact that those natural laws which govern the elementary processes (i.e., the laws of quantum mechanics) are of a statistical nature. (The causality of the macroscopic world can in any event be simulated by the leveling action which is manifest whenever many elementary processes operate simultaneously, i.e., by the "law of large numbers." Cf. the remarks at the end of Note 120 and Note 175.) Accordingly, we recognize **P.** (or **E.**.) as the most far reaching pronouncement on elementary processes.

This concept of quantum mechanics, which accepts its statistical expression as the actual form of the laws of nature, and which abandons the principle of causality, is the so-called statistical interpretation. It is due to M. Born,[122], and is the only consistently enforceable interpretation of quantum mechanics today -- i.e., of the sum of our experiences relative to the elementary processes.

---

[122] Z. Physik 37 (1926). The entire subsequent development (cf. Note 2) rests on this concept.

It is this interpretation to which we shall conform in the following (until we can proceed to a detailed and fundamental discussion of the situation).

3. SIMULTANEOUS MEASURABILITY AND MEASURABILITY IN GENERAL

The second circumstance which we had noted at the end of III.1., as surprising, was connected with the fact that **P**. gave information not only on those probabilities with which a quantity $\mathfrak{R}$ took on given numerical values, but also on the probability interrelations of several quantities $\mathfrak{R}_1, \ldots, \mathfrak{R}_l$. **P**. specified the probability that these quantities took on certain given values simultaneously (more precisely: that these quantities lay in certain intervals $I_1, \ldots, I_l$). (All in a given state ϕ.) But these quantities $\mathfrak{R}_1, \ldots, \mathfrak{R}_l$ were subjected to a characteristic limitation: their operators $R_1, \ldots, R_l$ had to commute pairwise. In the case of non-commuting $R_1, \ldots, R_l$ on the other hand, **P**. gave no information regarding the probability interrelations of the $\mathfrak{R}_1, \ldots, \mathfrak{R}_l$. In this case, **P**. could be used only to determine the probability distribution of each of these quantities by itself, without consideration of the others.

The most obvious step would be to assume that this is an incompleteness in **P**., and that there must exist a more general formula, containing this as a special case. Because even if quantum mechanics furnishes only statistical information regarding nature, the least we can expect from it is that it describe not only the statistics of individual quantities, but also the relations among several such quantities.

But, contrary to this concept, which appears reasonable at a first glance, we shall soon see that such a generalization of **P**. is not possible, and that in addition to the formal reasons (intrinsic in the structure of the mathematical tools of the theory), weighty physical grounds also suggest this type of a limitation. The necessity of this limitation and its physical meaning will give us an important insight into the nature of elementary processes.

In order to be clear on this point, we must investigate more precisely what the process of measurement of a quantity $\mathfrak{R}$ about which **P.** makes a (probability) statement, means for the quantum mechanical method of description.

First, let us refer to an important experiment which Compton and Simons carried out prior to the formulation of quantum mechanics.[123] In this experiment, light was scattered by electrons, and the scattering process was controlled in such a way that the scattered light and the scattered electrons were subsequently intercepted, and their energy and momenta measured. That is, there ensued collisions between light quanta and electrons, and the observer, since he measured the paths after collision, could prove whether or not the laws of elastic collision were satisfied. (We need consider only elastic collisions, since we do not believe that energy can be absorbed by electrons and light quanta in any form other than as kinetic energy. According to all experiments, both appear to have uniquely constituted structures. The collision calculation must naturally be carried out relativistically.[123]) Such a mathematical calculation was in fact possible, because the paths before collision were known, and those after the collision were observed. Therefore the collision problem was entirely determined. In order to determine the same process mechanically, two of these four paths, and the "central line" of the collision (the direction of the momentum transfer) suffices. In any case, therefore,

---

[123] Phys. Rev. 26 (1925). Cf. also the comprehensive treatment of W. Bothe in Handbuch der Physik, Vol. 23 ("Quanta") Berlin, 1926, Chapter 3, in particular § 73.

knowledge of 3 paths is sufficient, and the fourth acts as a check. The experiment gave complete confirmation to the mechanical laws of collision.

This result can also be formulated as follows, provided we admit the validity of the laws of collision, and regard the paths before the collision as known. The measurement of the path of either the light quantum or the electron after collision suffices to determine the position and the central line of the collision. The Compton-Simons experiment now shows that these two observations give the same result.

More generally, the experiment shows that the same physical quantity (namely, any coordinate of the place of collision or of the direction of the central line) is measured in two different ways (by capture of the light quantum and of the electron), and the result is always the same.

These two measurements do not occur entirely simultaneously. The light quantum and the electron do not arrive at once, and by suitable arrangement of the measuring apparatus either process may be observed first. The time difference is usually about 10^{-9} to 10^{-10} seconds. We call the first measurement M_1 and the second M_2. $\mathfrak{R}$ is the quantity measured. We then have the following situation. Although the entire arrangement is of such a type that, prior to the measurement, we can make only statistical statements regarding $\mathfrak{R}$, i.e., regarding M_1, M_2 (see the reference in Note 123), the statistical correlation between M_1 and M_2 is perfectly sharp (causal): the $\mathfrak{R}$ value of M_1 is certainly equal to that of M_2. Before the measurements M_1, M_2, therefore, both results are completely undetermined; after M_1 has been performed (but not M_2), the result of M_2 is already determined causally and uniquely.

We can formulate the principle that is involved as follows: by nature, three degrees of causality or noncausality may be distinguished. First, the $\mathfrak{R}$ value could be entirely statistical, i.e., the result of a measurement could be predicted only statistically; and if a second measurement were taken immediately after the first one, this would also have a dispersion, without regard to the value found initially -- for example, its dispersion might be equal to the original one.[124] Second, it is conceivable that the value of $\mathfrak{R}$ may have a dispersion in the first measurement, but that immediately subsequent measurement is constrained to give a result which agrees with that of the first. Third, $\mathfrak{R}$ could be determined causally at the outset.

The Compton-Simons experiment now shows that only the second case is possible in a statistical theory. Therefore, if the system is initially found in a state in which the values of $\mathfrak{R}$ cannot be predicted with certainty, then this state is transformed by a measurement M of $\mathfrak{R}$ (in the example above, M_1) into another state: namely, into one in which the value of $\mathfrak{R}$ is uniquely determined. Moreover, the new state, in which M places the system, depends not only on the arrangement of M, but also on the result of the measurement M (which could not be predicted causally in the original state) -- because the value of $\mathfrak{R}$ in the new state must actually be equal to this M-result.

Now let $\mathfrak{R}$ be a quantity whose operator R has a pure discrete spectrum $\lambda_1, \lambda_2, \ldots$ with the respective eigenfunctions $\phi_1, \phi_2, \ldots$ which then form a complete orthonormal set. In addition, let each eigenvalue be simple (i.e., of multiplicity 1, cf. II.6.), i.e., $\lambda_\mu \neq \lambda_\nu$ for $\mu \neq \nu$. Let us assume that we have measured $\mathfrak{R}$ and found a value λ^*. What is the state of the

---

[124]A statistical theory of elementary processes was erected by Bohr, Kramers and Slater on these basic concepts. Cf. Z. Physik 24 (1924), as well as the references cited in Note 123. The Compton-Simons experiment can be considered as a refutation of this view.

system after the measurement?

By virtue of the foregoing discussion, this state must be such, that a new measurement of $\mathfrak{R}$ gives the result λ^* with certainty (of course, this measurement must be made immediately, because after τ seconds, ϕ has changed to

$$e^{\frac{-2\pi i}{h}\tau . H}\phi$$

Cf. III.2., H is the energy operator.)

This question, as to when the measurement of $\mathfrak{R}$ in the state ϕ gives the value λ^* with certainty, we shall now answer in general, without limiting assumptions on the operator R.

Let $E(\lambda)$ be the resolution of the identity corresponding to R, I an interval $\{\lambda', \lambda''\}$. Our assumption can also be formulated in this way, that $\mathfrak{R}$ lies in I with the probability 0, if this does not contain λ^* -- or: with the probability 1 if λ^* belongs to I, i.e., if $\lambda' < \lambda^* \leq \lambda''$.

By **P.**, this means that $\|E(I)\phi\|^2 = 1$ or, since $\|\phi\| = 1$, $\|E(I)\phi\| = \|\phi\|$. Since $E(I)$ is a projection, and $1 - E(I)$ is also (THEOREM 13., II.4.),

$$\|\phi - E(I)\phi\| = \|\phi\|^2 - \|E(I)\phi\|^2 = 0$$

$$\phi - E(I)\phi = 0$$

$$E(\lambda'')\phi - E(\lambda')\phi = E(I)\phi = \phi$$

$\lambda' \to -\infty$ gives $E(\lambda'')\phi = \phi$; $\lambda'' \to +\infty$ gives $E(\lambda')\phi = 0$. (Cf. $\mathbf{S}_1$., II.7.) Therefore

$$E(\lambda)\phi = \begin{cases} \phi & \text{for } \lambda \geq \lambda^* \\ 0 & \text{for } \lambda < \lambda^* \end{cases}$$

But by II.8., this is characteristic for $R\phi = \lambda^*\phi$.

Another way of proving $R\phi = \lambda^*\phi$ rests on **E**$_1$. (i.e., **E**$_2$.). That $\mathfrak{R}$ has the value λ^* with certainty means that $(\mathfrak{R} - \lambda^*)^2$ has the expectation value 0. That is, the operator $F(R)$ with $F(\lambda) = (\lambda - \lambda^*)^2$, and therefore $(R - \lambda^*.1)^2$, has such an expectation value. We must then have

$$((R - \lambda^*.1)^2\phi,\phi) = ((R - \lambda^*.1)\phi,(R - \lambda^*.1)\phi) = ||(R - \lambda^*.1)\phi||^2$$

$$= ||R\phi - \lambda^*\phi||^2 = 0$$

i.e., $R\phi = \lambda^*\phi$.

For the special case that we considered originally, we thus have $R\phi = \lambda^*\phi$. As we discussed in II.6., this has the consequence that λ^* must be equal to a λ_μ (because $||\phi|| = 1$, $\phi \neq 0$) and $\phi = a\phi_\mu$. Since $||\phi|| = ||\phi_\mu|| = 1$, $|a|$ must $= 1$, and therefore a can be neglected without change in the state. Therefore: $\lambda^* = \lambda_\mu$, $\phi = \phi_\mu$ for some $\mu = 1,2,\ldots$. (The λ^*-assertion could also have been obtained directly from **P**., but not that regarding ϕ !)

Under the above assumptions on R, a measurement of $\mathfrak{R}$ then has the consequence of changing each state ψ into one of the states $\phi_1,\phi_2,\ldots$ which are connected with the respective results of measurement $\lambda_1,\lambda_2,\ldots$. The probabilities of these changes are therefore equal to the measurement probabilities for $\lambda_1,\lambda_2,\ldots$, and can therefore be calculated from **P**.

The probability that the value of $\mathfrak{R}$ lies in I is then $||E(I)\psi||^2$ by **P**. Hence, if we observe that by II.8., $E(I) = \sum_{\lambda_n \text{ in } I} P_{[\phi_n]}$,

$$P = ||E(I)\psi||^2 = (E(I)\psi,\ \psi) = \sum_{\lambda_n \text{ in } I} (P_{[\phi_n]}\psi,\ \psi)$$

$$= \sum_{\lambda_n \text{ in } I} |(\psi,\ \phi_n)|^2$$

One should therefore suspect that the probability for λ_n equals $|(\psi, \phi_n)|^2$. If we can so choose I that it contains a unique λ_m which is just λ_n, then this follows directly from the above formula. Otherwise (i.e., if the other λ_m are dense near λ_n), we can, for example, argue as follows: let $F(\lambda) = 1$ for $\lambda = \lambda_n$, otherwise 0. Then the desired probability of P_n is the expectation value of $F(\mathfrak{R})$, and hence by $\mathbf{E}_2$. (or $\mathbf{E}_1$.) is $(F(R)\psi, \psi)$. Now, by definition (II.8.)

$$(F(R)\psi, \psi) = \int_{-\infty}^{\infty} F(\lambda) d(\|E(\lambda)\psi\|^2)$$

and if we recall the definition of the Stieltjes integral, we can easily see that this is $= 0$ if $E(\lambda)\psi$ is continuous (in λ) for $\lambda = \lambda_n$, and in general the discontinuity of the (monotonic increasing) λ-function $\|E(\lambda)\psi\|^2$ at the point $\lambda = \lambda_n$. But this is equal to $\|P_{\mathfrak{M}}\psi\|^2$, where $\mathfrak{M}$ is the closed linear manifold spanned by all solutions of $R\psi = \lambda_n\psi$ (cf. II.8.). In the present case, $\mathfrak{M} = [\phi_n]$, and therefore

$$P_n = \|P_{[\phi_n]}\psi\|^2 = |(\psi, \phi_n)|^2$$

We have then answered the question as to what happens in the measurement of a quantity $\mathfrak{R}$, under the above assumptions for its operator R. To be sure, the "how" remains unexplained for the present. This discontinuous transition from ψ into one of the states $\phi_1, \phi_2, \ldots$ (which are independent of ψ, because ψ enters only into the respective probabilities $P_n = |(\psi, \phi_n)|^2$, $n = 1, 2, \ldots$ of this jump) is certainly not of the type described by the time dependent Schrödinger equation. This latter always results in a continuous change of ψ, in which the final result is uniquely determined and is dependent on ψ (cf. the discussion in III.2.).

We shall attempt to bridge this chasm later (cf. VI.).[125]

Let us still assume that R has a pure discrete spectrum, but no longer requires the simplicity of the eigenvalues. Then we can again form the $\phi_1, \phi_2, \ldots$ and $\lambda_1, \lambda_2, \ldots$, but duplications may now occur among the λ_n. After a measurement of $\mathfrak{R}$ a state ϕ with $R\phi = \lambda^*\phi$ is certainly present (λ^* is the result of the measurement.) As a consequence, λ^* is equal to one of the λ_n, but we can say only the following of ϕ: let those λ_n that equal λ^* be $\lambda_{n_1}, \lambda_{n_2}, \ldots$ (their number is finite or infinite). Then

$$\phi = \sum_\nu a_\nu \phi_{n_\nu}$$

(If there are infinitely many n_ν, then $\sum_\nu |a_\nu|^2$ must be finite.) Two such ϕ represent the same state only if they differ by no more than a numerical factor, i.e., if the ratio $a_1 : a_2 : \ldots$ is the same. Therefore, as soon as more than one n_ν exists, i.e., if the eigenvalue λ^* is multiple, then the state ϕ after the measurement is not uniquely determined by the knowledge of the result of the measurement.

We calculate the probability of λ^* (by **P.**, or **E$_1$.** or **E$_2$.**) exactly as before. It is

$$P(\lambda^*) = \sum_{\lambda_n = \lambda^*} |(\psi, \phi_n)|^2 = \sum_\nu |(\psi, \phi_{n_\nu})|^2$$

If R has no pure discrete spectrum, the situation is this:

All solutions f of $Rf = \lambda f$ span a closed

---

[125]That these jumps are related to the "quantum jumps"-concept of the older Bohr quantum theory was recognized by Jordan, Z. Physik 40 (1924).

linear manifold $\mathfrak{M}_\lambda$; all $\mathfrak{M}_\lambda$ together form an additional $\overline{\mathfrak{M}}$, and it is characteristic for the non-existence of a pure discrete spectrum that $\overline{\mathfrak{M}} \neq \mathfrak{R}_\infty$, i.e., that $\overline{\mathfrak{N}} = \mathfrak{R}_\infty - \overline{\mathfrak{M}} \neq (0)$. (Cf. II.8. for this, as well as for what follows.) $\mathfrak{M}_\lambda$ is at best $\neq (0)$ for a sequence of λ . These form the discrete spectrum of R . If we measure $\mathfrak{R}$ in the state ψ , then the probability that the result of the measurement will be λ^* is

$$P(\lambda^*) = ||P_{\mathfrak{M}_{\lambda^*}}\psi||^2 = (P_{\mathfrak{M}_{\lambda^*}}\psi, \psi)$$

This is best proved with the argumentation used above, which is based on $\mathbf{E}_2$. (or $\mathbf{E}_1$.) and on the function

$$F(\lambda) = \begin{cases} 1 , \text{ for } \lambda = \lambda^* \\ 0 , \text{ for } \lambda \neq \lambda^* \end{cases}$$

The probability that the value of $\mathfrak{R}$ will be some λ^* of the discrete spectrum $\mathfrak{P}$ of R is then

$$P = \sum_{\lambda^* \text{ in } \mathfrak{P}} (P_{\mathfrak{M}_{\lambda^*}}\psi, \psi) = (P_{\overline{\mathfrak{M}}}\psi, \psi) = ||P_{\overline{\mathfrak{M}}}\psi||^2$$

we can also see this directly with the aid of the function

$$F(\lambda) = \begin{cases} 1 , \text{ for } \lambda^* \text{ from } \mathfrak{P} \\ 0 , \text{ otherwise} \end{cases}$$

However, if we measure $\mathfrak{R}$ exactly, then afterwards a state ϕ with $R\phi = \lambda^*\phi$ must be present, and therefore the result of measurement λ^* must belong to $\mathfrak{P}$ -- the probability of obtaining an exact measurement is therefore (at most) $||P_{\overline{\mathfrak{M}}}\psi||^2$. But this number is not always 1 , and for the ψ of $\overline{\mathfrak{N}}$ it is in fact 0 --

therefore an exact measurement is not always possible.

We have seen that a quantity $\mathfrak{R}$ can always (i.e., for each state ψ) be measured exactly if and only if it possesses a pure discrete spectrum. If it possesses none, then it can be measured with only limited accuracy, i.e., the number continuum $-\infty < \lambda < +\infty$ can be divided into intervals $\ldots, I^{(-2)}, I^{(-1)}, I^{(0)}, I^{(1)}, I^{(2)}, \ldots$ (let the division points be $\ldots, \lambda^{(-2)}, \lambda^{(-1)}, \lambda^{(0)}, \lambda^{(1)}, \lambda^{(2)}, \ldots$, $I^{(n)} = \{\lambda^{(n)}, \lambda^{(n+1)}\}$; the maximum length of interval $\epsilon = \mathrm{Max}\,(\lambda^{(n+1)} - \lambda^{(n)})$, the spacing of the division points, is then the measurement accuracy), and the interval in which $\mathfrak{R}$ lies can be determined. This latter process can be pursued further mathematically. Namely, let $F(\lambda)$ be the following function (λ_n' some intermediate value from $I^{(n)}$, which is arbitrary for each $n = 0, \pm 1, \pm 2, \ldots$, but which is fixed in the following):

$$F(\lambda) = \lambda_n', \text{ if } \lambda \text{ lies in } I^{(n)}$$

Then the approximate measurement of $\mathfrak{R}$ is equivalent to the exact measurement of $F(\mathfrak{R})$. Now

$$F(R) = \int_{-\infty}^{\infty} F(\lambda)dE(\lambda) = \sum_{n=-\infty}^{+\infty} \int_{\lambda^{(n)}}^{\lambda^{(n+1)}} F(\lambda)dE(\lambda)$$

$$= \sum_{n=-\infty}^{+\infty} \lambda_n' \int_{\lambda^{(n)}}^{\lambda^{(n+1)}} dE(\lambda) = \sum_{n=-\infty}^{+\infty} \lambda_n' E(I^{(n)})$$

The equation $F(R)f = \lambda_n' f$ clearly holds for all f of the closed linear manifold belonging to $E(I^{(n)})$, i.e., for $F(R)$ $\mathfrak{M}_{\lambda_n'}$ contains the same. Consequently,

$$P_{\mathfrak{M}_{\lambda_n'}} \geq E(I^{(n)})$$

and therefore

$$P_{\overline{\mathfrak{M}}} \geq \sum_{n=-\infty}^{+\infty} P_{\mathfrak{M}_{\lambda'_n}} \geq \sum_{n=-\infty}^{\infty} E(I^{(n)})$$

$$= \sum_{n=-\infty}^{+\infty} (E(\lambda^{(n+1)}) - E(\lambda^{(n)})) = 1 - 0 = 1$$

From this it follows that

$$\sum_{n=-\infty}^{+\infty} P_{\mathfrak{M}_{\lambda'_n}} = P_{\overline{\mathfrak{M}}} = 1 \text{ , } P_{\mathfrak{M}_{\lambda'_n}} = E(I^{(n)})$$

i.e., $F(R)$ has a pure discrete spectrum, and this consists of the λ'_n .

Therefore $F(\mathfrak{R})$ is exactly measurable, and the probability that its value is λ'_n i.e., that the value of $\mathfrak{R}$ lies in $I^{(n)}$, is

$$||P_{\mathfrak{M}_{\lambda'_n}} \psi||^2 = ||E(I^{(n)})\psi||^2$$

in agreement with the statement of **P.** for $\mathfrak{R}$.

This result can also be interpreted physically, and it demonstrates that the theory is in good agreement with the ordinary intuitive physical point of view.

In the method of observation of classical mechanics (without any quantum conditions), we assign to each quantity $\mathfrak{R}$ in each state a completely determined value. At the same time, however, we recognize that each conceivable measuring apparatus, as a consequence of the imperfections of human means of observation (which result in the reading of the position of a pointer or of locating the blackening of a photographic plate with only limited accuracy), can furnish this value only with a certain (never vanishing) margin of error. This margin of error can, by sufficient refinement of the method of measurement, be made arbitrarily close to 0 -- but it is never exactly zero. One expects that this will also be true in quantum theory for those quantities $\mathfrak{R}$, which, according to the pictures that were customarily made of such things (especially before the discovery of quantum mechanics), are not quantized; for example, for the cartesian coordinates of an electron (which can take on every value between $-\infty$ and $+\infty$, and whose operators have continuous spectra). On the other hand, for those quantities which (according to our intuitive picture of them) are "quantized", the converse is true: since these are capable of assuming only discrete values, it suffices to observe them with just sufficient precision that no doubt can exist as to which one of these "quantized" values is occurring. That value is then as good as "observed" with absolute precision. For example, if we know of a hydrogen atom that it contains less energy than is necessary for the second lowest energy level, then we know its energy content with absolute precision: it is the lowest energy value.

This division into quantized and unquantized quantities corresponds, as the analysis of the matrix theory has already shown (cf. I.2. and II.6.), to the division into quantities $\mathfrak{R}$ with an operator R that has a pure discrete spectrum, and into such quantities for which this is not the case. And it was for the former, and only for these, that we found a possibility of an absolutely precise measurement -- while the latter could be observed only with arbitrarily good (but never absolute) precision.[126]

[126]In all such cases, we make the supposition that the structure of the observed system, and of the measuring apparatus -- i.e., all the force fields that operate, etc. -- is known exactly, and only the state, i.e., the instantaneous values of the coordinates, is sought. If these (idealized) assumptions do not prove to be correct, then additional sources of indeterminacy are of course present.

(In addition, it should be observed that the introduction of an eigenfunction which is "improper," i.e., which does not belong to Hilbert space -- mentioned in the preface as well as in I.3., cf. also II.8., especially the Notes 84, 86 -- gives a less good approach to reality than our treatment here. For such a method pretends the existence of such states in which quantities with continuous spectra take on certain values exactly, although this never occurs. Although such idealizations have often been advanced, we believe that it is necessary to discard them on these grounds, in addition to their mathematical untenability.) —

With this we have brought to a tentative conclusion the question as to the processes which occur in the measurement of a single quantity, and we can apply ourselves to the problem of the simultaneous measurement of several quantities.

First, let $\mathfrak{R}, \mathfrak{S}$ be two quantities with the respective operators R, S. We shall assume that they are simultaneously measurable. What follows from this?

We begin by assuming exact measurability so that R, S must both have pure discrete spectra: $\lambda_1, \lambda_2, \ldots$ and $\mu_1, \mu_2, \ldots$ respectively. Let the corresponding complete orthonormal sets of eigenfunctions be $\phi_1, \phi_2, \ldots$ and $\psi_1, \psi_2, \ldots$.

In order to discuss the simplest case first, we

There was a certain idealization even in our method of description of the inexact measurement. We assumed that these measurements consisted of distinguishing with absolute certainty whether or not a value lies in the interval $I = \{\lambda', \lambda''\}$, $\lambda' < \lambda''$. Actually, the boundaries λ', λ'' are indistinct, i.e., the necessary discriminations are valid only with a certain probability. Nevertheless, our method of description appears to be the most convenient one mathematically, at least for the present.

shall assume provisionally that one of the two operators, say, R, has simple eigenvalues, i.e., $\lambda_m \neq \lambda_n$ for $m \neq n$.

If we measure $\mathfrak{R}$, $\mathfrak{S}$ simultaneously, then a state is subsequently present in which $\mathfrak{R}$ as well as $\mathfrak{S}$ has the previously measured values with certainty. These values are say, $\lambda_{\overline{m}}$, $\mu_{\overline{n}}$. The state which then exists must satisfy the relations $R\psi = \lambda_{\overline{m}}\psi$, $S\psi = \mu_{\overline{n}}\psi$. From the first of these it follows that $\psi = \phi_{\overline{m}}$ (except for a numerical factor which we can neglect), while from the second, $\psi = \sum_\nu a_\nu \psi_{n_\nu}$ if $\mu_{n_1}, \mu_{n_2}, \ldots$ are all μ_n equal to $\mu_{\overline{n}}$. If the initial state was ϕ, then $\lambda_{\overline{m}}$, $\phi_{\overline{m}}$ has the probability $|(\phi, \phi_{\overline{m}})|^2$. For $\phi = \phi_m$ therefore, $\overline{m} = m$ is certain, so that we can say for each m that ϕ_m is a $\sum_\nu a_\nu \psi_{n_\nu}$ with equal μ_{n_ν}, i.e., $S\phi_m = \overline{\mu}\phi_m$ (with $\overline{\mu} = \mu_{n_1} = \mu_{n_2} = \ldots$). For $f = \phi_m$ consequently, $RSf = SRf$ (both are equal to $\lambda_m \overline{\mu} \cdot \phi_m$). Therefore this also holds for their linear aggregates, and, if R, S are continuous, for the limit points of these -- i.e., for all f. Therefore R, S commute.

If R, S are not continuous, then we argue as follows: The resolutions of the identity $E(\lambda)$, $F(\mu)$ corresponding to R, S are defined by

$$E(\lambda) = \sum_{\lambda_m \leq \lambda} P_{[\phi_m]}, \quad F(\mu) = \sum_{\mu_n \leq \mu} P_{[\psi_n]}$$

Consequently $F(\mu)\phi_m = \phi_m$ or 0, according to whether $\mu \geq$ or $<$ the above $\overline{\mu}$. Furthermore, $E(\lambda)\phi_m = \phi_m$ or 0, corresponding to $\lambda \geq$ or $< \lambda_m$. Therefore, in any case, $E(\lambda)F(\mu)\phi_m = F(\mu)E(\lambda)\phi_m$ for all ϕ_m. The commutativity of the $E(\lambda)$, $F(\mu)$ follows from this, just as above, and therefore (by II.10.) the commutativity of R, S also.

But according to II.10., there exists a complete orthonormal set of eigenfunctions common to R, S -- i.e., we may assume $\phi_m = \psi_m$. Since $\lambda_m \neq \lambda_n$ for $m \neq n$, we can set up a function $F(\lambda)$ with

$$F(\lambda) = \begin{cases} \mu_n, \text{ for } \lambda = \lambda_n, \; n = 1, 2, \ldots, \\ \text{arbitrary elsewhere} \end{cases}$$

and then $S = F(R)$, i.e., $\mathfrak{S} = F(\mathfrak{R})$. That is, $\mathfrak{R}$, $\mathfrak{S}$ are not only measurable simultaneously, but each measurement of $\mathfrak{R}$ is also one of $\mathfrak{S}$ since $\mathfrak{S}$ is a function of $\mathfrak{R}$, i.e., is determined causally by $\mathfrak{R}$.[127]

We now proceed to the more general case where nothing is assumed concerning the multiplicity of the eigenvalues of R, S. In this case we use an essentially different method.

First let us observe the quantity $\mathfrak{R} + \mathfrak{S}$. A simultaneous measurement of $\mathfrak{R}$, $\mathfrak{S}$ is also a measurement of $\mathfrak{R} + \mathfrak{S}$ because the addition of the results of the measurements gives the value of $\mathfrak{R} + \mathfrak{S}$. Consequently, the expectation value of $\mathfrak{R} + \mathfrak{S}$ in each state ψ is the sum of the expectation values of $\mathfrak{R}$ and of $\mathfrak{S}$. It should be noted that this holds independently of whether $\mathfrak{R}$, $\mathfrak{S}$ are statistically independent, or whether (and which) correlations exist between them -- because the law

Expectation value of the sum = Sum of the expectation values

holds in general, as is well known. Therefore, if T is the operator of $\mathfrak{R} + \mathfrak{S}$, then this expectation value is on the one hand $(T\psi, \psi)$, and on the other

---

[127] The latter proposition can be verified with the aid of **P.** The resolutions of the identity which belong to R and S may be formed by II.8.

$$(R\psi, \psi) + (S\psi, \psi) = ((R + S)\psi, \psi)$$

i.e., for all ψ

$$(T\psi, \psi) = ((R + S)\psi, \psi)$$

Therefore $T = R + S$. Consequently, $\mathfrak{R} + \mathfrak{S}$ has the operator $R + S$.[128] In the same way, we can show that $a\mathfrak{R} + b\mathfrak{S}$ (a, b real numbers) has the operator $aR + bS$. (This also follows from the first formula, if we substitute $\mathfrak{R}$, $\mathfrak{S}$ and R, S in the functions $F(\lambda) = a\lambda$, $G(\mu) = b\mu$.)

A simultaneous measurement of $\mathfrak{R}$, $\mathfrak{S}$ is also a measurement of

$$\frac{\mathfrak{R} + \mathfrak{S}}{2},\quad \left(\frac{\mathfrak{R} + \mathfrak{S}}{2}\right)^2,\quad \frac{\mathfrak{R} - \mathfrak{S}}{2},\quad \left(\frac{\mathfrak{R} - \mathfrak{S}}{2}\right)^2,\quad \left(\frac{\mathfrak{R} + \mathfrak{S}}{2}\right)^2 - \left(\frac{\mathfrak{R} - \mathfrak{S}}{2}\right)^2 = \mathfrak{R} \cdot \mathfrak{S}$$

Operators of these quantities (if we also make use of the fact that if T is the operator of $\mathfrak{T}$, $F(\mathfrak{T})$ has the operator $F(T)$ -- hence T^2 has the operator T^2) are therefore

$$\frac{R + S}{2},\quad \left(\frac{R + S}{2}\right)^2 = \frac{R^2 + S^2 + RS + SR}{4},\quad \frac{R - S}{2},\quad \left(\frac{R - S}{2}\right)^2$$

$$= \frac{R^2 + S^2 - RS - SR}{4},\quad \left(\frac{R + S}{2}\right)^2 - \left(\frac{R - S}{2}\right)^2 = \frac{RS + SR}{2}$$

That is, $\mathfrak{R} \cdot \mathfrak{S}$ has the operator $\frac{RS + SR}{2}$. This also holds for all $F(\mathfrak{R})$, $G(\mathfrak{S})$ (which are also measured), and therefore $F(\mathfrak{R}) \cdot G(\mathfrak{S})$ has the operator

[128] We have proved this law, according to which the operator of $\mathfrak{R} + \mathfrak{S}$ is the sum of the operators of $\mathfrak{R}$ and $\mathfrak{S}$, for simultaneously measurable $\mathfrak{R}$, $\mathfrak{S}$. Cf. what is said at the end of IV.1. and IV.2.

$$\frac{F(R)G(S) + G(S)F(R)}{2}$$

Now let $E(\lambda)$, $F(\mu)$ be the resolutions of the identity corresponding to R, S. Furthermore, let

$$F(\lambda) = \begin{cases} 1, & \text{for } \lambda \le \overline{\lambda} \\ 0, & \text{for } \lambda > \overline{\lambda} \end{cases} \quad \Bigg| \quad G(\mu) = \begin{cases} 1, & \text{for } \mu \le \overline{\mu} \\ 0, & \text{for } \mu > \overline{\mu} \end{cases}$$

As we know, $F(R) = E(\overline{\lambda})$, $G(S) = F(\overline{\mu})$, therefore $F(\mathfrak{R}) \cdot G(\mathfrak{S})$ has the operator $\frac{EF + FE}{2}$ (for brevity, we replace $E(\overline{\lambda})$, $F(\overline{\mu})$ by E, F). Since $F(\mathfrak{R})$ always $= 0, 1$, $F(\mathfrak{R})^2 = F(\mathfrak{R})$, and therefore

$$F(\mathfrak{R}) \cdot (F(\mathfrak{R}) \cdot G(\mathfrak{S})) = F(\mathfrak{R}) \cdot G(\mathfrak{S})$$

Let us now apply our multiplication formula to $F(\mathfrak{R})$ and $F(\mathfrak{R}) \cdot G(\mathfrak{S})$ (all these are simultaneously measurable). We then obtain the operator

$$\frac{E\,\frac{EF + FE}{2} + \frac{EF + FE}{2}\,E}{2} = \frac{E^2F + 2EFE + FE^2}{4} = \frac{EF + FE + 2EFE}{4}$$

for this product. This must equal $\frac{EF + FE}{2}$, from which it follows that

$$EF + FE = 2 \cdot EFE$$

Left hand multiplication with E gives

$$E^2F + EFE = 2E^2FE, \quad EF + EFE = 2 \cdot EFE, \quad EF = EFE$$

while right hand multiplication gives

$$EFE + FE^2 = 2 \cdot EFE^2, \quad EFE + FE = 2 \cdot EFE, \quad FE = EFE$$

Therefore $EF = FE$. That is, all $E(\overline{\lambda})$, $F(\overline{\mu})$ commute -- consequently R, S again commute.

By II.10., this condition: R, S commutative, means the same as the requirement that there exists a Hermitian operator T of which R and S are functions: $R = F(T)$, $S = G(T)$. If this operator belongs to the quantity $\mathfrak{T}$, then $\mathfrak{R} = F(\mathfrak{T})$, $S = G(\mathfrak{T})$. However, this condition is also sufficient for the simultaneous measurability, because a measurement of $\mathfrak{T}$ (an absolutely exact one, because T has a pure discrete spectrum, cf. II.10) measures simultaneously the functions $\mathfrak{R}$, $\mathfrak{S}$. The commutativity of R, S is therefore the necessary and sufficient condition.

If several variables $\mathfrak{R}, \mathfrak{S}, \ldots$ (but a finite number) are given, let their operators be $R, S, \ldots$, and if absolutely exact measurement is again required, then the situation with regard to simultaneous measurability is as follows. If all quantities $\mathfrak{R}, \mathfrak{S}, \ldots$ are simultaneously measurable, then all pairs formed from them must also be simultaneously measurable. That is, all operators R, S commute pair wise. Conversely, if $R, S, \ldots$ commute with each other, then by II.10., there exists an operator T, of which all are functions: $R = F(T), S = G(T), \ldots$ and therefore for the corresponding quantity $\mathfrak{T}$: $\mathfrak{R} = F(\mathfrak{T})$, $\mathfrak{S} = G(\mathfrak{T}), \ldots$. An exact measurement of $\mathfrak{T}$ ($\mathfrak{T}$ has again a pure discrete spectrum, cf. II.10.) is consequently a simultaneous measurement of $\mathfrak{R}, \mathfrak{S}, \ldots$. That is, the commutativity of $R, S, \ldots$ is necessary and sufficient for the simultaneous measurability of $\mathfrak{R}, \mathfrak{S}, \ldots$.

Now let us consider such measurements which are not absolutely exact, but only of (arbitrarily great) previously given accuracy. Then $R, S, \ldots$ no longer need have pure discrete spectra.

Since the limited accuracy measurements of $\mathfrak{R}, \mathfrak{S}, \ldots$ effect the same as absolutely exact measurements of $F(\mathfrak{R}), G(\mathfrak{S}), \ldots$ -- in which $F(\lambda), G(\lambda), \ldots$ are certain functions the manner of whose formation was described at the beginning of this section (in the discussion of a single measurement of limited accuracy, of course only $F(\lambda)$ was given, then $\mathfrak{R}, \mathfrak{S}, \ldots$ are certainly measurable simultaneously, if all the $F(\mathfrak{R}), G(\mathfrak{S}), \ldots$ are measurable simultaneously and, of course, with absolute accuracy. But the latter is equivalent to the commutativity of $F(R), G(S), \ldots$, and this follows from that of $R, S, \ldots$. Therefore the commutativity of R, S is in any case sufficient.

Conversely, if $\mathfrak{R}, \mathfrak{S}, \ldots$ are taken as simultaneously measurable, then we proceed as follows. A sufficiently exact measurement of $\mathfrak{R}$ permits us to distinguish whether its value is $> \bar{\lambda}$ or $\leq \bar{\lambda}$ (cf. our definition of "limited accuracy," discussed in Note 126). Then if $F(\lambda)$ is defined $= 1$, for $\lambda \leq \bar{\lambda}$ and $= 0$ for $\lambda > \bar{\lambda}$, respectively, $F(\mathfrak{R})$ is measurable with absolute accuracy. Correspondingly, if $G(\mu) = 1$ for $\mu \leq \bar{\mu}$, and $= 0$ for $\mu > \bar{\mu}$, then $G(\mathfrak{S})$ is measurable with absolute accuracy, and moreover, both quantities are measurable simultaneously. Therefore $F(R), G(S)$ commute. Now let $E(\lambda), F(\mu)$ be the resolutions of the identity belonging to R, S. Then $F(R) = E(\bar{\lambda}), G(S) = F(\bar{\mu})$, and therefore $E(\bar{\lambda}), F(\bar{\mu})$ commute -- for all $\bar{\lambda}, \bar{\mu}$. Consequently, R, S commute, and since the same must hold for each pair of R, S, all R, S must commute pairwise. Therefore this condition is also necessary.

We therefore see that the characteristic condition for the simultaneous measurability of an arbitrary (finite) number of quantities $\mathfrak{R}, \mathfrak{S}$ is the commutativity of their operators, $R, S, \ldots$. In fact, this holds for absolutely exact measurements as well as for arbitrarily exact measurements. In the first case, however, it is also required that the operators possess pure discrete spectra -- which is characteristic for the case of absolutely exact measurements.

We have now produced the mathematical proof that **P.** makes the most extensive statement which is in general possible in this theory (i.e., in one that includes **P.**). This is due to the fact that it presumes only the commutativity of the operators $R_1, \ldots, R_l$. Without this condition, nothing can be said regarding the results of simultaneous measurements for $\mathfrak{R}_1, \ldots, \mathfrak{R}_l$, since simultaneous measurements of these quantities are then in general not possible.

4. UNCERTAINTY RELATIONS

In the foregoing sections, we have obtained important information on the measuring process involving one quantity, or several, simultaneously measurable ones. We must now develop the procedure for quantities which are not simultaneously measurable, if we are interested in their statistics in the same system (in the same state ϕ).

Therefore, let two such quantities $\mathfrak{R}$, $\mathfrak{S}$, as well as their (non-commuting) operators R, S be given. In spite of this assumption, states ϕ may exist, in which both quantities have sharply defined values (i.e., dispersion 0) -- i.e., eigenfunctions common to both; only no complete orthogonal set can be formed from these, since then R, S would commute. (Cf. the construction given in II.8. for the corresponding resolutions of the identity $E(\lambda)$, $F(\lambda)$. If $\phi_1, \phi_2, \ldots$ is the complete orthogonal set mentioned, the $E(\lambda)$ as well as the $F(\lambda)$ are $P_{[\phi_\rho]}$ sums, and therefore commute, since the $P_{[\phi_\rho]}$ do.) What this means can easily be seen. The closed linear manifold $\mathfrak{M}$ spanned by these ϕ must be smaller than $\mathfrak{R}_\infty$ -- because were this equal to $\mathfrak{R}_\infty$, then the desired complete orthonormal set could be built up exactly as was done in the beginning of II.6., for the case of a single operator.

For the states of $\mathfrak{M}$, our $\mathfrak{R}$, $\mathfrak{S}$ are simultaneously measurable. This can be shown most readily by indicating a model for this simultaneous measurement. Since the common eigenfunctions of R, S span $\mathfrak{M}$, there is also an orthonormal set of such ϕ : $\phi_1, \phi_2, \ldots$ spanning $\mathfrak{M}$ (i.e., complete in $\mathfrak{M}$). (This is also obtained by the method of construction described previously in II.6.) We extend $\phi_1, \phi_2, \ldots$ to a complete set $\phi_1, \phi_2, \ldots, \psi_1, \psi_2, \ldots$ by the addition of an orthonormal set $\psi_1, \psi_2, \ldots$ which spans $\mathfrak{R}_\infty - \mathfrak{M}$. Now let $\lambda_1, \lambda_2, \ldots, \mu_1, \mu_2, \ldots$ be distinct numbers, and let T be defined by

$$T\left(\sum_n x_n \cdot \phi_n + \sum_n y_n \cdot \psi_n\right) = \sum_n \lambda_n x_n \cdot \phi_n + \sum_n \mu_n y_n \cdot \psi_n$$

where $\mathfrak{T}$ is the corresponding quantity.

A measurement of $\mathfrak{T}$ produces (as we know from III.3.) one of the states $\phi_1, \phi_2, \ldots, \psi_1, \psi_2, \ldots$. If a ϕ_n results (which can be sensed by observing that the result of the measurement is a λ_n), then we also know the values of $\mathfrak{R}$ and $\mathfrak{S}$: because $\mathfrak{R}$, $\mathfrak{S}$ have sharply defined values in ϕ_n by our assumptions, and we can predict with certainty that in a measurement of $\mathfrak{R}$ or $\mathfrak{S}$ immediately following, these respective values will be found. On the other hand, if ψ_n is the result, then we know nothing of the sort (ψ_n does not lie in $\mathfrak{M}$; therefore R, S are not sharply defined in ψ_n). The probability of finding ψ_n is, however, as we know, $(P_{[\psi_n]}\phi, \phi)$, and the probability of finding any ψ_n $(n = 1, 2, \ldots)$ is

$$\sum_n (P_{[\psi_n]}\phi, \phi) = (P_{\mathfrak{R}_\infty - \mathfrak{M}}\phi, \phi) = \|P_{\mathfrak{R}_\infty - \mathfrak{M}}\phi\|^2 = \|\phi - P_{\mathfrak{M}}\phi\|^2$$

If ϕ belongs to $\mathfrak{M}$, i.e., if $\phi = P_{\mathfrak{M}}\phi$, then this probability $= 0$, i.e., $\mathfrak{R}$, $\mathfrak{S}$ are measured simultaneously with certainty.[129]

---

[129] The further detailed discussion of the "simultaneous

Since we are now interested in non-simultaneously measurable quantities, we shall assume the existence of the extreme case $\mathfrak{M} = (0)$; i.e., we shall assume that $\mathfrak{R}$, $\mathfrak{S}$ are not simultaneously measurable in any state -- or: that no eigenfunctions common to R, S exist.

If R, S have the resolutions of the identity $E(\lambda)$, $F(\lambda)$, and are in the state ϕ , then, as we know from III.1., the expectation values of R, S are

$$\rho = (R\phi, \phi) \; , \; \sigma = (S\phi, \phi)$$

and their dispersions, i.e., the expectation values of $(\mathfrak{R} - \rho)^2$, $(\mathfrak{S} - \sigma)^2$ (cf. the discussion of the absolutely precise measurement in III.3.) are

$$\epsilon^2 = ((R - \rho.1)^2\phi, \phi) = ||(R - \rho.1)\phi||^2 = ||R\phi - \rho\phi||^2$$

$$\eta^2 = ((S - \sigma.1)^2\phi, \phi) = ||(S - \sigma.1)\phi||^2 = ||S\phi - \sigma\phi||^2$$

After a familiar transformation, this becomes:[130]

$$\epsilon^2 = ||R\phi||^2 - (R\phi, \phi)^2 \; , \quad \eta^2 = ||S\phi||^2 - (S\phi, \phi)^2$$

(because $||\phi|| = 1$ the Schwarz inequality, i.e., THEOREM

---

measurability for the ϕ of $\mathfrak{M}$ " for $\mathfrak{R}$, $\mathfrak{S}$ which are not measurable with absolute precision (continuous spectra) etc., is left to the reader. This can be carried out in the same way as in the treatment in III.3.

[130]The operator calculation is the following:

$$\epsilon^2 = ((R - \rho.1)^2\phi, \phi) = (R^2\phi, \phi) - 2\rho.(R\phi, \phi) + \rho^2$$

$$= ||R\phi||^2 - 2.(R\phi, \phi)^2 + (R\phi, \phi)^2 = ||R\phi||^2 - (R\phi, \phi)^2$$

and correspondingly for η^2 .

1., II.1., shows that the left sides are ≥ 0 .) There now arises the question: since ϵ and η cannot be jointly zero but ϵ alone can be made arbitrarily small, and η likewise ($\mathfrak{R}$, $\mathfrak{S}$ are measurable separately with arbitrary exactness, perhaps even absolutely exactly), must there be relations between ϵ, η which prevent their becoming arbitrarily small simultaneously, and what would be the form of such relations?

The existence of such relations was discovered by Heisenberg.[131] They are of great importance for the knowledge of the uncertainties of the description of nature produced by quantum mechanics. They are consequently known as uncertainty relations. We shall first derive the most important relation of this type mathematically, and then return to its fundamental meaning, and its connection with experiment.

In matrix theory, the operators P, Q with the commutation property

$$PQ - QP = \frac{h}{2\pi i} 1$$

played an important role: they were for example assigned to the coordinate and its conjugate momentum (cf. I.2.) -- or, more generally, to any two quantities which were canonically conjugate in classical mechanics (cf. for example the papers mentioned in Note 2). Let us examine any two such Hermitian operators, P, Q with

$$PQ - QP = a.1$$

(Since $(PQ - QP)^* = QP - PQ$, $(a.1)^* = \bar{a}.1 = -a.1$, $\bar{a} = -a$,

---

[131] Z. Physik 43 (1927). These considerations were extended by Bohr, Naturwiss. 16 (1928). The mathematical discussion that follows was first undertaken by Kennard, Z. Physik 44 (1926), and was given by Robertson in its present form.

i.e., a is pure imaginary. This operator equation is not necessarily understood to include the equality of the domains of definition of both sides: $PQ - QP$ need not have sense everywhere). For each ϕ then,

$$2\,\mathrm{Im}(P\phi, Q\phi) = -i[(P\phi, Q\phi) - (Q\phi, P\phi)] = -i[(QP\phi, \phi) - (PQ\phi, \phi)]$$

$$= (i\{PQ - QP\}\phi, \phi) = ia.\|\phi\|^2$$

Let $a \neq 0$, then we have (THEOREM 1., II.1.)

$$\|\phi\|^2 = \frac{-2i}{a}\,\mathrm{Im}(P\phi, Q\phi) \leq \frac{2}{|a|}\,|(P\phi, Q\phi)| \leq \frac{2}{|a|}\,\|P\phi\|.\|Q\phi\|$$

therefore, for $\|\phi\| = 1$

$$\|P\phi\|.\|Q\phi\| \geq \frac{|a|}{2}$$

Since $P - \rho.1$, $Q - \sigma.1$ also have the above commutation property, we have similarly

$$\|P\phi - \rho.\phi\|.\|Q\phi - \sigma.\phi\| \geq \frac{|a|}{2}$$

and if we introduce the mean values and the dispersions:

$$\rho = (P\phi, \phi), \qquad \epsilon^2 = \|P\phi - \rho.\phi\|^2$$

$$\sigma = (Q\phi, \phi), \qquad \eta^2 = \|Q\phi - \sigma.\phi\|^2$$

then this becomes:

$$(\mathbf{U}.) \qquad \epsilon\eta \geq \frac{|a|}{2}$$

In order that the equality sign hold, it is necessary and sufficient that in the $\leq$ inequalities used in the derivation, the $=$ sign always holds. With $P' = P - \rho.1$, $Q - Q - \sigma.1$ we then have

$$-\frac{i|a|}{a}\,\mathrm{Im}(P'\phi,\ Q'\phi) = |(P'\phi,\ Q'\phi)| = ||P'\phi||\cdot||Q'\phi||$$

By THEOREM 1., II.1., the second equation means that $P'\phi$, $Q'\phi$ differ only by a constant factor -- and since $||P'\phi||\cdot||Q'\phi|| \geq \frac{|a|}{2} > 0$ implies $P'\phi \neq 0$, $Q'\phi \neq 0$, there must be $P'\phi = c\cdot Q'\phi$, $c' \neq 0$. But the first equation means that $(P'\phi,\ Q'\phi) = c||Q'\phi||^2$ is pure imaginary, and in fact that its i-coefficient has the same sign as $-\frac{i|a|}{a}$ (real!), i.e., opposite to that of a. Therefore, $c = i\gamma$, γ real and $\gtrless 0$ for $\frac{a}{i} \lessgtr 0$, respectively. Consequently,

Eq.) $P'\phi = i\gamma\cdot Q'\phi$, γ real and $\lessgtr 0$ for $ia \lessgtr 0$, respectively

The definition of ρ, σ also requires that $(P'\phi) = 0$, $(Q'\phi,\ \phi) = 0$. Since $(P'\phi,\ \phi) = i\gamma(Q'\phi,\ \phi)$ follows from (**Eq.**), and in this equation a real quantity is on the left, and a purely imaginary quantity is on the right, therefore both sides must vanish, and so the desired equations hold automatically. We have yet to determine ϵ, η. We have the relations

$$\epsilon:\eta = ||P'\phi||:||Q'\phi|| = |c| = |\gamma|\ ,\quad \epsilon\eta = \frac{|a|}{2}$$

therefore, since ϵ, η are both positive,

$$\epsilon = \sqrt{\frac{|a||\gamma|}{2}}\ ,\quad \eta = \sqrt{\frac{|a|}{2|\gamma|}}$$

For the quantum mechanical case, $a = \frac{h}{2\pi i}$, we get from (**U.**)

(**U'.**) $$\epsilon\cdot\eta \geq \frac{h}{4\pi}$$

We can also discuss (**Eq.**) if P, Q are the operators of the Schrödinger theory, $P = \frac{h}{2\pi i}\frac{\partial}{\partial q}\cdots$, $Q = q\cdots$. (Cf.

I.2., we assume that a mechanical system with one degree of freedom is under consideration, and that its single coordinate is q.) Then (**Eq.**) becomes

$$\left(\frac{h}{2\pi i}\frac{\partial}{\partial q} - \rho\right)\phi = i\gamma(q - \sigma)\phi$$

and because $ia = \frac{h}{2\pi} > 0$, $\gamma > 0$. Therefore

$$\frac{\partial}{\partial q}\phi = \left\{-\frac{2\pi}{h}\gamma\cdot\dot{q} + \frac{2\pi}{h}\gamma\sigma + \frac{2\pi}{h}\rho\cdot i\right\}\phi$$

i.e.,

$$\phi = e^{\int^{q\cdot}\left\{-\frac{2\pi}{h}\gamma\cdot q + \frac{2\pi}{h}\gamma\cdot\sigma + \frac{2\pi}{h}\rho\cdot i\right\}dq}$$

$$= Ce^{\frac{-\pi\gamma}{h}q^2 + \frac{2\pi\gamma\sigma}{h}q + \frac{2\pi\rho}{h}iq} = C'e^{\frac{-\pi\gamma}{h}(q-\sigma)^2 + \frac{2\pi\rho}{h}iq}$$

Since $\gamma > 0$, $||\phi||^2 = \int_{-\infty}^{+\infty}|\phi(q)|^2dq$ is indeed finite, and C' is determined from $||\phi|| = 1$:

$$||\phi||^2 = \int_{-\infty}^{\infty}|\phi(q)^2|dq = |C'|^2\int_{-\infty}^{+\infty}e^{\frac{-2\pi\gamma}{h}(q-\sigma)^2}dq$$

$$= |C'|^2\sqrt{\frac{h}{2\pi\gamma}}\int_{-\infty}^{+\infty}e^{-x^2}dx$$

$$= |C'|^2\sqrt{\frac{h}{2\pi\gamma}}\sqrt{\pi} = |C'|^2\sqrt{\frac{h}{2\gamma}} = 1,$$

$$|C'| = \left(\frac{2\gamma}{h}\right)^{\frac{1}{4}}$$

Therefore, by neglect of the physically unimportant factor of absolute value 1,

$$C' = \left(\frac{2\gamma}{h}\right)^{\frac{1}{4}}$$

i.e.,

$$\phi = \phi(q) = \left(\frac{2\gamma}{h}\right)^{\frac{1}{4}} e^{\frac{-\pi\gamma}{h}(q-\sigma)^2 + \frac{2\pi\rho}{h} iq}$$

Then ϵ, η are given by

$$\epsilon = \sqrt{\frac{h\gamma}{4\pi}}, \quad \eta = \sqrt{\frac{h}{4\pi\gamma}}$$

Aside from the condition $\epsilon\eta = \frac{h}{4\pi}$, they are therefore arbitrary, since γ varies from 0 to $+\infty$. That is, each set of four quantities $\rho, \sigma, \epsilon, \eta$ satisfying $\epsilon\eta = \frac{h}{4\pi}$ is exactly realized by a ϕ. These ϕ were first investigated by Heisenberg, and applied to the interpretation of quantum mechanical situations. They are especially suitable for this because they represent the highest possible degree of approximation (in quantum mechanics) to classical mechanical relations (where p, q are both without dispersion!), and where ϵ and η can be prescribed without any restrictions. (Cf. reference in Note 131.)

With the foregoing considerations, we have comprehended only one phase of the uncertainty relations, that is, the formal one; for a complete understanding of these relations, it is still necessary to consider them from another point of view: from that of direct physical experience. For the uncertainty relations bear a more easily understandable and simpler relation to direct experience than many of the facts on which quantum mechanics was originally based, and therefore the above, entirely formal, derivation does not do them full justice. An intuitive discussion is all the more necessary, since one could obtain, at first glance, an impression that a contradiction exists here with the ordinary, intuitive point of view: it will not be clear to common sense without a further discussion, why the position and velocity (i.e., coordinate and momentum) of a material body cannot both be measured simultaneously and with arbitrarily high accuracy -- provided that sufficiently refined instruments of measurement were available. Therefore it is necessary to elucidate by an exact analysis of the finest processes of measurement (capable of execution perhaps only in the sense of ideal experiments) that this is not the case. Actually, the well known laws of wave optics, electrodynamics and elementary atomic processes place very great difficulties in the way of accurate measurement precisely where this is required by the uncertainty relations. And in fact, this can already be recognized if the processes mentioned are investigated purely classically (not quantum theoretically). This is an important point of principle. It shows that the uncertainty relations, although apparently paradoxical, do not conflict with classical experience (i.e., with the area in which the quantum phenomena do not call for an essential correction of the earlier ways of thinking) -- and classical experience is the only kind which is valid independently of the correctness of quantum mechanics, indeed the only kind directly accessible to our ordinary, intuitive way of thinking.[132]

We are then to show that if p, q are two canonically conjugate quantities, and a system is in a state in which the value of p can be given with the accuracy ϵ (i.e., by a p measurement with an error range ϵ), then q can be known with no greater accuracy

---

[132] The fundamental meaning of this circumstance was emphasized by Bohr. See the reference in Note 131. In addition, the method of description followed below is not entirely classical at one point: the existence of light quanta will be assumed, i.e., the fact that light of frequency ν never appears in quantities of energy smaller than $h\nu$.

than $\eta = \frac{h}{2\pi} : \epsilon$. Or: a measurement of p with the accuracy ϵ must bring about an indeterminancy $\eta = \frac{h}{4\pi} : \epsilon$ in the value of q .

Naturally, in these very qualitative considerations, we cannot expect to recover each detail with perfect exactness. Thus, instead of showing that $\epsilon\eta = \frac{h}{4\pi}$ we shall be able to show only that $\epsilon\eta \sim h$, for the most precise measurement possible (i.e., it is of the same order of magnitude as h) . As a typical example, we shall consider the conjugate pair position (coordinate)-momentum of a particle **T** .[133]

First let us investigate the determination of position. This results when one looks at **T** , i.e., when **T** is illuminated and the scattered light is absorbed in the eye. Therefore, a light quantum **L** is emitted from the light source **L** in the direction of **T** , is deflected from its straight line path $\beta\beta_1$ into $\beta\beta_2$ by collision with **T** , and at the end of its path is annihilated by absorption at the screen Sc (which represents the eye or a photographic plate)(Fig. 1). The measurement takes place by the determination that **L** hits the screen not at 1 (at the end of its undeflected path $\beta\beta_1$) , but at 2 (at the end of $\beta\beta_2$) . But in order to be able to furnish the position of the collision (i.e., of **T**) from this, the directions of β and β_2 must also be known (i.e., **L**'s direction before and after collision): we achieve this by the interposition of the slit systems ss and s's' . (In this way we are actually not performing a measurement of the coordinate of **T** but we obtain only an answer to the question of whether or not this coordinate has a certain value, corresponding to the intersection of the directions β and β_2 . This value, however, can be selected at will by appropriate arrangement of the slits.

---

[133]The subsequent discussion is due to Heisenberg and Born. See references in Note 131.

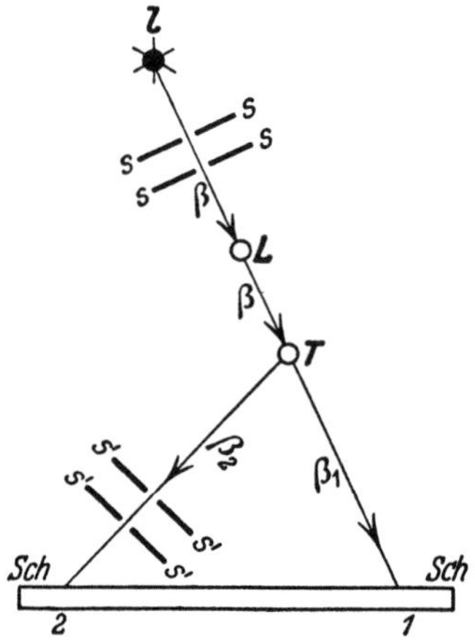

The superposition of several such determinations, i.e., the use of additional slits s's' , is equivalent to the complete coordinate measurement.) Now what is the accuracy of this measurement of position?

This measurement has a fundamental limitation in the laws of optical image formation. Indeed, it is impossible, with the light of wavelength λ , to picture sharply objects which are smaller than λ , or even to reduce the scattering to such an extent that one can speak of a (distorted) image. To be sure, we did not require a faithful optical image, since the mere fact of the deviation of **L** suffices to determine the position of **T** . Nevertheless, the slits ss and s's' cannot be narrower than λ , since otherwise **L** cannot pass without appreciable diffraction. Rather, a bundle of interference lines will then occur -- so that nothing further can be deduced from the direction of the line connecting the successive slits ss and s's' , concerning the directions β and β_2 of the light ray. As a consequence it is never possible, with this projectile **L** , to aim and hit with an accuracy greater than λ .

The wavelength λ is then the measure of the error in measuring the coordinate: $\lambda \sim \epsilon$. Further characteristics of **L** are: its frequency ν , its energy E , its momentum $\bar{p}$, and there exist the well known relations

$$\nu = \frac{c}{\lambda}, \; E = h\nu = \frac{hc}{\lambda}, \; \bar{p} = \frac{E}{c} = \frac{h\nu}{c} = \frac{h}{\lambda}$$

(c is the velocity of light).[134] Consequently, $\bar{p} \sim \frac{h}{\epsilon}$. Now there is a momentum change in the (not exactly known) collision process between **L** and **T** , which is clearly of the order of magnitude of $\bar{p}$, i.e., of the same order as $\frac{h}{\epsilon}$. Hence there results an uncertainty $\eta \sim \frac{h}{\epsilon}$ in the momentum.

This would show that $\epsilon\eta \sim h$, if one detail had not been overlooked. The collision process is really not so unknown. We actually know the directions of motion of **L** before and after (β and β_2) , and therefore its momentum also -- and the amount of momentum transferred to **T** can be obtained from this. Consequently, $\bar{p}$ is not a measure of η , but rather the possible uncertainty of direction of the rays β and β_2 will furnish such a measure. Now in order to be able to establish more precisely the relations between the "aiming" at the small object **T** and the uncertainty of direction which is associated with it, it is appropriate to use a better focusing device than the slit ss -- namely, a lens. Consequently, the well-known theory of the microscope must be considered. This asserts the following: in order to illuminate an element of surface with the linear extension ϵ (i.e., to hit **T** with **L** with the precision ϵ) , a wavelength λ and a lens aperture ϕ are necessary, between which the relation

$$\frac{\lambda}{2 \sin \frac{\phi}{2}} \sim \epsilon$$

---

[134] Cf., for example, Einstein's original paper, Ann. Physik 14 (1905); or any modern text.

exists (Fig. 2).[135] The uncertainty of the tt-component of the momentum of L therefore rests on the fact that its direction lies between $-\frac{\phi}{2}$ and $+\frac{\phi}{2}$, but is otherwise unknown. Consequently, the error amounts to

$$2 \sin \frac{\phi}{2} \cdot \overline{p} = \frac{\lambda}{\epsilon} \cdot \frac{h}{\lambda} = \frac{h}{\epsilon}$$

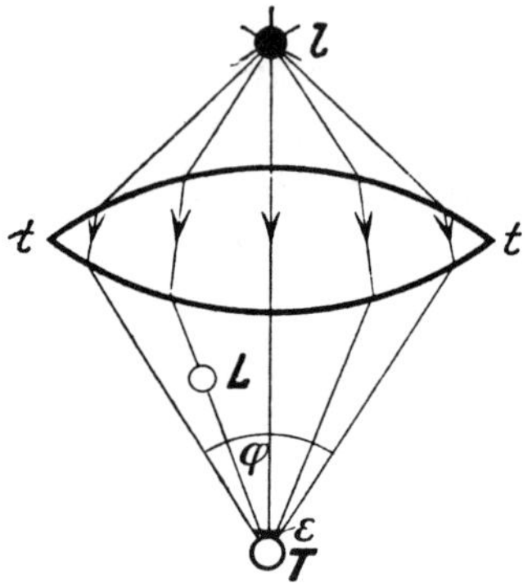

But this is the correct measure for η, therefore $\eta \sim \frac{h}{\epsilon}$ again, i.e.,

This example shows the mechanism of the uncertainty principle very clearly: in order to aim more accurately, we need a large eye (large aperture ϕ), and very short wavelength radiation -- i.e., very uncertain

---

[135] For the theory of the microscope, cf., for example, Handbuch der Physik, Berlin, 1927, vol. 18, Chapter 2.G. In very precise measurements, ϵ, and therefore λ, is very small, i.e., γ rays or still shorter lengths are to be used. A normal lens fails under such circumstances. The only type which could be used would be one whose molecules are neither shattered by these γ rays nor knocked out of their positions by them. Since the existence of such molecules, or particles, encroaches on no known natural law, their use is possible for the purposes of an ideal experiment.

(and large) momentum for the light quantum, which produces collisions (Compton effect) with the observed object **T**, that are out of control by a wide margin. In this way they cause the dispersion in **T**'s momentum.

Let us also consider the complementary measurement process: the measurement of the velocity (momentum). It should first be noted that the natural method of the measurement of the velocity of **T** is to measure its position at two different times, say 0 and t, and divide the change in coordinates by t. In this case, however, the velocity in the time interval $0, t$ must be constant; if it changes, then this change is a measure of the deviation between the above calculated mean velocity, and the actual velocity (say instantaneously at t), i.e., a measure of the uncertainty in the measurement. The same holds for a measurement of the momentum. Now if the coordinate measurements are obtained with the precision ϵ this does not actually affect the precision of measurement of the mean momentum, since t can be chosen arbitrarily large. Nevertheless, it does produce momentum changes of the order of $\frac{h}{\epsilon}$, and therefore an uncertainty relative to the final momentum (in its relationship to the mean momentum mentioned above) of $\eta \sim \frac{h}{\epsilon}$. Hence this procedure gives $\epsilon\eta \sim h$. A different (more favorable) result can therefore only obtain -- if at all -- from such momentum measurements which are not connected with position measurement. Such measurements are entirely possible, and are frequently used in astronomy. They rely on the Doppler effect, and we shall now consider this effect.

The Doppler effect is the following, as is well known. Light which is emitted from a body **T** moving with velocity v, and which is emitted with the frequency ν_0 (measured on the moving body), is actually measured by the observer at rest as having a different frequency ν, which can be calculated from the relation $(\nu - \nu_0)/\nu_0 = \frac{v}{c} \cos \phi$. ($\phi$ is the angle between the direction of motion and the direction of emission. This formula is non-relativistic, i.e., it is valid only for small values of $\frac{v}{c}$; however, this could easily be corrected.) The determination of the velocity is therefore possible if ν is observed and ν_0 is known -- perhaps because it is a particular spectral line of a known element. More exactly, the component of the velocity in the direction of the observation (light emission direction)

$$v \cos \phi = \frac{c(\nu - \nu_0)}{\nu}$$

is measured, or equivalently the corresponding component of momentum

$$p' = p \cos \phi = \frac{mc(\nu - \nu_0)}{\nu_0}$$

(m the mass of the body $\mathbf{T}$). The dispersion of p' evidently depends on the dispersion $\Delta\nu$ of ν. Therefore

$$\eta \sim \frac{mc\Delta\nu}{\nu_0} \sim mc \frac{\Delta\nu}{\nu}$$

The momentum of $\mathbf{T}$ is of course changed by the fact that $\mathbf{T}$ emits a light quantum of frequency ν (and therefore of momentum $\bar{p} = \frac{h\nu}{c}$), but the uncertainty of this quantity $\frac{h\Delta\nu}{c}$ can ordinarily be neglected in comparison with $mc \frac{\Delta\nu}{\nu}$.[136]

ν is measured by any interference method, but this type of measurement will give an absolutely sharp

---

[136] $\frac{mc\Delta\nu}{\nu}$ large in comparison to $\frac{h\Delta\nu}{c}$ means that ν is small in comparison with $\frac{mc^2}{h}$, i.e., $E = h\nu$ is small compared with mc^2. That is, the energy of the light quantum $\mathbf{L}$ is small in comparison to the relativistic rest mass of $\mathbf{T}$ -- an assumption which is unavoidable for a non-relativistic calculation.

ν-value only with a purely monochromatic light wave train. Such a wave train has the form $a \sin [2\pi(\frac{q}{\lambda} - \nu t) + \alpha]$ (q the coordinate, t the time, a the amplitude, α the phase -- this expression may stand for any component of the electric or magnetic field strength -- and is therefore extended infinitely in time and space. To avoid this, we must replace this expression (which can also be written as $a \sin [2\pi\nu(\frac{q}{c} - t) + \alpha]$ since $\lambda = \frac{c}{\nu}$) by another, $F(\frac{q}{c} - t)$, which is $\neq 0$ only in a finite interval of its argument. If the light source has this form, then the well-known Fourier analysis may be carried out:

$$F(x) = \int_0^{+\infty} a_\nu \sin (2\pi\nu x + \alpha_\nu) d\nu$$

Then the interference picture shows all frequencies ν for which $a_\nu \neq 0$. In fact, the frequency interval $\nu, \nu + d\nu$ has the relative intensity $a_\nu{}^2 d\nu$. The dispersion of ν , i.e., $\Delta\nu$, is to be calculated from this distribution.

If our wave train has the length τ in x , i.e., in t and q the respective lengths τ and $c\tau$, then it can be seen that the ν dispersion is $\sim \frac{1}{\tau}$.[137] An indeterminacy of the position now results from this

[137] Let, for example, $F(x)$ be a finite monochromatic wave train of frequency ν_0 extending from 0 to x :

$$F(x) = \begin{cases} \alpha \sin 2\pi\nu_0 x , & \text{for } 0 \leq x \leq \tau , \\ 0 , & \text{otherwise} \end{cases}$$

(Because of the continuity at the junction, $\sin 2\pi\nu_0\tau$ must be $= 0$ there, i.e., $\nu_0 = \frac{n}{2\tau}$, $n = 1,2,3,\ldots$.) Then, on the basis of the known inversion formulae of the Fourier integral (cf. the reference in Note 87) $a_\nu{}^2 = b_\nu{}^2 + c_\nu{}^2$ with

method of measurement, by the fact that **T** undergoes recoil $\frac{h\nu}{c}$ (in the direction of observation) for the

---

$$\left.\begin{matrix} b_\nu \\ c_\nu \end{matrix}\right\} = 2\int_{-\infty}^{\infty} F(x) \begin{matrix}\cos\\ \sin\end{matrix} 2\pi\nu x \cdot dx = 2a\int_0^{\tau} \sin 2\pi\nu_0 x \begin{matrix}\cos\\ \sin\end{matrix} 2\pi\nu x\cdot dx$$

$$= \pm a \int_0^{\tau}\left(\begin{matrix}\sin\\ \cos\end{matrix}\pi(\nu+\nu_0)x - \begin{matrix}\sin\\ \cos\end{matrix}\pi(\nu-\nu_0)x \right)\cdot dx$$

$$= -a\left[\frac{\begin{matrix}\cos\\ \sin\end{matrix}\pi(\nu+\nu_0)x}{\pi(\nu+\nu_0)} - \frac{\begin{matrix}\cos\\ \sin\end{matrix}\pi(\nu-\nu_0)x}{\pi(\nu-\nu_0)}\right]_0^{\tau}$$

$$= \begin{cases} -a\left[\dfrac{(-1)^n\cos\pi\nu\tau-1}{\pi(\nu+\nu_0)}\right] - \dfrac{(-1)^n\cos\pi\nu\tau-1}{\pi(\nu-\nu_0)}\Bigg] = \dfrac{-2a\nu_0(1-(-1)^n\cos\pi\nu t}{\pi(\nu^2-\nu_0^2)} \\[2ex] -a\left[\dfrac{(-1)^n\sin\pi\nu t}{\pi(\nu+\nu_0)} - \dfrac{(-1)^n\sin\pi\nu t}{\pi(\nu-\nu_0)}\right] = \dfrac{2a\nu_0(-1)^n\sin\pi\nu t}{\pi(\nu^2-\nu_0^2)} \end{cases}$$

therefore

$$a_\nu = \frac{2a\nu_0\sqrt{2-2(-1)^n\cos\pi\nu\tau}}{\pi(\nu^2-\nu_0^2)} = \frac{4a\nu_0\left|\begin{matrix}\sin\\ \cos\end{matrix}\frac{1}{2}\pi\nu\tau\right|}{\pi(\nu^2-\nu_0^2)} = \frac{4a\nu_0|\sin\pi(\nu-\nu_0)\tau|}{\pi(\nu^2-\nu_0^2)}$$

As we see, the frequencies in the neighborhood of $\nu = \nu_0$ are most strongly represented, and the greatest part of the energy of the wave train falls in that frequency interval in which $\pi(\nu - \nu_0)\tau$ has moderate values. Therefore the dispersion of $\nu - \nu_0$ (or, which is the same thing, that of ν) has the order of magnitude of $\frac{1}{\tau}$. Exact calculation of the expression

$$\frac{\int_0^{\infty} a_\nu^{\,2}(\nu - \nu_0)^2 d\nu}{\int_0^{\infty} a_\nu^{\,2} d\nu}$$

gives the same result.

individual light emission, i.e., a velocity change $\frac{h\nu}{mc}$ results. Since the emission process takes the time τ, we cannot localize the time of this change in velocity more accurately than τ. Hence an indeterminacy of position $\epsilon \sim \frac{h\nu}{mc}\tau$ results. Therefore:

$$\epsilon \backsim \frac{h\nu}{mc}\tau , \quad \eta \sim mc \frac{\Delta\nu}{\nu} = \frac{mc}{\nu}\frac{1}{\tau} , \quad \epsilon\eta \backsim h$$

So we have $\epsilon\eta \backsim h$.

If **T** is not self-luminous, as was assumed here, but scatters other light (i.e., if it is illuminated), the calculation proceeds in a similar fashion.

5. PROJECTIONS AS PROPOSITIONS

As in III.1., let us consider a physical system **S** with k degrees of freedom, the configuration space of which is described by the k coordinates $q_1, \dots, q_k$ (cf. also I.2.). All physical quantities $\mathfrak{R}$ which can be formed in the system S are, in the manner of classical mechanics, functions of $q_1, \dots, q_k$ and the conjugate momenta $p_1, \dots, p_k$: $\mathfrak{R} = \mathfrak{R}(q_1, \dots, q_k, p_1, \dots, p_k)$ (for example, the energy is the Hamiltonian function $H(q_1, \dots, q_k, p_1, \dots, p_k)$). In quantum mechanics on the other hand, as we already pointed out in III.1., the quantities $\mathfrak{R}$ correspond one-to-one to the hypermaximal Hermitian operators R; in particular, $q_1, \dots, q_k$ correspond to the operators $Q_1 = q_1 \cdot \dots, Q_k = q_k \cdot \dots,$ and $p_1, \dots, p_k$ to the operators $P_1 = \frac{h}{2\pi i}\frac{\partial}{\partial q_1} \dots, P_k = \frac{h}{2\pi i}\frac{\partial}{\partial q_k} \dots$. It has already been noted in the case of the Hamiltonian function (I.2.) that it is not possible in general to define

$$R = \mathfrak{R}(Q_1, \dots, Q_k, P_1, \dots, P_k)$$

because of the non-commutativity of the Q_l, P_l. Nevertheless, without being able to give any final and complete rules regarding the relationship between the

$$\mathfrak{R}(q_1, \ldots, q_k, p_1, \ldots, p_k)$$

and the R , we stated the following special rules in III.1. and III.3.:

L. If the operators R, S correspond to the simultaneously observable quantities $\mathfrak{R}$, $\mathfrak{S}$, then the operator aR + bS (a, b real numbers) corresponds to the quantity $a\mathfrak{R} + b\mathfrak{S}$.

F. If the operator R corresponds to the quantity $\mathfrak{R}$, then the operator F(R) corresponds to the quantity $F(\mathfrak{R})$ [$F(\lambda)$ an arbitrary real function].

L., **F**. permit a certain generalization. That of **F**. is rigorously implied, and runs as follows:

F*. If the operators R, S, ... correspond to the simultaneously measurable quantities $\mathfrak{R}$, $\mathfrak{S}$, ... (which are consequently commutative; we assume that their number is finite), then the operator F(R, S, ...) corresponds to the quantity $F(\mathfrak{R}, \mathfrak{S}, \ldots)$.

In this case, we shall assume that $F(\lambda, \mu, \ldots)$ is a real polynomial in $\lambda, \mu, \ldots$, so that the meaning of F(R, S, ...) may be clear (R, S, ... commutative) although **F***. could also be defined for arbitrary $F(\lambda, \mu, \ldots)$ [for the definition of the general F(R, S, ...) , see the reference in Note 94]. Now since each polynomial is obtained by repetition of the three operations $a\lambda$, $\lambda + \mu$, $\lambda\mu$, it suffices to consider these, and since $\lambda\mu = \frac{1}{4}[(\lambda + \mu)^2 - (\lambda - \mu)^2]$, i.e., is equal to

$$\frac{1}{4}\cdot(\lambda + \mu)^2 + (-\tfrac{1}{4})\cdot(\lambda + (-1)\cdot\mu)^2$$

we can also replace these three operations by $a\lambda$, $\lambda + \mu$, λ^2. But the first two fall under **L**., and the latter under **F**. Consequently, **F***. is proved.

On the other hand, **L**. is extended in quantum mechanics even to the case where $\mathfrak{R}$, $\mathfrak{S}$ are not simultaneously measurable. We shall discuss this question later (in IV.1.), and at present shall limit ourselves to the observation that even the meaning of $a\mathfrak{R} + b\mathfrak{S}$ for $\mathfrak{R}, \mathfrak{S}$ not simultaneously measurable is not yet clear.

Apart from the physical quantities $\mathfrak{R}$, there exists another category of concepts that are important objects of physics -- namely the properties of the states of the system S. Some such properties are: that a certain quantity $\mathfrak{R}$ takes the value λ -- or that the value of $\mathfrak{R}$ is positive -- or that the values of two simultaneously observable quantities $\mathfrak{R}, \mathfrak{S}$ are equal to λ and μ respectively -- or that the sum of the squares of these values is > 1, etc. We denote the quantities by $\mathfrak{R}, \mathfrak{S}, \ldots$ and the properties by $\mathfrak{E}, \mathfrak{F}, \ldots$. The hypermaximal Hermitian operators R, S, ... correspond to the quantities, as was discussed above. What now corresponds to the properties?

To each property $\mathfrak{E}$ we can assign a quantity which we define as follows: each measurement which distinguished between the presence or absence of $\mathfrak{E}$ is considered as a measurement of this quantity, such that its value is 1 if $\mathfrak{E}$ is verified, and zero in the opposite case. This quantity which corresponds to $\mathfrak{E}$ will also be denoted by $\mathfrak{E}$.

Such quantities take on only the values 0 and 1, and conversely, each quantity $\mathfrak{R}$ which is capable of these two values only, corresponds to a property $\mathfrak{E}$ which is evidently this: "the value of $\mathfrak{R}$ is $\neq 0$." The quantities $\mathfrak{E}$ that correspond to the properties are therefore characterized by this behavior.

That $\mathfrak{E}$ takes on only the values 0, 1 can also be formulated as follows: Substituting $\mathfrak{E}$ into the polynomial $F(\lambda) = \lambda - \lambda^2$ makes it vanish identically. If $\mathfrak{E}$ has the operator E, then $F(\mathfrak{E})$ has the operator $F(E) = E - E^2$, i.e., the condition is that $E - E^2 = 0$ or $E = E^2$. In other words: the operator E of $\mathfrak{E}$ is a projection.

The projections E therefore correspond to the

properties $\mathfrak{E}$ (through the agency of the corresponding quantities $\mathfrak{E}$ which we just defined). If we introduce, along with the projections E, the closed linear manifolds $\mathfrak{M}$ belonging to them $(E = P_{\mathfrak{M}})$, then the closed linear manifolds correspond equally to the properties of $\mathfrak{E}$.

The calculation with the corresponding $\mathfrak{E}$, E, $\mathfrak{M}$ will now be investigated in detail.

If in a state ϕ we want to determine whether or not a property $\mathfrak{E}$ is verified, then we must measure the quantity $\mathfrak{E}$, and ascertain whether its value is 1 or 0 (these processes are identical, by definition.) The probability of the former, i.e., that $\mathfrak{E}$ is verified, is consequently equal to the expectation value of $\mathfrak{E}$, i.e.,

$$(E\phi, \phi) = \|E\phi\|^2 = \|P_{\mathfrak{M}}\phi\|^2$$

and that of the latter, i.e., that $\mathfrak{E}$ is not verified, is equal to the expectation value of $1 - \mathfrak{E}$, i.e.,

$$((1 - E)\phi, \phi) = \|(1 - E)\phi\|^2 = \|\phi - P_{\mathfrak{M}}\phi\|^2$$

(The sum is of course equal to (ϕ, ϕ), i.e., to 1). Consequently, $\mathfrak{E}$ is certainly present or certainly absent, if the second or first probability respectively is equal to zero, i.e., for $P_{\mathfrak{M}}\phi = \phi$ or $= 0$. That is, if ϕ belongs to $\mathfrak{M}$ or is orthogonal to $\mathfrak{M}$ respectively; or if ϕ belongs to $\mathfrak{M}$ or to $\mathfrak{R}_\infty - \mathfrak{M}$.

$\mathfrak{M}$ can therefore be defined as the set of all ϕ which possess the property $\mathfrak{E}$ with certainty. (Actually, only the subset of $\mathfrak{M}$ lying on the surface $\|\phi\| = 1$. $\mathfrak{M}$ itself is obtained by multiplication of these ϕ with positive constants and the adjunction of the zero.)

If we call the property opposite that of $\mathfrak{E}$ (the denial of $\mathfrak{E}$) "not $\mathfrak{E}$," then it follows immediately from the above that if E, $\mathfrak{M}$ belong to $\mathfrak{E}$, then $1 - E$, $\mathfrak{R}_\infty - \mathfrak{M}$ belong to "not $\mathfrak{E}$."

As with quantities, there also arises the question of simultaneous measurability (or rather decidability of properties. It is clear that $\mathfrak{E}$, $\mathfrak{F}$ are simultaneously decidable if and only if the corresponding quantities $\mathfrak{E}$, $\mathfrak{F}$ are simultaneously measurable (whether with arbitrarily great or with absolute accuracy is unimportant, since they are capable of the values 0, 1 only), i.e., if E, F commute. The same holds for several properties $\mathfrak{E}$, $\mathfrak{F}$, $\mathfrak{G}$, ...

From properties $\mathfrak{E}$, $\mathfrak{F}$ which are simultaneously decidable we can form the additional properties " $\mathfrak{E}$ and $\mathfrak{F}$ " and " $\mathfrak{E}$ or $\mathfrak{F}$." The quantity corresponding to " $\mathfrak{E}$ and $\mathfrak{F}$ " is 1 if those corresponding to $\mathfrak{E}$ and to $\mathfrak{F}$ are both 1 , and it is 0 if one of these is 0 . Hence it is the product of these quantities. By $\mathbf{F}^*$., its operator is then the product of the operators of $\mathfrak{E}$ and $\mathfrak{F}$, i.e., EF . By THEOREM 14., II.4., the corresponding closed linear manifold is the set $\mathfrak{P}$ common to $\mathfrak{M}$, $\mathfrak{N}$.

On the other hand, " $\mathfrak{E}$ or $\mathfrak{F}$ " can be written as

"not [(not $\mathfrak{E}$) and (not $\mathfrak{F}$)] "

and therefore its operator is

$$1 - (1 - E)(1 - F) = E + F - EF$$

(because of its origin, this is also a projection). Since F - EF is a projection, the linear manifold belonging to E + F - EF is $\mathfrak{M} + (\mathfrak{N} - \mathfrak{P})$ (THEOREM 14., II.4.). It is a subset of $\{\mathfrak{M}, \mathfrak{N}\}$ and evidently embraces $\mathfrak{M}$, and by symmetry, $\mathfrak{N}$ also, and therefore $\{\mathfrak{M}, \mathfrak{N}\}$. Consequently it is equal to $\{\mathfrak{M}, \mathfrak{N}\}$, and this, since it is closed, is itself equal to $[\mathfrak{M}, \mathfrak{N}]$.

If $\mathfrak{E}$ is a property which is always present (i.e., empty), then the corresponding quantity is identically 1 , i.e., $E = 1$, $\mathfrak{M} = \mathfrak{R}_\infty$. On the other hand, if $\mathfrak{E}$

is never present (i.e., impossible) then the corresponding quantity is identically 0, i.e., $E = 0$, $\mathfrak{M} = (0)$. If two properties $\mathfrak{E}$, $\mathfrak{F}$ are incompatible, then they must at any rate be simultaneously decidable, and " $\mathfrak{E}$ and in addition $\mathfrak{F}$ " must be impossible: i.e., E, F commute, $EF = 0$. But since $EF = 0$ implies commutativity (THEOREM 14., II.4.), this by itself is characteristic. If E, F are presumed to be commutative, then $EF = 0$ means merely that the subset common to $\mathfrak{M}$ and $\mathfrak{N}$ consists only of the 0. However, the commutativity of E, F does not follow from this alone. Indeed $EF = 0$ is equivalent to all $\mathfrak{M}$ being orthogonal to all $\mathfrak{N}$ (THEOREM 14., II.4.).

If $\mathfrak{R}$ is a quantity with the operator R, to which may belong the resolution of the identity $E(\lambda)$, then the operator of the property " $\mathfrak{R}$ lies in the interval $I = \{\lambda', \mu'\}$ " $(\lambda' \leq \mu')$ is $E(\mu') - E(\lambda')$. To say this it suffices to observe that the probability of the above proposition is $((E(\mu') - E(\lambda'))\phi, \phi)$ (cf. **P.** in II.,1.). Another way: the quantity belonging to the property in question is $\mathfrak{E} = F(\mathfrak{R})$, with

$$F(\lambda) = \begin{cases} 1, & \text{for } \lambda' < \lambda \leq \mu', \\ 0, & \text{otherwise} \end{cases}$$

and $F(R) = E(\mu') - E(\lambda')$ (cf. II.8. or III.1.). We called this operator $E(I)$ in III.1.

Summarizing, we have thus obtained the following information on the relations between properties $\mathfrak{E}$, their projections E and the closed linear manifolds $\mathfrak{M}$ of these projections:

α) In the state ϕ the property $\mathfrak{E}$ is or is not present with the respective probabilities

$$(E\phi, \phi) = ||E\phi||^2 = ||P_{\mathfrak{M}}\phi||^2$$

and

$$((1 - E)\phi, \phi) = ||(1 - E)\phi||^2 = ||\phi - P_{\mathfrak{M}}\phi||^2$$

β) $\mathfrak{E}$ is certainly present or certainly absent for the ϕ of $\mathfrak{M}$ and $\mathfrak{R}_\infty - \mathfrak{M}$ respectively, and only for these.

γ) For the simultaneous decidability of several properties $\mathfrak{E}, \mathfrak{F}, \ldots$, the commutativity of their operators $E, F, \ldots$ is characteristic.

δ) If $E, \mathfrak{M}$ belong to $\mathfrak{E}$, then $1 - E, \mathfrak{R}_\infty - \mathfrak{M}$ belong to "not $\mathfrak{E}$."

ε) If $E, \mathfrak{M}$ belong to $\mathfrak{E}$ and $F, \mathfrak{N}$ to $\mathfrak{F}$ and if $\mathfrak{E}$, $\mathfrak{F}$ can be decided simultaneously, then EF, and the common part of $\mathfrak{M}$, $\mathfrak{N}$ belong to "$\mathfrak{E}$ and $\mathfrak{F}$," and $E + F - EF$, $\{\mathfrak{M}, \mathfrak{N}\}$ (this is equal to $[\mathfrak{M}, \mathfrak{N}]$) belong to "$\mathfrak{E}$ or $\mathfrak{F}$."

η) $\mathfrak{E}$ always holds if $E = 1$, or also if $\mathfrak{M} = \mathfrak{R}_\infty$; it never holds if $E = 0$, or also if $\mathfrak{M} = (0)$.

θ) $\mathfrak{E}$, $\mathfrak{F}$ are incompatible if $EF = 0$, or also if all $\mathfrak{M}$ is orthogonal to all $\mathfrak{N}$.

ζ) Let $\mathfrak{R}$ be a quantity and R its operator, I an interval. Let $E(\lambda)$ be the resolution of the identity belonging to R, $I = \{\lambda', \mu'\}$ $(\lambda' \leq \mu')$, $E(I) = E(\mu') - E(\lambda')$ (cf. III.1.). Then the operator $E(I)$ belongs to the property "$\mathfrak{R}$ lies in I."

From α) - ζ) we can derive the earlier probability statements **P**., **E**$_1$., **E**$_2$., as well as the statements of III.3. on simultaneous measurability. It is clear that the latter are equivalent to γ); **P**. follows from α), ε), ζ), and has **E**$_1$., **E**$_2$. as its consequence.

As can be seen, the relation between the properties of a physical system on the one hand, and the projections on the other, makes possible a sort of logical calculus with these. However, in contrast to the concepts of ordinary logic, this system is extended by the concept of "simultaneous decidability" which is characteristic for quantum mechanics.

Moreover, the calculus of these propositions, based on projections, has the advantage over the calculus of quantities, which is based on the totality of (hypermaximal) Hermitian operators, that the concept of "simultaneous decidability" represents a refinement of the concept of "simultaneous measurability." For example, in order that the questions "does $\mathfrak{R}$ lie in I ?" and "does $\mathfrak{S}$ lie in J ?" ($\mathfrak{R}$, $\mathfrak{S}$ have the respective operators R and S, and these the resolutions of the identity $E(\lambda)$ and $F(\mu)$ respectively, $I = \{\lambda', \lambda''\}$, $J = \{\mu', \mu''\}$) be simultaneously decidable, we require (by γ), ζ)) only that the operators $E(I) = E(\lambda'') - E(\lambda')$ and $F(J) = F(\mu'') - F(\mu')$ commute. For simultaneous measurability of $\mathfrak{R}$, $\mathfrak{S}$, however, the commutativity of all $E(\lambda)$ with all $F(\mu)$ is necessary.

6. RADIATION THEORY

We have obtained once again the statistical statements of quantum mechanics developed in I.2., substantially generalized and systematically arranged -- with one single exception. We are lacking the Heisenberg expression for the transition probability from one stationary state of a quantized system into another -- although this played an important role in the development of quantum mechanics (cf. the comments in I.2.). Following the method of Dirac,[138] we shall now show how these transition probabilities can be derived from the ordinary statistical statements of quantum mechanics, i.e., from the theory just now developed. This is all the more important, since such a derivation will give us a deeper insight into the

[138] Proc. Roy. Soc. 114 (1927). Cf. also the presentation in Weyl, Gruppentheorie und Quantenmechanik, 2nd. ed., p. 91 ff. Leipzig 1931.

mechanism of transitions of the stationary states, and into the relations of the Einstein-Bohr energy-frequency conditions. The radiation theory advanced by Dirac is one of the most beautiful achievements in the quantum mechanical field.

Let **S** be a system (say, a quantized atom) with an energy to which corresponds the Hermitian operator H_o. We represent the coordinates which describe the configuration space of **S** by the single symbol ξ (if, for example, **S** consists of l particles, then there are $3l$ cartesian coordinates: $x_1 = q_1$, $y_1 = q_2$, $z_1 = q_3$, ..., $x_l = q_{3l-2}$, $y_l = q_{3l-1}$, $z_l = q_{3l}$ -- ξ stands for all of these together); furthermore, for simplicity, let H_o have a pure discrete spectrum: Eigenvalues $W_1, W_2, \ldots$, eigenfunctions $\phi_1(\xi), \phi_2(\xi), \ldots$ (several W_m may coincide). An arbitrary state of S, i.e., a wave function $\phi(\xi)$, is expanded according to the time dependent Schrödinger equation (cf. III.2.):

$$\frac{h}{2\pi i} \frac{\partial}{\partial t} \phi_t(\xi) = - H_o \phi_t(\xi)$$

i.e., if

$$\phi_t(\xi) = \phi(\xi) = \sum_{k=1}^{\infty} a_k \phi_k(\xi)$$

for $t = t_o$, then in general,

$$\phi_t(\xi) = \sum_{k=1}^{\infty} a_k e^{-\frac{2\pi i}{h} W_k (t-t_o)} \phi_k(\xi)$$

A state $\phi(\xi) = \phi_k(\xi)$ therefore changes into

$$e^{-\frac{2\pi i}{h} W_k (t-t_o)} \phi_k(\xi)$$

i.e., into itself (since the factor

$$e^{-\frac{2\pi i}{h} W_k(t-t_o)}$$

is irrelevant). Hence the $\phi_k(\xi)$ are stationary. Thus we find in general no transitions from one into another. How is it that we do nevertheless speak of such transitions? The explanation is simple. We have disregarded the agent which causes these transitions -- radiation. The stationary quantum orbits break down, on the basis of the original Bohr theory, only under the emission of radiation (Cf. reference in Note 5), but if this is neglected (as in the set-up just given), then it is quite reasonable that an absolute and permanent stability results. We must therefore extend the system to be investigated, so that we include the radiation which may be emitted by **S** -- i.e., we must include, in general, all radiation which can interact with **S** under any circumstances. If we denote by **L** the system which is formed by the radiation (i.e., the electromagnetic field of the classical theory, less the stationary field resulting from the electronic and nuclear charges), then we investigate **S** + **L** .

We must then do the following:

1. A set-up for the quantum mechanical description of **L** , i.e., the configuration space of **L** is to be given.

2. The energy operator of **S** + **L** must be found. This can be divided into three parts.

α) The energy of **S** , which is present independently of **L** , i.e., the unperturbed **S** energy: For the system **S** it was the operator H_0 .

β) The energy of **L** , which is present independently of **S** , i.e., the unperturbed **L** energy. Call its operator H_1 .

γ) The remainder of the energy, i.e., the interaction energy of **S** and **L** . Call its operator H_I

Clearly we have questions here which, in accord with the fundamental principles of quantum mechanics, must first be answered classically. The results so obtained can then be translated into operator form (cf. I.2.). We therefore adopt (at first) a purely classical point of view relative to the nature of the radiation: We consider it (in the sense of the electromagnetic theory of radiation) as an oscillatory state of the electromagnetic field.[139]

In order to avoid unnecessary complications (the loss of the radiation in infinite space, etc.) we consider **S** and **L** enclosed in a very large cavity **H** of volume $\mathcal{V}$, which shall have perfectly reflecting walls. As is well known, the state of the electromagnetic field in **H** is described by the electric and magnetic field strengths $\mathfrak{E} = \{\mathfrak{E}_x, \mathfrak{E}_y, \mathfrak{E}_z\}$, $\mathfrak{H} = \{\mathfrak{H}_x, \mathfrak{H}_y, \mathfrak{H}_z\}$. All quantities $\mathfrak{E}_x, \ldots, \mathfrak{H}_z$ are functions of the cartesian coordinates x, y, z of the general point in **H** and of the time t. It should also be pointed out that we shall now frequently consider real space vectors $\mathfrak{a} = \{\mathfrak{a}_x, \mathfrak{a}_y, \mathfrak{a}_z\}$, $\mathfrak{b} = \{\mathfrak{b}_x, \mathfrak{b}_y, \mathfrak{b}_z\}$, etc. (e.g. $\mathfrak{E}$, $\mathfrak{H}$), and for these, concepts such as the inner or scalar product

$$[\mathfrak{a}, \mathfrak{b}] = \mathfrak{a}_x \mathfrak{b}_x + \mathfrak{a}_y \mathfrak{b}_y + \mathfrak{a}_z \mathfrak{b}_z$$

This will not be confused with the inner product (ϕ, ψ) in $\mathfrak{R}_\infty$. We denote the differential operator

$$\frac{\partial^2}{\partial x^2} + \frac{\partial^2}{\partial y^2} + \frac{\partial^2}{\partial z^2}$$

by Δ, and the well-known vector operations by div, grad,

---

[139]The interested reader will find treatments of the electromagnetic theory of radiation in any textbook of electrodynamics, for example, Abraham und Becker: Theorie der Elektrizität, Berlin 1930. Cf. these also for the following developments which belong to the framework of the Maxwell theory.

curl. The vectors $\mathfrak{E}$, $\mathfrak{H}$ satisfy the Maxwell equations in the empty space **H** :

$$\operatorname{div} \mathfrak{H} = 0 \ , \ \operatorname{curl} \mathfrak{E} + \frac{1}{c}\frac{\partial}{\partial t} \mathfrak{H} = 0 \ ,$$

$$\operatorname{div} \mathfrak{E} = 0 \ , \ \operatorname{curl} \mathfrak{H} - \frac{1}{c}\frac{\partial}{\partial t} \mathfrak{E} = 0$$

The first equation of the first row is satisfied by $\mathfrak{H} = \operatorname{curl} \mathfrak{A}$ ($\mathfrak{A} = \{\mathfrak{A}_x, \mathfrak{A}_y, \mathfrak{A}_z\}$ is the so-called vector potential; its components also depend on x, y, z, t) , and the second by $\mathfrak{E} = -\frac{1}{c}\frac{\partial}{\partial t}\mathfrak{A}$. The equations of the second row now become:

(A.) $$\operatorname{div} \mathfrak{A} = 0 \ , \ \Delta\mathfrak{A} - \frac{1}{c^2}\frac{\partial^2}{\partial t^2}\mathfrak{A} = 0$$

(The vector potential is usually introduced in a somewhat different way, in order to improve the symmetry in space and time. That the present set-up for $\mathfrak{A}$ furnishes a general solution of Maxwell's equations -- where it is to be noted particularly that the first equation of the second row actually gives only $\frac{\partial}{\partial t}\operatorname{div}\mathfrak{A} = 0$, i.e. $\operatorname{div}\mathfrak{A} = f(x, y, z)$ -- is shown in most treatments of the Maxwell theory. Hence it will not be developed to any extent here. Cf. the reference in Note 139.) (A.) is our starting point for the following. The fact that the walls of **H** are reflecting is expressed by the condition that $\mathfrak{A}$ must be perpendicular to the walls at the boundaries of **H** . The well-known method of finding all such $\mathfrak{A}$'s is this: Since t nowhere enters explicitly into the problem, the most general $\mathfrak{A}$ is a linear combination of all those solutions which are products of an x, y, z dependent vector with a time dependent scalar:

$$\mathfrak{A} = \mathfrak{A}(x, y, z, t) = \overline{\mathfrak{A}}(x, y, z)\cdot\tilde{q}(t)$$

Therefore (A.) gives

(A_1.) $$\operatorname{div}\overline{\mathfrak{A}} = 0 \ , \ \Delta\overline{\mathfrak{A}} = \eta\,\overline{\mathfrak{A}}$$

$\overline{\mathfrak{A}}$ perpendicular to the boundary of **H** on the boundary.

$$(A_2.) \qquad \frac{\partial^2}{\partial t^2}\tilde{q}(t) = c^2\eta\tilde{q}(t)$$

Because of $(A_1.)$, η depends on x, y, z only, because of $(A_2.)$ it depends on t only. Hence η is constant.

$(A_1.)$ is therefore an eigenvalue problem in which η is the eigenvalue parameter, and $\overline{\mathfrak{A}}$ the general eigenfunction. The theory of this problem is fully known and we shall give only the results here:[140] $(A_1.)$ has a pure discrete spectrum, all eigenvalues $\eta_1, \eta_2, \dots$ (let the corresponding $\overline{\mathfrak{A}}$ be $\overline{\mathfrak{A}}_1, \overline{\mathfrak{A}}_2, \dots$) are negative, and $\eta_n \rightarrow -\infty$ as $n \rightarrow \infty$. We can normalize a complete set $\overline{\mathfrak{A}}_1$, $\overline{\mathfrak{A}}_2$ by

$$\iiint\limits_{\mathbf{H}} [\overline{\mathfrak{A}}_m, \overline{\mathfrak{A}}_n]dxdydz = \begin{cases} 4\pi c^2 & \text{for } m = n \\ 0 & \text{for } m \neq n \end{cases}$$

(We choose $4\pi c^2$ instead of the customary 1, because it will prove to be somewhat more practical later.) If we denote η_n (< 0) by

$$-\frac{4\pi^2\rho_n^2}{c^2}$$

then $(A_2.)$ gives

$$\tilde{q}_n(t) = \gamma \cos 2\pi\rho_n(t - \tau) \qquad (\gamma, \tau \text{ arbitrary})$$

Therefore the general solution $\mathfrak{A}$ is equal to

$$\mathfrak{A} = \mathfrak{A}(x, y, z, t) = \sum_{n=1}^{\infty} \overline{\mathfrak{A}}_n(x, y, z)\cdot\tilde{q}_n(t)$$

$$= \sum_{n=1}^{\infty} \overline{\mathfrak{A}}_n(x, y, z)\cdot\gamma_n \cos 2\pi\rho_n(t - \tau_n)$$

---

[140] Cf. R. Courant und D. Hilbert, Methoden der mathematischen Physik I, pp. 358-363. Berlin 1924.

($\gamma_1, \gamma_2, \ldots, \tau_1, \tau_2, \ldots$ arbitrary constants)

The energy of the arbitrary field

$$\mathfrak{A} = \sum_{n=1}^{\infty} \overline{\mathfrak{A}}_n(x, y, z) \cdot \tilde{q}_n(t)$$

[$\mathfrak{A}$ is now not assumed to be a solution of (A.), i.e., the $\tilde{q}_n(t)$ are arbitrary] amounts to

$$E = \frac{1}{8\pi} \iiint_H ([\mathfrak{E}, \mathfrak{E}] + [\mathfrak{H}, \mathfrak{H}])dxdydz$$

$$= \frac{1}{8\pi} \iiint_H \left(\frac{1}{c^2}[\frac{\partial}{\partial t}\mathfrak{A}, \frac{\partial}{\partial t}\mathfrak{A}] + [\text{curl}\ \mathfrak{A}, \text{curl}\ \mathfrak{A}]\right) dxdydz$$

$$= \frac{1}{8\pi} \sum_{m,n=1}^{\infty} \iiint_H \left(\frac{1}{c^2} \frac{\partial}{\partial t}\tilde{q}_m(t) \frac{\partial}{\partial t}\tilde{q}_n(t)[\overline{\mathfrak{A}}_m, \overline{\mathfrak{A}}_n] + \right.$$

$$\left. + \tilde{q}_m(t)\tilde{q}_n(t)[\text{curl}\ \overline{\mathfrak{A}}_m, \text{curl}\ \overline{\mathfrak{A}}_n]\right) dxdydz$$

Upon integrating by parts, we find:[141]

$$\iiint_H [\text{curl}\ \overline{\mathfrak{A}}_m, \text{curl}\ \overline{\mathfrak{A}}_n]dxdydz = \iiint_H [\text{curl curl}\ \overline{\mathfrak{A}}_m, \overline{\mathfrak{A}}_n]dxdydz$$

$$= \iiint_H [-\Delta \overline{\mathfrak{A}}_m + \text{grad div}\ \overline{\mathfrak{A}}_m, \overline{\mathfrak{A}}_n]dxdydz$$

$$= \frac{4\pi^2 \rho_m^2}{c^2} \iiint_H [\overline{\mathfrak{A}}_m, \overline{\mathfrak{A}}_n]dxdydz$$

---

[141]We have

$$\iiint_H [\mathfrak{a}, \text{curl}\ \mathfrak{b}]dxdydz = \iiint_H [\text{curl}\ \mathfrak{a}, \mathfrak{b}]dxdydz$$

because of

$$[\mathfrak{a}, \text{curl}\ \mathfrak{b}] - [\text{curl}\ \mathfrak{a}, \mathfrak{b}] = \text{grad}\ [\mathfrak{a} \times \mathfrak{b}]$$

($\mathfrak{a} \times \mathfrak{b}$ is the so-called outer or vector product of $\mathfrak{a}$, $\mathfrak{b}$), if the normal components of $\mathfrak{a} \times \mathfrak{b}$ vanish on the boundary

Therefore

$$E = \frac{1}{8\pi} \sum_{m,n=1}^{\infty} \left(\frac{1}{c^2} \frac{\partial}{\partial t} \tilde{q}_m(t) \frac{\partial}{\partial t} \tilde{q}_n(t) + \frac{4\pi^2 \rho_m^{\ 2}}{c^2} \tilde{q}_m(t)\tilde{q}_n(t) \right) \iiint_{\mathbf{H}} [\overline{\mathfrak{A}}_m, \overline{\mathfrak{A}}_n] dxdydz$$

$$= \frac{1}{2} \sum_{m=1}^{\infty} \left[\left(\frac{\partial}{\partial t} \tilde{q}_m(t) \right)^2 + 4\pi^2 \rho_m^{\ 2} (\tilde{q}_m(t))^2 \right]$$

But we can regard the $\tilde{q}_1, \tilde{q}_2, \cdots$ as the coordinates describing the instantaneous state of the field, i.e., as the coordinates of the configuration space of $\mathbf{L}$. The conjugate momenta $\tilde{p}_m$ (classical mechanical) are obtained from the formula

$$E = \frac{1}{2} \sum_{n=1}^{\infty} \left((\frac{\partial}{\partial t} \tilde{q}_n)^2 + 4\pi^2 \rho_n^{\ 2} \tilde{q}_n^{\ 2} \right)$$

This gives

$$\tilde{p}_n = \frac{\partial}{\partial (\frac{\partial}{\partial t} \tilde{q}_n)} E = \frac{\partial}{\partial t} \tilde{q}_m, \quad E = \frac{1}{2} \sum_{n=1}^{\infty} (\tilde{p}_n^{\ 2} + 4\pi^2 \rho_n^{\ 2} \tilde{q}_n^{\ 2})$$

(cf. I.2.). This yields the classical mechanical equations of motion

$$\frac{\partial}{\partial t} \tilde{p}_n = - \frac{\partial}{\partial \tilde{q}_n} E = - 4\pi^2 \rho_n^{\ 2} \tilde{q}_n, \quad \frac{\partial}{\partial t} \tilde{q}_n = \frac{\partial}{\partial \tilde{p}_n} E = \tilde{p}_n$$

This is precisely the (A_2.) following from Maxwell's equations. Consequently Jean's theorem holds:

The radiation field $\mathbf{L}$ can be described purely

---

of $\mathbf{H}$. Since $\mathfrak{a} \times \mathfrak{b}$ is perpendicular to $\mathfrak{a}$ and to $\mathfrak{b}$, this is certainly the case if $\mathfrak{a}$ or $\mathfrak{b}$ is perpendicular to $\mathbf{H}$. We have $\mathfrak{a} = \text{curl}\, \mathfrak{A}_n$, $\mathfrak{b} = \mathfrak{A}_n$, so that the former indeed occurs.

classically by the coordinates $\tilde{q}_1, \tilde{q}_2, \cdots$ -- which are connected through

$$\mathfrak{A} = \mathfrak{A}(x, y, z) = \sum_{n=1}^{\infty} \tilde{q}_n \overline{\mathfrak{A}}_n(x, y, z)$$

with the instantaneous vector potential $\mathfrak{A}$ describing the field -- with the aid of the energy (Hamiltonian function)

$$E = \frac{1}{2} \sum_{n=1}^{\infty} (\tilde{p}_n{}^2 + 4\pi^2 \rho_n{}^2 \tilde{q}_n{}^2)$$

A point of mass 1 constrained on a straight line which has the coordinate $\tilde{q}$, and which lies in the potential field $C\tilde{q}^2$, $C = 2\pi^2\rho^2$, has the energy $\frac{1}{2}[(\frac{\partial}{\partial t}\tilde{q})^2 + 4\pi^2\rho^2\tilde{q}^2]$. Or, since $\tilde{p} = \frac{\partial}{\partial t}\tilde{q}$ again results, this has the value $\frac{1}{2}(\tilde{p}^2 + 4\pi^2\rho^2\tilde{q}^2)$. The equation of motion of such a particle is consequently

$$\frac{\partial^2}{\partial t^2}\tilde{q} + 4\pi^2\rho^2\tilde{q} = 0$$

whose solution is $\tilde{q} = \gamma \cos 2\pi\rho(t - \tau)$ (γ, τ arbitrary). Because of the form of its motion, this sequence is called "a linear oscillator of frequency ρ." $\mathbf{L}$ may consequently be regarded as the combination of a sequence of linear oscillators, whose frequencies are the eigenfrequencies of the space $\mathbf{H}$: $\rho_1, \rho_2, \cdots$.

This "mechanical" description of the electromagnetic field is important because it can immediately be reinterpreted in the sense of the usual method of quantum mechanics. The configuration space of $\mathbf{L}$ is described by $\tilde{q}_1, \tilde{q}_2, \cdots$, and in the expression for E, the $\tilde{p}_n$, $\tilde{q}_n$ are to be replaced by $\frac{h}{2\pi i}\frac{\partial}{\partial q_n} \cdots$ and $\tilde{q}_n \cdots$ respectively. We call these operators $\tilde{P}_n$ and $\tilde{Q}_n$. Then questions 1. and 2. $\boldsymbol{\beta}$) are answered, and in particular,

$$H_1 = \frac{1}{2} \sum_{n=1}^{\infty} (\tilde{P}_n{}^2 + 4\pi^2 \rho_n{}^2 Q_n{}^2)$$

is the operator sought in the sense of 2.β). 2.α) was solved previously, since we assumed H_0 to be known. There then remains only 2.γ), but this now also causes no additional difficulties.

By classical electrodynamics, the interaction of **S** with **L** is to be calculated in the following way: Let **S** consist of l particles (perhaps protons and electrons) with the respective charges and masses $e_1, m_1, \ldots, e_l, m_l$, and the cartesian coordinates $x_1 = q_1, y_1 = q_2, z_1 = q_3, \ldots;$ $x_l = q_{3l-2}, y_l = q_{3l-1}, z_l = q_{3l}$ (these make up the symbol ξ used above), and let the corresponding momenta be $p_1^x, p_1^y, p_1^z, \ldots, p_l^x, p_l^y, p_l^z$. The interaction energy is then (in sufficient approximation)

$$\sum_{\nu=1}^{l} \frac{e_\nu}{cm_\nu} \left(p_\nu^x \mathfrak{A}_x(x_\nu, y_\nu, z_\nu) + p_\nu^y \mathfrak{A}_y(x_\nu, y_\nu, z_\nu) + p_\nu^z \mathfrak{A}_z(x_\nu, y_\nu, z_\nu)\right) \text{ [142]}$$

The corresponding operator of quantum mechanics is obtained if we replace $p_\nu^x, p_\nu^y, p_\nu^z, x_\nu, y_\nu, z_\nu$ $(\nu = 1, \ldots, l)$ by the operators

$$\frac{h}{2\pi i}\frac{\partial}{\partial x_\nu} \cdots, \frac{h}{2\pi i}\frac{\partial}{\partial y_\nu} \cdots, \frac{h}{2\pi i}\frac{\partial}{\partial z_\nu} \cdots, \; x_\nu \cdots, y_\nu \cdots, z_\nu \cdots$$

which we call $P_\nu^x, P_\nu^y, P_\nu^z, Q_\nu^x, Q_\nu^y, Q_\nu^z$. If we consider

$$\mathfrak{A}(x, y, z) = \sum_{n=1}^{\infty} \tilde{q}_n \overline{\mathfrak{A}}_n(x, y, z)$$

then we have the desired H_I:

$$H_I = \sum_{n=1}^{\infty} \sum_{\nu=1}^{l} \frac{e_\nu}{cm_\nu} \tilde{Q}_n \{P_\nu^x \overline{\mathfrak{A}}_{n,x}(Q_\nu^x, Q_\nu^y, Q_\nu^z) + P_\nu^y \overline{\mathfrak{A}}_{n,y}(Q_\nu^x, Q_\nu^y, Q_\nu^z) + P_\nu^z \overline{\mathfrak{A}}_{n,z}(Q_\nu^x, Q_\nu^y, Q_\nu^z)\}$$

---

[142] For example, see the reference in Note 138.

It should be observed here that we replaced the products $p_\nu^x \overline{\mathfrak{A}}_{n,x}(x, y, z)$ by operators using an arbitrary order of factors; $P_\nu^x \overline{\mathfrak{A}}_{n,x}(Q_\nu^x, Q_\nu^y, Q_\nu^z), \ldots$ -- although we could just as well have taken $\overline{\mathfrak{A}}_{n,x}(Q_\nu^x, Q_\nu^y, Q_\nu^z)P_\nu^x, \ldots$. (In order to be certain of the Hermitian character of the resulting operator, a symmetric form, such as $\frac{1}{2}(P_\nu^x \overline{\mathfrak{A}}_{n,x}(Q_\nu^x, Q_\nu^y, Q_\nu^z) + \overline{\mathfrak{A}}_{n,x}(Q_\nu^x, Q_\nu^y, Q_\nu^z)P_\nu^x), \ldots$ would actually be necessary). Fortunately, all this makes no difference, because

$$[P_\nu^x \overline{\mathfrak{A}}_{n,x}(Q_\nu^x, Q_\nu^y, Q_\nu^z) + \ldots] - [\overline{\mathfrak{A}}_{n,x}(Q_\nu^x, Q_\nu^y, Q_\nu^z)P_\nu^x + \ldots]$$

$$= [P_\nu^x \overline{\mathfrak{A}}_{n,x}(Q_\nu^x, Q_\nu^y, Q_\nu^z) - \overline{\mathfrak{A}}_{n,x}(Q_\nu^x, Q_\nu^y, Q_\nu^z)P_\nu^x] + \ldots \quad [143]$$

$$= \frac{h}{2\pi i}\frac{\partial}{\partial x}\overline{\mathfrak{A}}_{n,x}(Q_\nu^x, Q_\nu^y, Q_\nu^z) + \ldots = \frac{h}{2\pi i}\operatorname{div}\overline{\mathfrak{A}}_n(Q_\nu^x, Q_\nu^y, Q_\nu^z) =$$

Thus the total energy of our system **S** + **L** , that is, its operator

$$H = H_0 + H_1 + H_i$$

is now completely specified. But before we transform H further, let us note the following: **S** + **L**'s configuration space is described by the coordinates ξ (i.e., $q_1, \ldots, q_{3l}$, or also $x_1, y_1, z_1, \ldots, x_l, y_l, z_l$) and

---

[143] Since P_ν^x commutes with Q_ν^y, Q_ν^z , but not with Q_ν^x we have to prove the following relation, in order to justify the subsequent transformation (we neglect superfluous indices, and replace A by F) :

$$PF(Q) - F(Q)P = \frac{h}{2\pi i} F'(Q)$$

if $P = \frac{h}{2\pi i}\frac{\partial}{\partial q} \ldots Q = q \ldots$. This relation, which is of especial importance in matrix theory, can be verified most easily by direct calculation.

$\tilde{q}_1, \tilde{q}_2, \ldots$. Therefore the wave function depends on these. Now it is inconvenient formally and of doubtful validity to admit systems with infinitely many degrees of freedom, or wave functions with infinitely many arguments. Our original directions were always based on a finite number of coordinates. We shall therefore begin by considering only the first N of the $\tilde{q}_1, \tilde{q}_2, \ldots$, the $\tilde{q}_1, \ldots, \tilde{q}_N$ (i.e., we shall limit $\mathfrak{A}$ to the linear combinations of $\overline{\mathfrak{A}}_1, \ldots, \overline{\mathfrak{A}}_N$), and then only after we have obtained a complete result based on these, shall we carry out the necessary transition to the limit $N \to \infty$.

Then

$$H = H_0 + \frac{1}{2} \sum_{n=1}^{N} (\tilde{P}_n{}^2 + 4\pi^2 \rho_n{}^2 \tilde{Q}_n{}^2)$$

$$+ \sum_{n=1}^{N} \sum_{\nu=1}^{l} \frac{e_\nu}{cm_\nu} \tilde{Q}_n \{ P_\nu{}^x \overline{\mathfrak{A}}_{n,x}(Q_\nu{}^x, Q_\nu{}^y, Q_\nu{}^z) +$$

$$+ P_\nu{}^y \overline{\mathfrak{A}}_{n,y}(Q_\nu{}^x, Q_\nu{}^y, Q_\nu{}^z) + P_\nu{}^z \overline{\mathfrak{A}}_{n,z}(Q_\nu{}^x, Q_\nu{}^y, Q_\nu{}^z) \}$$

It is convenient to introduce the (non-Hermitian) operator $\tilde{R}_n$ and its adjoint $\tilde{R}_n^*$ in place of the $\tilde{P}_n, \tilde{Q}_n$:

$$\tilde{R}_n = \frac{1}{\sqrt{2h\rho_n}} (2\pi\rho_n \tilde{Q}_n + i\tilde{P}_n) \ , \ \tilde{R}_n^* = \frac{1}{\sqrt{2h\rho_n}} (2\pi\rho_n \tilde{Q}_n - i\tilde{P}_n)$$

Then

$$\tilde{Q}_n = \frac{1}{2\pi} \sqrt{\frac{h}{2\rho_n}} (\tilde{R}_n + \tilde{R}_n^*)$$

and because $\tilde{P}_n \tilde{Q}_n - \tilde{Q}_n \tilde{P}_n = \frac{h}{2\pi i} 1$

$$\tilde{R}_n \tilde{R}_n^* = \frac{1}{2h\rho_n} (\tilde{P}_n{}^2 + 4\pi^2 \rho_n{}^2 \tilde{Q}_n{}^2) + \frac{1}{2} \cdot 1 \ ,$$

$$\tilde{R}_n^* \tilde{R}_n = \frac{1}{2h\rho_n} (\tilde{P}_n{}^2 + 4\pi^2 \rho_n{}^2 \tilde{Q}_n{}^2) - \frac{1}{2} \cdot 1$$

Therefore, in particular, $\tilde{R}_n\tilde{R}_n^* - \tilde{R}_n^*\tilde{R}_n = 1$. Then the energy formula becomes

$$H = H_0 + \sum_{n=1}^{N} h\rho_n \cdot \tilde{R}_n^*\tilde{R}_n + \sum_{n=1}^{N}\sum_{\nu=1}^{1} \frac{e_\nu}{2\pi c m_\nu}\sqrt{\frac{h}{2\rho_\nu}}\,(\tilde{R}_n + \tilde{R}_n^*)$$

$$+ [P_\nu^{\,x}\overline{\mathfrak{A}}_{n,x}(Q_\nu^{\,x}, Q_\nu^{\,y}, Q_\nu^{\,z}) + P_\nu^{\,y}\overline{\mathfrak{A}}_{n,y}(Q_\nu^{\,x}, Q_\nu^{\,y}, Q_\nu^{\,z})$$

$$+ P_\nu^{\,z}\overline{\mathfrak{A}}_{n,z}(Q_\nu^{\,x}, Q_\nu^{\,y}, Q_\nu^{\,z})] + C$$

in which $C = \text{constant} = \frac{1}{2}\sum_{n=1}^{N} h\rho_n$. Since an additive constant is meaningless in the expression for the energy, we can neglect C. This is all the more desirable, since C becomes infinite for $N \to \infty$, and therefore would upset the proper completion of the theory.

The Hermitian operator $\tilde{R}_n^*\tilde{R}_n$ is hypermaximal, and in fact it has a pure discrete spectrum, consisting of the numbers $0, 1, 2, \ldots$. The corresponding eigenfunctions are called $\psi_0^{\,n}(\tilde{q}_n), \psi_1^{\,n}(\tilde{q}_n), \psi_2^{\,n}(\tilde{q}_n), \ldots$.

[If we write

$$\frac{1}{2\pi}\sqrt{\frac{h}{\rho_n}}\,q$$

in place of $\tilde{q}_n$, then

$$\frac{1}{\sqrt{2h\rho_n}}\,2\pi\rho_n\tilde{q}_n = 2\pi\sqrt{\frac{\rho_n}{2h}}\,\tilde{q}_n \quad\text{and}\quad \frac{1}{\sqrt{2h\rho_n}}\,\frac{h}{2\pi i}\frac{\partial}{\partial\tilde{q}_n} = \frac{1}{2\pi}\sqrt{\frac{h}{2\rho_n}}\,\frac{1}{i}\frac{\partial}{\partial\tilde{q}_n}$$

go over into $\frac{1}{\sqrt{2}}\,q$ and $\frac{1}{\sqrt{2}}\frac{1}{i}\frac{\partial}{\partial q}$ respectively, so that

$$\tilde{R}_n = \frac{1}{\sqrt{2}}\left(q + \frac{\partial}{\partial q}\right), \quad \tilde{R}_n^* = \frac{1}{\sqrt{2}}\left(q - \frac{\partial}{\partial q}\right)$$

$$\tilde{R}_n\tilde{R}_n^* = -\frac{1}{2}\frac{\partial^2}{\partial q^2} + \frac{1}{2}q^2 + \frac{1}{2}$$

$$\tilde{R}_n^*\tilde{R}_n = -\frac{1}{2}\frac{\partial^2}{\partial q^2} + \frac{1}{2}q^2 - \frac{1}{2}$$

The eigenvalue theory of these operators can be found in many treatises, for example, Courant-Hilbert, p. 261, formulas (42), (43) and the related subject matter, as well as p. 76, formulas (60), (61); or Weyl, Gruppentheorie und Quantenmechanik, p. 74 and thereafter.]

Since the $\psi_1(\xi), \psi_2(\xi), \ldots$ form a complete orthogonal set in the ξ space, and the $\psi_0{}^n(\tilde{q}_n), \psi_1{}^n(\tilde{q}_n), \ldots$ form one in the $\tilde{q}_n$ space, the $\Phi_{kM_1\ldots M_N}(\xi, \tilde{q}_1, \ldots, \tilde{q}_N) = \psi_k(\xi)\cdot\psi^1_{M_1}(\tilde{q}_1)\ldots\psi^N_{M_N}(\tilde{q}_N)$, $k = 1, 2, \ldots$; $M_1, \ldots, M_N = 0, 1, 2, \ldots$ form a complete orthogonal set in the $\xi, \tilde{q}_1, \ldots, \tilde{q}_N$-space -- i.e., in the configuration space. We can then expand $\Phi = \Phi(\xi, \tilde{q}_1, \ldots, \tilde{q}_N)$ as follows:

$$\Phi(\xi, \tilde{q}_1, \ldots, \tilde{q}_N) = \sum_{k=1}^{\infty} \sum_{M_1=0}^{\infty} \cdots \sum_{M_N=0}^{\infty} a_{kM_1\ldots M_N} \Phi_{kM_1\ldots M_N}(\xi, \tilde{q}_1, \ldots, \tilde{q}_N)$$

$$= \sum_{k=1}^{\infty} \sum_{M_1=0}^{\infty} \cdots \sum_{M_N=0}^{\infty} a_{kM_1\ldots M_N} \psi_k(\xi)\cdot\psi^1_{M_1}(\tilde{q}_1)\ldots\psi^N_{M_N}(\tilde{q}_N)$$

It is of no significance that we enumerate the complete orthogonal set and the expansion coefficients with $N + 1$ indices $k, M_1, \ldots, M_N$ instead of with one. Indeed, the considerations of II.2. justify this conclusion. The Hilbert space of the wave functions Φ can also be interpreted as that one of the $(N + 1$ fold) sequences $a_{kM_1\ldots M_N}$ (with finite $\sum_{k=1}^{\infty} \sum_{M_1=0}^{\infty} \cdots \sum_{M_N=0}^{\infty} |a_{kM_1\ldots M_N}|^2$) .

Now which operator is H in this conception of the Hilbert space? In order to answer this, let us first calculate the $H\Phi_{kM_1\ldots M_N}$. Since H_0 operates on ξ alone, and $\psi_k(\xi)$ is an eigenfunction of H_0 with the eigenvalue W_k, and moreover, since $R_n^*R_n$ operates on q_n alone, and $\psi^n_{M_n}(q_n)$ is the eigenfunction of $R_n^*R_n$ with eigenvalue M_n, then

$$H\Phi_{kM_1\ldots M_N} = \left(W_k + \sum_{n=1}^{N} h\rho_n \cdot M_n\right)\Phi_{kM_1\ldots M_N} + \sum_{n=1}^{N}\sum_{\nu=1}^{l}\frac{e_\nu}{2\pi c m_\nu}\sqrt{\frac{h}{2\rho_n}}$$

$$\times [P_\nu^x \mathfrak{A}_{n,x}(Q_\nu^x, Q_\nu^y, Q_\nu^z) + P_\nu^y \mathfrak{A}_{n,y}(Q_\nu^x, Q_\nu^y, Q_\nu^z)$$

$$+ P_\nu^z \mathfrak{A}_{n,z}(Q_\nu^x, Q_\nu^y, Q_\nu^z)]\psi_k(\xi)$$

$$\times \psi_{M_1}^1(\tilde{q}_1)\ldots(\tilde{R}_n + \tilde{R}_n^*)\psi_{M_n}^n(\tilde{q}_n)\ldots\psi_{M_N}^N(\tilde{q}_N)$$

For all operators A which (like the $[\ldots]$ expression) affect only the variable ξ, we can employ the expansion

$$A\psi_k(\xi) = \sum_{j=1}^{\infty}(A\psi_k, \psi_j)\cdot\psi_j(\xi) = \sum_{j=1}^{\infty}(A)_{kj}\cdot\psi_j(\xi)$$

in which $(A)_{kj} = (A\psi_k, \psi_j)$. Furthermore, as was shown in the place mentioned above,

$$\tilde{R}_n\psi_M^n(\tilde{q}_n) = \sqrt{M}\ \psi_{M-1}^n(\tilde{q}_n)\ ,\quad \tilde{R}_n^*\psi_M^n(\tilde{q}_n) = \sqrt{M+1}\ \psi_{M+1}^n(\tilde{q})$$

(for $M = 0$, the right side of the first equation is to be equated to zero without regard to the meaningless ψ_{-1}^n appearing in it). Consequently

$$H\Phi_{kM_1\ldots M_N} = \left(W_k + \sum_{n=1}^{N} h\rho_n \cdot M_n \quad \Phi_{kM_1\ldots M_n}\right) + \sum_{j=1}^{\infty}\sum_{n=1}^{N}\sqrt{\frac{h}{2\rho_n}}$$

$$\times\left(\sum_{\nu=1}^{l}\frac{e_\nu}{2\pi c m_\nu}\left(P_\nu^x \mathfrak{A}_{n,x}(Q_\nu^x, Q_\nu^y, Q_\nu^z) + \ldots\right)_{kj}\right)$$

$$\times\left(\sqrt{M_n + 1}\ \Phi_{kM_1\ldots M_n+1\ldots M_N} + \sqrt{M_n}\ \Phi_{kM_1\ldots M_n-1\ldots M_N}\right)$$

We can now give the form of H as an $a_{kM_1\ldots M_N}$ operator. It is

$$\mathsf{H} \sum_{kM_1 \ldots M_N} a_{kM_1 \ldots M_N} \Phi_{kM_1 \ldots M_N} = \sum_{kM_1 \ldots M_N} a'_{kM_1 \ldots M_N} \Phi_{kM_1 \ldots M_N}$$

with

$$\mathsf{H} a_{kM_1 \ldots M_N} = a'_{kM_1 \ldots M_N} = \left(W_k + \sum_{n=1}^{N} h\rho_n \cdot M_n \right) a_{kM_1 \ldots M_N}$$
$$+ \sum_{j=1}^{\infty} \sum_{n=1}^{N} \sqrt{\frac{h}{2\rho_n}} \left(\sum_{\nu=1}^{l} \frac{e_\nu}{2\pi c m_\nu} \left(P_\nu^x \mathfrak{A}_{n,x}(Q_\nu^x, Q_\nu^y, Q_\nu^z) + \ldots \right)_{kj} \right)$$
$$\times \left(\sqrt{M_n}\, a_{jM_1 \ldots M_n - 1 \ldots M_N} + \sqrt{M_n + 1}\, a_{jM_1 \ldots M_n + 1 \ldots M_N} \right)$$

The discussion of H has now been carried far enough for us to undertake the transition to the limit $N \to \infty$. Since the system of indexing of the $a_{kM_1 \ldots M_N}$ changes in the process, there arises an entirely new H operator. We must introduce components $a_{kM_1 M_2 \ldots}$ with infinitely many indices $M_1, M_2, \ldots$. However, we must limit ourselves to such $M_1, M_2, \ldots$ sequences in which only a finite number of elements is different from zero -- if for no other reason, then in order to make sure of the finiteness of the sum

$$\sum_{n=1}^{\infty} h\rho_n \cdot M_n$$

appearing in H . From now on therefore, the Hilbert space of all sequences $a_{kM_1 M_2 \ldots}$ with finite

$$\sum_{k=1}^{\infty} \sum_{M_1=0}^{\infty} \sum_{M_2=0}^{\infty} \cdots |a_{kM_1 M_2 \ldots}|^2$$

will be used in which the indices $k, M_1, M_2, \ldots$ run over the following region: $k = 1, 2, \ldots;\ M_1, M_2, \ldots = 0, 1, 2, \ldots$;

with only a finite (but arbitrary) number of $M_n \neq 0$.[144] The final form of H is then

$$H a_{kM_1M_2\ldots} = a'_{kM_1M_2\ldots} = \left(W_k + \sum_{n=1}^{\infty} h\rho_n \cdot M_n \right) \cdot a_{kM_1M_2\ldots}$$

$$+ \sum_{j=1}^{\infty} \sum_{n=1}^{\infty} w_{kj}^n \left(\sqrt{M_n + 1}\, a_{jM_1M_2\ldots M_n+1\ldots} + \sqrt{M_n}\, a_{jM_1M_2\ldots M_n-1\ldots} \right)$$

in which w_{kj}^n is defined by

$$w_{kj}^n = \sqrt{\frac{h}{2\rho_n}} \sum_{\nu=1}^{l} \frac{e_\nu}{2\pi c m_\nu} \left(P_\nu{}^x \mathfrak{A}_{n,x}(Q_\nu{}^x, Q_\nu{}^y, Q_\nu{}^z) + \ldots \right)_{kj}$$

Before we draw from this result the physical conclusions that interest us, we should recall that it was obtained on the basis of the electrodynamic theory of radiation. We now want to determine whether or not the standard quantum mechanical transformation which we performed, suffices to account for the deviations of radiation from the wave model -- for its discrete-corpuscular nature. (Note, that it would be quite reasonable to expect that in order to achieve this one would have to start directly from

---

[144] That the totality of all these index systems $k, M_1, M_2, \ldots$ actually forms a sequence can be shown most simply as follows. Let $\pi_1, \pi_2, \pi_3, \ldots$ be the series of prime numbers $2, 3, 5, \ldots$. The products $\pi_1^{k-1} \cdot \pi_2^{M_1} \cdot \pi_3^{M_2} \ldots$ are in reality finite, because all $M_n = 0$, with a finite number of exceptions, i.e., $\pi_{n+1}^{M_n} = 0$. Then if $k, M_1, M_2, \ldots$ run through our entire set of indices, the $\pi_1^{k-1} \cdot \pi_2^{M_1} \cdot \pi_3^{M_2} \ldots$ run through all numbers $1, 2, 3, \ldots$, and assume each value once. We can therefore use the $\pi_1^{k-1} \cdot \pi_2^{M_1} \cdot \pi_3^{M_2} \ldots$ to obtain a simple running indexing for the $a_{kM_1M_2\ldots}$.

a corpuscular model for light, instead of quantizing the electromagnetic field, as we did here).

It can immediately be seen in our expression for that something like the corpuscular light quantum is included in it. Suppose that we neglect the second term, which produces a sort of perturbation, and which, as we shall see later, gives rise to the quantum jumps of the system S from one "stationary state" into another. (This latter is the phenomenon which is actually of interest to us, but which is nevertheless a less striking entity than the material system S itself, and that is already present. As we shall see, these entities are represented by the first term of H.) We then have left

$$H_1 a_{kM_1M_2\ldots} = \left(W_k + \sum_{n=1}^{\infty} h\rho_n \cdot M_n\right) \cdot a_{kM_1M_2\ldots}$$

But this expression for the energy can be interpreted as follows: It is the energy W_k of the system S, increased by the amounts $h\rho_n \cdot M_n$ ($n = 1, 2, \ldots$). Hence it is plausible to interpret the numbers $M_n = 0, 1, 2, \ldots$ as the numbers of particles with the respective energies $h\rho_n$. But $h\rho_n$ is precisely that energy which, according to Einstein, has to be assigned to a light quantum of frequency ρ_n (cf. Note 134). Consequently the structure of H_1 justifies the view that the electromagnetic field existing in H (less the electrostatic part), i.e., L, actually consists of light quanta with the frequencies $\rho_1, \rho_2, \ldots$, and with the energies $h\rho_1, h\rho_2, \ldots$. Furthermore, the numbers of such particles are then given by the indices $M_1, M_2, \ldots$ ($= 0, 1, 2, \ldots$). The fact that no other frequencies than $\rho_1, \rho_2, \ldots$ occur can be made plausible by saying that these are the eigen-frequencies of the cavity H. Indeed, the vector potentials

$$\overline{\mathfrak{A}}_n(x, y, z) \cdot \gamma \cos 2\pi\rho_n(t - \tau)$$

represent the only possible stationary electromagnetic oscillations possible in **H**.

While these speculations and interpretations are of only heuristic value, a fully satisfactory and final answer to our question is obtained only if it is possible for us to arrive at the energy expression H, starting out from the light quantum model for the radiation **L**. That we first carried out the classical treatment is due to the fact that the pre-quantum mechanical light quantum hypothesis furnished no expression for the interaction energy of a light quantum with matter (the re-interpretation of classical electrodynamics never succeeded at this point). Now, however, we shall be able to determine this interaction term by comparison of coefficients if our result (to be derived by use of a non-specialized expression for the interaction energy) coincides in form with H

What is the state space of **L** (question 1.) on the basis of the light quantum hypothesis? A single light quantum (in the space **H**) may be characterized by certain coordinates, whose totality we shall represent by the symbol u.[145] Its stationary states may have the wave

---

[145]As coordinates of the light quantum, we may want to use, for example, its momenta p_x, p_y, p_z, as well as a coordinate π describing its state of polarization. p_x, p_y, p_z determine the direction of the light quantum i.e., its direction cosines $\alpha_x, \alpha_y, \alpha_z$ $(\alpha_x^2 + \alpha_y^2 + \alpha_z^2 = 1)$, as well as its frequency ν, its wavelength λ, and energy; because, by Einstein, the momentum vector has the magnitude $\frac{h\nu}{c}$ (cf. Note 134). Therefore

$$p_x = \frac{h\nu}{c}\alpha_x, \quad p_y = \frac{h\nu}{c}\alpha_y, \quad p_z = \frac{h\nu}{c}\alpha_z$$

i.e.,

$$\nu = \frac{c}{h}\sqrt{p_x^2 + p_y^2 + p_z^2}, \quad \lambda = \frac{c}{\nu}, \quad \text{Energy} = h\nu,$$

functions $\psi_1(u), \psi_2(u), \ldots$ (which form an orthonormal set)

$$\alpha_x = \frac{cp_x}{h\nu}, \quad \alpha_y = \frac{cp_y}{h\nu}, \quad \alpha_z = \frac{cp_z}{h\nu}$$

It is disturbing to observe that our eigen-oscillations $\overline{\mathfrak{A}}_n(x, y, z) \cdot \gamma \cos 2\pi\rho_n(t - \tau)$ here are standing waves, as it could not be otherwise possible in the cavity **H** because of the reflecting walls -- so that $\overline{\mathfrak{A}}_n$ can be related with no unique "ray-direction" $\alpha_x, \alpha_y, \alpha_z$. We can see immediately that, along with $\alpha_x, \alpha_y, \alpha_z$, at least the opposite direction $-\alpha_x, -\alpha_y, -\alpha_z$ is also present, and the same holds for the momentum. As a consequence, we must use other coordinates in **H** than p_x, p_y, p_z, π.

In some recent expositions of the subject this inconvenience has been overcome by the following artifice: Let **H** be the parallelepipedon

$$-A < x < A, \quad -B < y < B, \quad -C < z < C$$

whose boundary surfaces $x = \pm A$, $y = \pm B$, $z = \pm C$ are not treated as reflecting walls. Rather, $x = +A$ is identified with $x = -A$, $y = B$ with $y = -B$, $z = C$ with $z = -C$. That is, radiation that impinges upon the wall $x = A$ at A, y, z, resumes at $-A, y, z$ its progress in the same direction (back again into **H**) as if nothing had happened, etc. [Cf., for example, a treatment of L. Landau and R. Peierls, Z. Physik <u>62</u> (1930)]. We can also say that the space is taken periodically in the x, y, z directions with the respective periods $2A, 2B, 2C$.

The analytical treatment remains the same, but the boundary condition is now

$$\mathfrak{A}(A, y, z) = \mathfrak{A}(-A, y, z), \quad \mathfrak{A}(x, B, z) = \mathfrak{A}(x, -B, z),$$

$$\mathfrak{A}(x, y, C) = \mathfrak{A}(x, y, -C)$$

and the energies $E_1, E_2, \ldots$. These correspond to the electromagnetic eigen-oscillations $\overline{\mathfrak{A}}_1, \overline{\mathfrak{A}}_2, \ldots$ with the frequencies $\rho_1, \rho_2, \ldots$. (In the sense of the Einstein concept $E_n = h\rho_n$, which we shall also prove.) In this way, the following is to be observed: In the electromagnetic consideration, we had so normalized the energy of the light that its minimum value was 0 , which corresponded to the indices $M_1 = M_2 = \ldots = 0$. Consequently, we have recognized non-existence as a possible state of the light, which is in fact justified. In actuality, light quanta are emitted and absorbed, i.e., light is produced and annihilated. Yet to quantum mechanics in general, such a concept is entirely foreign: each particle contributes coordinates to the state space of the system, and therefore enters so intimately into the formal description of the total system that it appears in effect indestructible. After the annihilation then, we must ascribe to the particle a sort of latent existence, in which its coordinates are still part of the description of the configuration space. Consequently, one of the states $\psi_n(u)$, with an energy $E_n = 0$, must correspond to the non-existence of the light quantum. We prefer to represent this with $\psi_0(u) (E_0 = 0)$ so that $\psi_1(u), \psi_2(u), \ldots$ correspond to the existing light

---

(instead of $\frac{\partial}{\partial n}\mathfrak{A} = 0$ at the boundary), and the "elementary solutions," with which we perform the expansions, are the

$$\begin{matrix}\cos\\ \sin\end{matrix}\ [2\pi\nu(t - c(\alpha_z x + \alpha_y y + \alpha_z z))]$$

(instead of the $\overline{\mathfrak{A}}(x, y, z)\cdot\tilde{\rho}(t)$) . We can easily determine the

$$\nu = \rho_n,\ \alpha_x = \alpha_{n,x},\ \alpha_y = \alpha_{n,y},\ \alpha_z = \alpha_{n,z} \quad (n = 1, 2, \ldots)$$

belonging to the eigen-solutions, and the further development of the theory coincides with that in the text.

quantum, but $\psi_0(u), \psi_1(u), \psi_2(u), \dots$ form the complete orthogonal set.

We now proceed to the consideration of **L**, the system of all light quanta. Since we also count non-existent light quanta, all light quanta that can ever be represented in **L** right from the start, i.e., of an infinite number. But since it is inconvenient to operate with infinitely many constituents in **L**, we first consider the existence of only S light quanta ($S = 1, 2, \dots$), and at the end pass to the limit $S \to +\infty$.[146] We designate these light quanta by the numbers $1, \dots, S$, and call their coordinates $u_1, \dots, u_S$. The configuration space of **L** is therefore described by $u_1, \dots, u_S$, and that of **S** + **L** by $\xi, u_1, \dots, u_S$. The most general wave function for S + L is then $f(\xi, u_1, \dots, u_S)$, and the $\phi_k(\xi) \cdot \psi_{n_1}(u_1) \cdots \psi_{n_S}(u_S)$, $k = 1, 2, \dots$, $u_1, \dots, u_S = 0, 1, 2, \dots$ form a complete orthogonal set.

The light quanta have now the fundamental property of being exactly identical, i.e., there is no conceivable way of distinguishing between two light quanta with the same coordinate u. Or, in another way, a state in which the light quanta m and n have the corresponding u values $u_m = u'$, $u_n = u''$ is not distinguishable from the state in which $u_m = u''$, $u_n = u'$. (This is the classical, not quantum mechanical, method of description, since we give the value of u, and not of the wave function $\phi(u)$.)

---

[146]This transition to the limit $S \to +\infty$ is different from the limit transition $N \to +\infty$ taken in the electromagnetic theory. For, if we interpret the $M_1, M_2, \dots$ there as numbers of light quanta, then N is a limit for the number of incoherent light quanta (i.e., of light quanta not of the same frequency and direction -- these together make up its momentum -- and polarization; cf. Note 143), while S is a limit for the total number of light quanta.

Quantum mechanically, this means that the states belonging to the wave functions $f(\xi, u_1, \ldots, u_m, \ldots, u_n, \ldots, u_S)$ and $f(\xi, u_1, \ldots, u_n, \ldots, u_m, \ldots, u_S)$ are indistinguishable. That is, each physical quantity $\mathfrak{R}$ has the same expectation value in them (therefore, since this also holds for the $F(\mathfrak{R})$, each physical quantity has the same statistics also -- cf. the discussion of $\mathbf{E}_1$. and $\mathbf{E}_2$. in III.1.). If we denote the functional operation which permutes u_m, u_n by O_{mn} (O_{mn} is simultaneously Hermitian and unitary, $O_{mn}^2 = 1$, as can be seen immediately), then this means that $\mathfrak{R}$ has the same expectation value for f as for $O_{mn}f$ i.e.,

$$(Rf, f) = (RO_{mn}f, O_{mn}f) = (O_{mn}RO_{mn}f, f)$$

therefore

$$R = O_{mn}RO_{mn}, \text{ or also } O_{mn}R = RO_{mn}$$

This means that in the present case only such operators R are admissible which commute with all O_{mn} ($m, n = 1, \ldots, S$, $m \neq n$) i.e., (with reference to the definition of the O_{mn}) into which all the coordinates $u_1, \ldots, u_S$ enter symmetrically.

A wave function f, which is symmetrical in all variables $u_1, \ldots, u_S$ i.e., for which $O_{mn}f = f$ holds ($m, n = 1, \ldots, S$, $m \neq n$), is transformed by such an operator R into one of the same kind: $O_{mn}Rf = RO_{mn}f = Rf$. These f form a closed linear manifold, and therefore a Hilbert sub-space $\overline{\mathfrak{R}}_\infty^{(S)}$ in the Hilbert space $\mathfrak{R}_\infty^{(S)}$ of all f -- and the R map elements of $\overline{\mathfrak{R}}_\infty^{(S)}$ on the same space, i.e., they can be regarded as operators in the Hilbert space $\overline{\mathfrak{R}}_\infty^{(S)}$. Consequently $\overline{\mathfrak{R}}_\infty^{(S)}$ is just as useful for the purposes of quantum mechanics as the $\mathfrak{R}_\infty^{(S)}$ originally considered, and in view of the symmetry of $\mathbf{L}$ with respect to the exchanges of the light quanta, the question arises as to whether one cannot limit oneself to symmetric wave functions, i.e., whether $\mathfrak{R}_\infty^{(S)}$ should be replaced by $\overline{\mathfrak{R}}_\infty^{(S)}$. We shall do this, and the result, i.e., the complete agreement which will be achieved with the H-expression derived electromagnetically, will ultimately justify our step.[147]

The $\phi_k(\xi)\cdot\psi_{n_1}(u_1)\cdots\psi_{n_S}(u_S)$ formed a complete orthonormal set in $\mathfrak{R}_\infty^{(S)}$. Making use of this, we shall now form one in $\overline{\mathfrak{R}}_\infty^{(S)}$. Let $M_0, M_1, \ldots$ be any numbers $= 0, 1, 2, \ldots$, with $M_0 + M_1 + \cdots = S$ (therefore only a finite number of them differ from zero). We denote by $[M_1, M_2, \ldots]$ the totality of all index systems $n_1, \ldots, n_S$ in which 0 appears M_0 times, 1 appears M_1 times, ... There are exactly $M_0!\cdot M_1!\cdots$ such systems. We set

$$\Phi_{M_0 M_1 \cdots}(u_1 \cdots u_S) = \sum_{n_1 \cdots n_S \text{ in } [M_0, M_1, \ldots]} \psi_{n_1}(u_1)\cdots\psi_{n_S}(u_S)$$

Since $\Phi_{M_0 M_1 \cdots}$ is the sum of $M_0!\cdot M_1!\cdots$ pairwise orthogonal summands of magnitude 1 , its square is the sum of $M_0!\cdot M_1!\cdots$ terms 1 , and its magnitude is then

$$\sqrt{M_0!\cdot M_1!\cdots}$$

Two different $\Phi_{M_0 M_1 \cdots}$ have pairwise orthogonal summands, and are therefore orthogonal. The

$$\psi_{M_0 M_1 \cdots}(u_1, \ldots, u_S) = \frac{1}{\sqrt{M_0!\cdot M_1!\cdots}} \Phi_{M_0 M_1 \cdots}(u_1, \ldots, u_S)$$

then form an orthonormal set. An $f(\xi, u_1, \ldots, u_S)$ symmetric

---

[147]This introduction of $\overline{\mathfrak{R}}_\infty^{(S)}$ in place of $\mathfrak{R}_\infty^{(S)}$ is equivalent to replacing ordinary statistics by the so-called Bose-Einstein statistics, if we consider the consequences of this relation without reference to quantum mechanics. Cf. Dirac in the reference cited in Note 138.

in the $u_1, \ldots, u_S$ has the same inner product with all summands of $\phi_k(\xi)\Phi_{M_0 M_1 \ldots}(u_1, \ldots, u_S)$, hence it is orthogonal to each of them if it is orthogonal to $\phi_k(\xi)\Phi_{M_0 M_1 \ldots}(u_1, \ldots, u_S)$; i.e., to $\phi_k(\xi)\psi_{M_0 M_1 \ldots}(u_1, \ldots, u_S)$. Therefore, if it is orthogonal to all $\phi_k(\xi)\psi_{M_0 M_1 \ldots}(u_1, \ldots, u_S)$, it is also orthogonal to all $\phi_k(\xi)\psi_{n_1}(u_1)\ldots\psi_{n_S}(u_S)$, and therefore it is $= 0$. Consequently, the $\phi_k(\xi)\psi_{M_0 M_1 \ldots}(u_1, \ldots, u_S)$ (which themselves belong to $\overline{\mathfrak{R}}_\infty^{(S)}$) form a complete orthonormal set in $\overline{\mathfrak{R}}_\infty^{(S)}$.

Let us now consider the set of energies entering into **S** + **L**. First, there is the energy of **S** [2α)], whose operator for **S** is defined by $H_0\phi_k(\xi) = W_k\phi_k(\xi)$, and therefore for **S** + **L** by

$$H_0\phi_k(\xi)\psi_{M_0 M_1 \ldots}(u_1, \ldots, u_S) = W_k\phi_k(\xi)\psi_{M_0 M_1 \ldots}(u_1, \ldots, u_S)$$

Second [2β)], each light quantum 1' has the energy $H_{1'}\psi_n(u) = E_n\psi_n(u)$. Therefore the mth quantum in **S** + **L** $(m = 1, \ldots, S)$ has the energy

$$H_{1_m}\phi_k(\xi)\cdot\psi_{n_1}(u_1)\ldots\psi_{n_m}(u_m)\ldots\psi_{n_S}(u_S)$$

$$= E_{n_m}\phi_k(\xi)\cdot\psi_{n_1}(u_1)\ldots\psi_{n_m}(u_m)\ldots\psi_{n_S}(u_S)$$

and $H_1 = H_{1_1} + \ldots + H_{1_S}$ is to be formed. Finally [2γ)], let the interaction energy of a light quantum 1' with **S** be described by an operator V, at present not known exactly, which we identify by its matrix

$$V_{1'}\phi_k(\xi)\psi_n(u) = \sum_{j=1}^{\infty}\sum_{p=0}^{\infty} v_{kn|jp}\phi_j(\xi)\psi_p(u)$$

In **S** + **L**, for the mth light quantum, we then have

$$V_{1_m}\phi_k(\xi)\cdot\psi_{n_1}(u_1)\cdots\psi_{n_m}(u_m)\cdots\psi_{n_S}(u_S)$$

$$= \sum_{j=1}^{\infty}\sum_{p=0}^{\infty} V_{kn_m|jp}\phi_j(\xi)\cdot\psi_{n_1}(u_1)\cdots\psi_p(u_m)\cdots\psi_{n_S}(u_S)$$

$$= \sum_{j=1}^{\infty}\sum_{p_1\cdots p_m\cdots p_S=0}^{\infty} \delta(n_1 - p_1)\cdots V_{kn_m|jp_m}\cdots\delta(n_S - p_S)$$

$$\times\ \phi_j(\xi)\cdot\psi_{p_1}(u_1)\cdots\psi_{p_m}(u_m)\cdots\psi_{p_S}(u_S)$$

($\delta(n)$ is 1 for $n = 0$, and it is 0 for $n \neq 0$), and we must form

$$H_i = V_{1_1} + \cdots + V_{1_S}$$

Altogether we then have

$$H\phi_k(\xi)\cdot\psi_{n_1}(u_1)\cdots\psi_{n_S}(u_S)$$

$$= (W_k + E_{n_1} + \cdots + E_{n_S})\phi_k(\xi)\cdot\psi_{n_1}(u_1)\cdots\psi_{n_S}(u_S)$$

$$+ \sum_{j=1}^{\infty}\sum_{p_1\cdots p_S=0}^{\infty}\sum_{m=1}^{S} \delta(n_1 - p_1)\cdots V_{kn_m|jp_m}\cdots\delta(n_S - p_S)$$

$$\times\ \phi_j(\xi)\cdot\psi_{p_1}(u_1)\cdots\psi_{p_S}(u_S)$$

By a simple transformation, this becomes

$$\phi_k(\xi)\Phi_{M_0M_1\cdots}(u_1,\ldots,u_S) = (W_k + \sum_{n=0}^{\infty} M_nE_n)\phi_k(\xi)\Phi_{M_0M_1\cdots}(u_1,\ldots,u_S)$$

$$+ \sum_{j=1}^{\infty}\sum_{n,p}^{\infty} M_nV_{kn|jp}\phi_j(\xi)\Phi_{M_0M_1\cdots M_n-1\cdots M_p+1\cdots}(u_1,\ldots,u_S)$$

(for $n = p$; $\ldots M_n-1\ldots M_p+1\ldots$ is to be replaced by

$\ldots M_n \ldots)$, and therefore, for the orthonormal functions,

$$H\phi_k(\xi)\psi_{M_0 M_1 \ldots}(u_1, \ldots, u_S) = (W_k + \sum_{n=0}^{\infty} M_n E_n)\phi_k \psi_{M_0 M_1 \ldots}(u_1, \ldots, u_S)$$

$$+ \sum_{j=1}^{\infty} \sum_{n,p}^{\infty} \sqrt{M_n(M_p+1-\delta(n-p))}\, V_{kn|jp}\phi_j(\xi)\psi_{M_0 M_1 \ldots M_n-1 \ldots M_p+1 \ldots}(u_1, \ldots u$$

We can expand the general $f(\xi, u, \ldots, u_S)$ of $\overline{\mathfrak{R}}_\infty^{(S)}$ in terms of these orthonormal functions:

$$f(\xi, u_1, \ldots, u_S) = \sum_{k=1}^{\infty} \sum_{\substack{M_0 M_1 \ldots = 0 \\ (M_0+M_1+\ldots=S)}}^{\infty} a_{kM_0M_1\ldots}\phi_k(\xi)\psi_{M_0M_1\ldots}(u_1, \ldots, u_S)$$

Therefore $\overline{\mathfrak{R}}_\infty^{(S)}$ may also be conceived as the Hilbert space of the sequences $a_{kM_0M_1\ldots}$ with $k = 1, 2, \ldots;\ M_0, M_1 \ldots = 0, 1, 2, \ldots;\ M_0 + M_1 + \ldots = S$; with finite

$$\sum_{kM_0M_1\ldots} |a_{kM_0M_1\ldots}|^2$$

In this case H is defined by $a_{kM_0M_1\ldots} = a'_{kM_0M_1\ldots}$

$$H \sum_{k=1}^{\infty} \sum_{\substack{M_0, M_1 \ldots = 0 \\ (M_0+M_1+\ldots=S)}}^{\infty} a_{kM_0M_1\ldots}\phi_k(\xi)\psi_{M_0M_1\ldots}(u_1, \ldots, u_S)$$

$$= \sum_{k=1}^{\infty} \sum_{\substack{M_1, M_2, \ldots = 0 \\ (M_1+M_2+\ldots=S)}}^{\infty} a'_{kM_0M_1\ldots}\phi_k(\xi)\psi_{M_0M_1\ldots}(u_1, \ldots, u_S)$$

and therefore

$$Ha_{kM_0M_1\ldots} = a'_{kM_0M_1\ldots} = (W_k + \sum_{n=0}^{\infty} M_n E_n)a_{kM_0M_1\ldots}$$

$$+ \sum_{j=1}^{\infty} \sum_{n,p=0}^{\infty} \sqrt{M_n(M_p+1-\delta(n-p))}\, \overline{V_{kn|jp}}\, a_{jM_0M_1\ldots M_n-1\ldots M_p+1\ldots}$$

(k, j, and n, p have exchanged their roles in comparison to the $\phi_k(\xi)\psi_{M_0M_1\ldots}(u_1,\ldots,u_S)$ formula; in place of $V_{jp|kn}$ we have written $\overline{V}_{kn|jp}$, keeping in mind the Hermitian nature of V .)

We proceed now to prepare for the transition to the limit $S \to +\infty$. Since M_0 is determined by $M_1, M_2, \ldots$, according to $M_0 = S - M_1 - M_2 - \ldots$, we can write $a_{kM_1M_2\ldots}$ instead of $a_{kM_0M_1\ldots}$. In this way the indices are limited by the conditions

$$k = 1,2,\ldots;\ M_1,M_2,\ldots = 0,1,2,\ldots;\ M_1 + M_2 + \ldots \leq S$$

If we consider $E_0 = 0$ and introduce the notation $SV_{ko|jo} = V_{k|j}$, $\sqrt{S}V_{ko|jn} = V_{k|jn}$, $\sqrt{S}V_{kn|jo} = \overline{V}_{j|kn}$ ($V_{kn|jp}$ is Hermitian), then

$$\mathsf{H}a_{kM_1M_2\ldots} = a'_{kM_1M_2\ldots} = (W_k + \sum_{n=1}^{\infty} M_nE_n)a_{kM_1M_2\ldots}$$

$$+ \sum_{j=1}^{\infty} V_{k|j}a_{jM_1M_2\ldots}$$

$$+ \sum_{j=1}^{\infty}\sum_{n=1}^{\infty} \sqrt{M_n}\sqrt{\frac{S-M_1-M_2-\ldots+1}{S}}\, V_{j|kn}a_{jM_1M_2\ldots M_n-1\ldots}$$

$$+ \sum_{j=1}^{\infty}\sum_{n=1}^{\infty} \sqrt{M_n+1}\sqrt{\frac{S-M_1-M_2-\ldots}{S}}\, \overline{V}_{k|jn}a_{jM_1M_2\ldots M_n+1\ldots}$$

$$+ \sum_{j=1}^{\infty}\sum_{n,p=1}^{\infty} \sqrt{M_n(M_p+1)}\, \overline{V}_{kn|jp}a_{jM_1M_2\ldots M_n-1\ldots M_p+1\ldots}$$

Now let $S \to +\infty$. The $a_{kM_1M_2\ldots}$ are again defined over all sequences $kM_1M_2\ldots$ with $k = 1,2,\ldots;$ $M_1,M_2,\ldots = 0,1,2,\ldots$; with only a finite (but arbitrary) number of $M_n \neq 0$ (cf. Note 144), and for H we obtain

$$Ha_{kM_1M_2\cdots} = a'_{kM_1M_2\cdots} = (W_k + \sum_{n=1}^{\infty} M_nE_n)a_{kM_1M_2\cdots}$$

$$+ \sum_{j=1}^{\infty} V_{k|j}a_{jM_1M_2\cdots}$$

$$+ \sum_{j=1}^{\infty}\sum_{n=1}^{\infty}(V_{j|kn}\sqrt{M_n+1}\, a_{jM_1M_2\cdots M_n+1\cdots} + \bar{V}_{k|jn}\sqrt{M_n}\, a_{jM_1M_2\cdots M_n-1\cdots}$$

$$+ \sum_{j=1}^{\infty}\sum_{n,p=1}^{\infty} \bar{V}_{kn|jp}\sqrt{M_n(M_p+1)}\, a_{jM_1M_2\cdots M_n-1\cdots M_p+1\cdots}$$

The similarity with the equation derived from the electromagnetic theory of radiation is now apparent; to make the two relations identical, we need only set

$$E_n = h_{\rho_n},\ V_{k|j} = 0,\ V_{k|jn} = w^n_{jk} = \bar{w}^n_{kj},\ V_{kn|jp} = 0$$

We then see that the light quanta concept proves to be identical to the classical electromagnetic concept if we observe these rules:

1. The latter is rewritten according to the general quantum mechanical scheme;

2. The energy of each light quantum is equated to the h-fold frequency, corresponding to the Einstein rule;

3. The interaction energy of the light quantum with matter is appropriately defined (cf. the above expressions for V) .

In this way one of the most difficult paradoxes of the earlier form of the quantum theory, the dual nature of light (electromagnetic waves and discrete corpuscules or light quanta) is brilliantly resolved.[148] To be sure,

---

[148]The reader will find further discussion on how this "dual nature" was conceived, and how paradoxical it was

it is difficult to find a direct, clear-cut interpretation of the interaction energy V which has just been calculated. This is even more difficult since the individual matrix elements $V_{kn|jp}$ different from zero (those with $n \neq 0$, $p = 0$ or $n \neq$, $p = 0$) depend upon the number of all possible light quanta S (they are proportional to $\frac{1}{\sqrt{S}}$) -- although in the end one has to effect $S \to +\infty$. Nevertheless, we can accept this with the interpretation that each model-description is only an approximation, while the exact content of the theory is furnished solely by the expression for the H operator.

We now return to our actual task: the determination of the transition probabilities. In the sense of the time dependent Schrödinger equation, the changes in the

$$a_{kM_1M_2\cdots} = a_{kM_1M_2\cdots}(t)$$

are determined by

---

considered, in the contemporary literature. For example, see the works listed in Note 6.

It has often been said that the quantum mechanics involves the same dual nature, since the discrete particles (electrons, protons) are also described by wave functions, and exhibit typical wave properties, i.e., diffraction by a grating. [Cf., the experiments of Davison-Germer, Phys. Rev. 50 (1927), Prox. Mat. Acad. Sci. U.S.A. 15 (1928); also C. F. Thompson, Proc. Roy. Soc. 117 (1928), and Rupp, Ann. Physik 85 (1928)]. In contrast with this, however, it is to be noted that quantum mechanics derives both "natures" from a single unified theory of the elementary phenomena. The paradox of the earlier quantum theory lay in the circumstance that one had to draw alternately on two contradictory theories (electromagnetic theory of radiation of Maxwell-Hertz, light quantum theory of Einstein) for the explanation of the experience.

$$\frac{h}{2\pi i}\frac{\partial}{\partial t}a_{kM_1M_2\cdots} = -Ha_{kM_1M_2\cdots} = -(W_k + \sum_{n=1}^{\infty} h\rho_n\cdot M_n)a_{kM_1M_2\cdots}$$
$$-\sum_{j=1}^{\infty}\sum_{n=1}^{\infty} w_{kj}^{n}\cdot(\sqrt{M_n+1}\; a_{jM_1M_2\cdots M_n+1\cdots} + \sqrt{M}\; a_{jM_1M_2\cdots M_n-1\cdots})$$

Since the chief change of the $a_{kM_1M_2\cdots}$ is caused by the first term in this expression, it is appropriate to separate this by the substitution

$$a_{kM_1M_2\cdots}(t) = e^{-\frac{2\pi i}{h}(W_k + \sum_{n=1}^{\infty} h\rho_n\cdot M_n)t}\cdot b_{kM_1M_2\cdots}(t)$$

Then

$$\frac{\partial}{\partial t} b_{kM_1M_2\cdots}$$
$$= \frac{2\pi i}{h}\sum_{j=1}^{\infty}\sum_{n=1}^{\infty} w_{kj}^{n}\cdot\Big(e^{-\frac{2\pi i}{h}(W_j - W_k + h\rho_n)t}\sqrt{M_n+1}\; b_{jM_1M_2\cdots M_n+1\cdots}$$
$$- e^{-\frac{2\pi i}{h}(W_j - W_k - h\rho_n)t}\sqrt{M_n}\; b_{jM_1M_2\cdots M_n-1\cdots}\Big)$$

The physical meaning of the $a_{kM_1M_2\cdots}$ and the $b_{kM_1M_2\cdots}$ may be seen from their origin: for finite $\bar{M}_0 + \bar{M}_1 + \bar{M}_2 + \cdots = S$, $\phi_{\bar{k}}(\xi)\psi_{\bar{M}_0\bar{M}_1}\cdots(u_1,\ldots,u_S)$ was that state in which S is in the kth quantum orbit, and $\bar{M}_0, \bar{M}_1, \bar{M}_2, \ldots$ light quanta of the respective states $\psi_0, \psi_1, \psi_2, \ldots$ are present -- i.e., $\bar{M}_0$ in the state of "non-existence," and $\bar{M}_1, \bar{M}_2, \ldots$ in the states belonging to the corresponding characteristic oscillations $\bar{\mathfrak{A}}_1, \bar{\mathfrak{A}}_2, \ldots$.

The $a_{kM_1M_2\cdots}$ belonging to this wave function are then

$$a_{kM_1M_2\cdots} = \delta(k - \overline{k})\cdot\delta(M_1 - \overline{M}_1)\cdot\delta(M_2 - \overline{M}_2)\cdots$$

(Only a finite number of factors are different from 1, since $M_n = \overline{M}_n = 0$ with only a finite number of exceptions.) This, of course, remains valid after $S \to +\infty$, too. For an arbitrary state $a_{kM_1M_2\cdots}$ of **S + L**, therefore, the configuration mentioned (if this is measured, cf. the comments in III.3. on the non-degenerate pure discrete spectrum) has the probability

$$\Big|\sum_{kM_1M_2\cdots} a_{kM_1M_2\cdots}\delta(k - \overline{k})\delta(M_1 - \overline{M}_1)\delta(M_2 - \overline{M}_2)\cdots\Big|^2$$
$$= |a_{\overline{k}\overline{M}_1\overline{M}_2\cdots}|^2 = |b_{\overline{k}\overline{M}_1\overline{M}_2\cdots}|^2 .$$

In particular, the total probability that S be found in the kth quantum orbit is

$$\theta_{\overline{k}} = \sum_{\overline{M}_1\overline{M}_2\cdots} |b_{\overline{k}\overline{M}_1\overline{M}_2\cdots}|^2$$

Let the atom be initially $(t = 0)$ in the $\overline{k}$th state, and let $\overline{M}_1, \overline{M}_2, \cdots$ light quanta of the respective states $\overline{\mathfrak{A}}_1, \overline{\mathfrak{A}}_2, \cdots$ be present, i.e.,

$$b_{kM_1M_2\cdots} = a_{kM_1M_2\cdots} = \delta(k - \overline{k})\cdot\delta(M_1 - \overline{M}_1)\cdot\delta(M_2 - \overline{M}_2)\cdots$$

In the sense of the differential equation above, and as a first approximation (i.e., for such short times t that the right sides can still be considered constant), only those $\frac{\partial}{\partial t} b_{kM_1M_2\cdots}$ will differ from zero for which an $M_1, M_2, \cdots, M_n + 1, \cdots$ or an $M_1, M_2, \cdots, M_n - 1, \cdots$ coincides with $\overline{M}_1, \overline{M}_2, \cdots$: i.e., all $k, \overline{M}_1, \overline{M}_2, \cdots, \overline{M}_n \pm 1, \cdots$. If we integrate in this case,

$$b_{kM_1M_2\ldots M_n+1} = w^n_{k\bar{k}} \frac{1-e^{-\frac{2\pi i}{h}(W_{\bar{k}} - W_k - h\rho_n)t}}{W_{\bar{k}} - W_k - h\rho_n} \sqrt{M_n + 1} ,$$

$$b_{kM_1M_2\ldots M_n-1\ldots} = w^n_{k\bar{k}} \frac{1-e^{-\frac{2\pi i}{h}(W_{\bar{k}} - W_k + h\rho_n)t}}{W_{\bar{k}} - W_k - h\rho_n} \sqrt{M_n}$$

All other $b_{kM_1M_2\ldots}$ are equal to zero in this approximation. (Except for $b_{\bar{k}\bar{M}_1M_2\ldots}$ this would be equal to initial value 1 in this approximation, i.e., except for t^2 terms. Yet, the conclusion $\frac{\partial}{\partial t} b_{\bar{k}M_1M_2\ldots} = 0$ becomes doubtful due to the fact that the right side of our differential equation in this case contains an infinite number of terms $b_{k\bar{M}_1\bar{M}_2\ldots\bar{M}_n\pm 1\ldots}$ which do not vanish in our approximation. Hence we may not argue from the smallness of each of these summands -- for small t -- to the smallness of their sum. Actually, the calculation of the next order approximation would show that the deviation of $b_{\bar{k}M_1M_2\ldots}$ from 1 is proportional to t and not to t^2 .[149] However, since

$$\sum_{kM_1M_2} |b_{kM_1M_2\ldots}|^2 = \sum_{kM_1M_2} |a_{kM_1M_2\ldots}|^2 = 1 ,$$

$$|b_{\bar{k}M_1M_2\ldots}|^2 = 1 - \sum_{kM_1M_2\ldots \neq \bar{k}M_1M_2\ldots} |b_{kM_1M_2\ldots}|^2$$

the direct determination of this $b_{\bar{k}M_1M_2\ldots}$ is not actually necessary.)

---

[149]The exact solution of this differential equation was given by Weisskopf and Wigner, [Z. Physik 63 (1930)], and from it the validity of these statements can be recognized.

The qualitative nature of the process is clearly recognizable here: $b_{kM_1M_2\ldots M_n+1\ldots}$, which corresponds to the emission of an $\mathfrak{N}_n$ light quantum (of frequency ρ_n), becomes larger as the denominator $W_{\overline{k}} - W_k - h\rho_n$ becomes smaller, i.e., the closer the light frequency ρ_n lies to the "Bohr frequency" $(W_{\overline{k}} - W_k)/h$;[150] in the same way, the $b_{kM_1M_2\ldots M_n-1\ldots}$ which corresponds to absorption, increases as ρ_n becomes closer to $(W_k - W_{\overline{k}})/h$. We then see that the Bohr frequency relation does not hold exactly (of course, not all frequencies are at one's disposal in the ρ_n), but with large probability only -- if the time t is short and the ρ_n are very dense (which is the case for a large cavity $\mathbf{H}$). In addition, the $w^n_{k\overline{k}}$ increase the frequency of occurrence of this process. We shall soon identify them with the transition probabilities.

From our $b_{kM_1M_2\ldots M_n\pm1\ldots}$-formulas, it follows that

$$|b_{kM_1M_2\ldots M_n+1\ldots}|^2 = \frac{2}{h^2}(M_n+1)|w^n_{k\overline{k}}|^2 \frac{1-\cos 2\pi\left(\rho_n - \frac{W_{\overline{k}} - W_k}{h}\right)t}{\left(\rho_n - \frac{W_{\overline{k}} - W_k}{h}\right)^2}$$ [151]

---

[150] N. Bohr, as is well-known, stated the fundamental principle in 1913 (cf. the reference cited in Note 5) that in transitions from a stationary state of energy $W^{(1)}$ into a stationary state of energy $W^{(2)}$, the atom emits radiation of the frequency $(W^{(1)} - W^{(2)})/h$ $(W^{(1)} > W^{(2)})$. In our case, this corresponds to the $(W_{\overline{k}} - W_k)/h$.

[151] We have

$$|e^{ix} - 1|^2 = (e^{ix} - 1)\overline{(e^{ix} - 1)} = (e^{ix} - 1)(e^{-ix} - 1)$$

$$= 2 - e^{ix} - e^{-ix} = 2 - 2\cos x = 2(1 - \cos x)$$

$$|b_{kM_1M_2\ldots M_n-1\ldots}|^2 = \frac{2}{h^2} M_n |w^n_{k\bar{k}}|^2 \frac{1-\cos 2\pi(\rho_n - \frac{W_k - W_{\bar{k}}}{h})t}{(\rho_n - \frac{W_k - W_{\bar{k}}}{h})^2}$$

and

$$|b_{kM_1M_2\ldots}|^2 = 0$$

for $kM_1M_2 \ldots \neq \bar{k}M_1M_2 \ldots$, and $\neq k\bar{M}_1M_2 \ldots M_n \pm 1 \ldots$

From this we get, for θ_k, $k \neq \bar{k}$,

$$\theta_k = \sum_{n=1}^{\infty} \frac{2}{h^2}(M_n + 1)|w^n_{k\bar{k}}|^2 \frac{1-\cos 2\pi(\rho_n - \frac{W_{\bar{k}} - W_k}{h})t}{(\rho_n - \frac{W_{\bar{k}} - W_k}{h})^2}$$

$$+ \sum_{n=1}^{\infty} \frac{2}{h^2} M_n |w^n_{k\bar{k}}|^2 \frac{1-\cos 2\pi(\rho_n - \frac{W_k - W_{\bar{k}}}{h})t}{(\rho_n - \frac{W_k - W_{\bar{k}}}{h})^2}$$

(The first $\sum_{n=1}^{\infty}$ corresponds to emission, the second $\sum_{n=1}^{\infty}$ to absorption.) In order to be able to give these θ_k in closed form, we must now make simplifying assumptions, in which on the one hand we take **H** as very large (i.e., its volume $\mathcal{V} \to \infty$), and on the other consider the characteristic oscillations $\bar{\mathfrak{A}}_n$ of **H** statistically. For this purpose, we combine all terms in each of the above summations which belong to ρ_n between ρ and $\rho + d\rho$ (for $w^n_{k\bar{k}}$ we introduce its value, and assume $d\rho \ll \rho$):

$$\frac{1}{4\pi^2c^2h\rho}\left[\sum_{\substack{n\\ \rho\leq\rho_n<\rho+d\rho}} \left|\sum_{\nu=1}^{l} \frac{e_\nu}{m_\nu}(P^x_\nu \mathfrak{A}_{n,x}(Q^x_\nu, Q^y_\nu, Q^z_\nu)+\ldots)_{k\bar{k}}\right|^2 (M_n+1)\right]$$

$$\times \frac{1-\cos 2\pi(\rho - \frac{W_{\bar{k}} - W_k}{h})t}{(\rho - \frac{W_{\bar{k}} - W_k}{h})^2}$$

We then repeat this procedure, but with $\overline{M}_n$ in place of $M_n + 1$, and $(W_k - W_{\overline{k}})/h$ in place of $(W_{\overline{k}} - W_k)/h$. The expression in brackets [...] is now to be evaluated.

The customary method of describing the $M_1, M_2, \ldots$ is not the enumeration of their values, but much less, namely, the listing of their intensities, i.e., the radiation energy $I(\rho)d\rho$ that lies in the spectrum interval from ρ to $\rho + d\rho$ and in the unit volume. This means that

$$\sum_{\substack{n \\ \rho \leq \rho_n < \rho + d\rho}} h\rho_n \cdot \overline{M}_n \approx h\rho \sum_{\substack{n \\ \rho \leq \rho_n < \rho + d\rho}} \overline{M}_n = \mathscr{V} \cdot I(\rho)d\rho ,$$

$$\sum_{\substack{n \\ \rho \leq \rho_n < \rho + d\rho}} \overline{M}_n = \frac{\mathscr{V} I \cdot (\rho) d\rho}{h\rho}$$

The number of the ρ_n in the interval $\rho \leq \rho_n < \rho + d\rho$ is

$$\frac{8\pi \mathscr{V} \rho^2}{c^2} d\rho$$

according to an asymptotic formula of Weyl (cf. the reference in Note 140) which is valid generally, and therefore,

$$\sum_{\substack{n \\ \rho \leq \rho_n < \rho + d\rho}} (\overline{M}_n + 1) \approx \frac{\mathscr{V}\left(I(\rho) + \frac{8\pi h\rho^3}{c^3}\right)}{h\rho} d\rho$$

For the [...] above, we obtain the following expression, if

$$|\sum_{\nu=1}^{l} \frac{e_\nu}{m_\nu} (P_\nu^{x} \overline{\mathscr{M}}_{n,x} (Q_\nu^{x}, Q_\nu^{y}, Q_\nu^{z}) + \ldots)_{k\overline{k}}|^2$$

executes (sufficiently rapid) fluctuations in the interval

$\rho \leq \rho_n < \rho + d\rho$ (about a mean value which may be called $w_{k\bar{k}}(\rho)$):

$$w_{k\bar{k}}(\rho) \frac{\mathcal{V}\left(I(\rho) + \frac{8\pi h\rho^3}{c^3}\right)}{h\rho} d\rho \quad \text{and} \quad w_{k\bar{k}}(\rho) \frac{\mathcal{V} I(\rho)}{h\rho} d\rho$$

If in addition we write $\nu_{\bar{k}k}$ for $(W_{\bar{k}} - W_k)$, and $\nu_{k\bar{k}}$ for $(W_{\bar{k}} - W_k)/h$, then we get for our sums

$$\theta_k = \frac{\mathcal{V}}{4\pi^2 c^2 h^2} \int_0^\infty \left\{ \left(I(\rho) + \frac{8\pi h\rho^3}{c^3} \right) \frac{1 - \cos 2\pi(\rho - \nu_{\bar{k}k})t}{(\rho - \nu_{\bar{k}k})^2} + I(\rho) \frac{1 - \cos 2\pi(\rho - \nu_{k\bar{k}})t}{(\rho - \nu_{k\bar{k}})^2} \right\} \frac{w_{k\bar{k}}(\rho)}{\rho^2} d\rho$$

For small t , this integral evidently has the order of magnitude of t^2 (because $1 - \cos 2\pi ct$ does), except in that portion of the region of integration in which the denominators $(\rho - \nu_{\bar{k}k})^2$ or $(\rho - \nu_{k\bar{k}})^2$ are small. Here, contributions can arise which are large in comparison with t^2 , and if this is the case, then these contributions are the asymptotic expression for θ_k . It will actually be shown that this is the case, because we shall obtain contributions of the order of t .

Since $\nu_{\bar{k}k} = - \nu_{k\bar{k}} = (W_{\bar{k}} - W_k)/h$, for $W_{\bar{k}} > W_k$, only the denominator of the first term is small, for $W_{\bar{k}} < W_k$, only that of the second term -- we then get for $W_{\bar{k}} >$ or $< W_k$ respectively only the first or second term respectively. Moreover, since the ρ lying further away from $\nu_{\bar{k}k}$ and $\nu_{k\bar{k}}$ -- for brevity, we call $\bar{\nu}_{k\bar{k}} = |W_{\bar{k}} - W_k|/h$ -- give only t^2 contributions to the integral, we can replace the integrand by its value for $\rho = \bar{\nu}_{k\bar{k}}$, i.e., $I w_{k\bar{k}}(\bar{\nu}_{k\bar{k}})$ in which

$$I = I(\bar{\nu}_{k\bar{k}}) + \frac{8\pi h}{c^3} \bar{\nu}_{k\bar{k}}^3$$

or $= I(\bar{\nu}_{k\bar{k}})$. Therefore

$$\theta_k = \frac{\mathcal{V} I w_{k\bar{k}}(\bar{\nu}_{k\bar{k}})}{4\pi^2 c^2 h^2 \bar{\nu}^2_{k\bar{k}}} \int_0^\infty \frac{1 - \cos 2\pi(\rho - \bar{\nu}_{k\bar{k}})t}{(\rho - \bar{\nu}_{k\bar{k}})^2} d\rho$$

Since this again leads only to a t^2 contribution, we can replace $\int_0^\infty$ by $\int_{-\infty}^\infty$ and then introduce the new variable $x = 2\pi(\rho - \bar{\nu}_{k\bar{k}})t$. Since

$$\int_{-\infty}^\infty \frac{1 - \cos 2\pi(\rho - \bar{\nu}_{k\bar{k}})t}{(\rho - \bar{\nu}_{k\bar{k}})^2} d\rho = 2\pi t \int_{-\infty}^\infty \frac{1 - \cos x}{x^2} dx = 2\pi^2 t$$ [152]

we have finally,

$$\theta_k = \frac{\mathcal{V} I w_{k\bar{k}}(\bar{\nu}_{k\bar{k}})}{2h^2 \bar{\nu}^2_{k\bar{k}}} t$$

with which it is also proved that θ_k is of the order of t .

In order to evaluate $w_{k\bar{k}}(\bar{\nu}_{k\bar{k}})$, we must find an expression for

$$\left| \sum_{\nu=1}^{1} \frac{e_\nu}{m_\nu} (P_\nu{}^x \mathfrak{A}_{n,x}(Q_\nu{}^x, Q_\nu{}^y, Q_\nu{}^z) + \cdots)_{k\bar{k}} \right|^2$$

which is free of $\mathfrak{A}_n$. This can be obtained if we replace $\mathfrak{A}_n$ (considering its rapid fluctuations) by an irregularly oriented vector of constant length -- because of its constancy in space, i.e., of its independence from the

---

[152]We obtain (cf. Courant-Hilbert, p. 49)

$$\int_{-\infty}^\infty \frac{1 - \cos x}{x^2} dx = 2\int_0^\infty \frac{1 - \cos x}{x^2} dx$$

$$= \int_0^\infty \frac{1 - \cos(2y)}{y} dy = 2\int_0^\infty \frac{\sin^2 y}{y^2} dy = \pi$$

$Q_\nu{}^x, Q_\nu{}^y, Q_\nu{}^z$, it is a number vector times the matrix 1 -- and the constant length γ_n can be obtained from the normalization

$$\iiint_{\mathbf{H}} [\overline{\mathfrak{A}}_n, \overline{\mathfrak{A}}_n] dx\, dy\, dz = 4\pi c^2$$

Therefore

$$\mathcal{V}\gamma_n^2 = 4\pi c^2 , \quad \gamma_n^2 = \frac{4\pi c^2}{\mathcal{V}}$$

On the average, 1/3 of the $[\overline{\mathfrak{A}}_n, \overline{\mathfrak{A}}_n] = \overline{\mathfrak{A}}_{n,x}{}^2 + \overline{\mathfrak{A}}_{n,y}{}^2 + \overline{\mathfrak{A}}_{n,z}{}^2 = \gamma_n^2$ contributes to the x component $\overline{\mathfrak{A}}_{n,x}{}^2$. Hence $\frac{1}{3}\gamma_n^2 = \frac{4\pi c^2}{3\mathcal{V}}$ and similarly for $\overline{\mathfrak{A}}_{n,y}{}^2$ and $\overline{\mathfrak{A}}_{n,z}{}^2$. Consequently, we have

$$w_{k\overline{k}}(\rho) = \underset{\rho \leq \rho_n < \rho + d\rho}{\text{mean}} \left| \sum_{\nu=1}^{l} \frac{e_\nu}{m_\nu} (P_\nu{}^x \overline{\mathfrak{A}}_{n,x}(Q_\nu{}^x, Q_\nu{}^y, Q_\nu{}^z) + \ldots)_{k\overline{k}} \right|^2$$

$$\approx \frac{4\pi c^2}{3\mathcal{V}} \left(\left| \left(\sum_{\nu=1}^{l} \frac{e_\nu}{m_\nu} P_\nu{}^x \right)_{k\overline{k}} \right|^2 + \cdots \right)$$

Since H_0, the energy of the system **S** alone, is equal to the kinetic energy plus the potential energy, and is therefore of the form

$$H_0 = \sum_{\nu=1}^{l} \frac{1}{2m_\nu} ((P_\nu^x)^2 + (P_\nu^y)^2 + (P_\nu^z)^2) + V(Q_1^x, Q_1^y, Q_1^z, \ldots, Q_l^x, Q_l^y, Q_l^z)$$

it is true[153] that

---

[153] $P_\nu{}^x$ commutes with all $Q_\mu{}^x, Q_\mu{}^y, Q_\mu{}^z, P_\mu{}^x, P_\mu{}^y, P_\mu{}^z$ except $Q_\nu{}^x$. In fact,

$$P_\nu{}^x Q_\nu{}^x - Q_\nu{}^x P_\nu{}^x = \frac{h}{2\pi i} 1$$

$$H_oQ_\nu^x - Q_\nu^x H_o = \frac{h}{2\pi i}\frac{1}{m_\nu} P_\nu^x$$

and since H_o is a diagonal matrix with the diagonal elements $W_1, W_2, \ldots$ $[(H_o)_{kj} = W_k\delta_{kj}]$, it follows from this for the matrix elements that

$$(P_\nu^x)_{k\bar{k}} = \frac{2\pi i m_\nu}{h}(H_oQ_\nu^x - Q_\nu^x H_o)_{k\bar{k}} = \frac{2\pi i m_\nu}{h}(W_k - W_{\bar{k}})(Q_\nu^x)_{k\bar{k}}$$

$$= \pm\, i\cdot 2\pi m_\nu \bar{\nu}_{k\bar{k}}(Q_\nu^x)_{k\bar{k}}$$

Therefore

$$w_{k\bar{k}}(\rho) = \frac{16\pi^3c^2}{3\nu h^2}\bar{\nu}^2_{k\bar{k}}\left(\left|\left(\sum_{\nu=1}^{1} e_\nu Q_\nu^x\right)_{k\bar{k}}\right|^2 + \cdots\right)$$

Substitution in the θ_k results in

$$\theta_k = \frac{8\pi^3}{3h^2}\left(\left|\left(\sum_{\nu=1}^{1} e_\nu Q_\nu^x\right)_{k\bar{k}}\right|^2 + \cdots\right) It$$

If we set

$$w_{k\bar{k}} = \left|\left(\sum_{\nu=1}^{1} e_\nu Q_\nu^x\right)_{k\bar{k}}\right|^2 + \cdots$$

this result is evidently to be interpreted as follows: The atom **S** in the kth state carries out the following transitions (quantum jumps):

1. A transition occurs into a higher state

Therefore

$$H_oQ_\nu^x - Q_\nu^x H_o = \frac{1}{2m_\nu}(P_\nu^x)^2 Q_\nu^x - Q_\nu^x \frac{1}{2m_\nu}(P_\nu^x)^2 = \frac{h}{2\pi i}\frac{1}{m_\nu}P_\nu^x$$

cf. Note 143.

$\bar{k}$ $(W_{\bar{k}} > W_k)$, $\frac{8\pi^3}{3h^2} w_{k\bar{k}} I \left(\frac{W_{\bar{k}} - W_k}{h} \right)$ times per second -- i.e., proportional to the intensity of the radiation field of the corresponding Bohr frequency $(W_{\bar{k}} - W_k)/h$.

2. A transition occurs into a lower state $\bar{k}$ $(W_{\bar{k}} < W_k)$ $\frac{8\pi^3}{3h^2} w_{k\bar{k}} I \left(\frac{W_k - W_{\bar{k}}}{h} \right)$ times per second -- i.e., proportional to the intensity of the radiation field of the corresponding Bohr frequency $(W_k - W_{\bar{k}})/h$.

3. A transition also occurs into a lower state $\bar{k}$ $(W_{\bar{k}} < W_k)$ $\frac{64\pi^4}{3hc^3} w_{k\bar{k}} \left(\frac{W_k - W_{\bar{k}}}{h} \right)^3$ times per second -- i.e., in complete independence of the radiation field present.

1. corresponds to absorption from the radiation field; 2. to emission which is induced by the radiation field; 3. however corresponds to spontaneous emission which the atom will always undergo, so long as it has not obtained complete stability in its lowest stationary state (minimum W_k) .

The three transition mechanisms 1-3 had already been found thermodynamically by Einstein before the discovery of quantum mechanics,[154] only the value of the "transition probability" $w_{k\bar{k}}$ was lacking. The value

$$w_{k\bar{k}} = \left| \left(\sum_{\nu=1}^{l} e_\nu Q_\nu^{\,x} \right)_{k\bar{k}} \right|^2 + \left| \left(\sum_{\nu=1}^{l} e_\nu Q_\nu^{\,y} \right)_{k\bar{k}} \right|^2 + \left| \left(\sum_{\nu=1}^{l} e_\nu Q_\nu^{\,z} \right)_{k\bar{k}} \right|^2$$

above is, as we have already mentioned, contained in the first interpretation contributed by Heisenberg. We have now obtained it again (by Dirac's method) from the general theory.

---

[154] Physik. Z. 18 (1917).

CHAPTER IV

DEDUCTIVE DEVELOPMENT OF THE THEORY

1. THE FUNDAMENTAL BASIS OF THE STATISTICAL THEORY

In Chapter III we succeeded in reducing all assertions of quantum mechanics to the statistical formula (called there $\mathbf{E}_2$.)

$$(\overline{\mathbf{E}}.)\ \mathrm{Exp}\ (\mathfrak{R}, \phi) = (R\phi, \phi)$$

(Exp $(\mathfrak{R}, \varphi)$ is the expectation value of the quantity $\mathfrak{R}$ in the state ϕ, R is the operator of $\mathfrak{R}$). In the following, we shall show how this formula can itself be derived from a few general qualitative assumptions, and simultaneously we shall check the entire structure of quantum mechanics, as it was developed in III. Before we do this, however, a further remark is necessary.

In the state ϕ the quantity $\mathfrak{R}$ has the expectation value $\rho = (R\phi, \phi)$ and has as its dispersion ϵ^2 the expectation value of the quantity $(\mathfrak{R} - \rho)^2$, i.e., $((R - \rho.1)^2\phi, \phi) = ||R\phi||^2 - (R\phi, \phi)^2$ (cf. Note 130; all these are calculated with the aid of $\overline{\mathbf{E}}.$!) which is in general > 0 (and $= 0$ only for $R\phi = \rho.\phi$, cf. III.3.) -- therefore there exists a statistical distribution of $\mathfrak{R}$, even though ϕ is one individual state -- as we have repeatedly noted.) But the statistical character may become even more prominent, if we do not even know what state is actually present -- for example, when several states $\phi_1, \phi_2, \ldots$ with the respective probabilities $w_1, w_2, \ldots$ $(w_1 \geq 0, w_2 \geq 0, \ldots, w_1 + w_2 + \ldots = 1)$ constitute the description. Then the expectation value of the quantity $\mathfrak{R}$, in the sense of the generally valid rules of the calculus of probabilities is $\rho' = \sum_n w_n \cdot (R\phi_n, \phi_n)$.

Now, in general $(R\phi, \phi) = \operatorname{Tr}(P_{[\phi]} \cdot R)$. Indeed, if we choose a complete orthonormal set $\psi_1, \psi_2, \ldots$, so that $\psi_1 = \phi$ (and therefore $\psi_2, \psi_3, \ldots$ are orthogonal to ϕ), then

$$P_{[\phi]}\psi_n \begin{cases} = \phi, & \text{for } n = 1 \\ = 0, & \text{otherwise} \end{cases} \Bigg| ,$$

therefore

$$\operatorname{Tr}(P_{[\phi]} \cdot R) = \sum_{m,n} (P_{[\phi]}\psi_n, \psi_m)(R\psi_m, \psi_n)$$

$$= \sum_m (\phi, \psi_m)(R\psi_m, \phi) = (R\phi, \phi).$$

Therefore our $\rho' = \operatorname{Tr}\left(\left\{ \sum_n w_n P_{[\phi_n]} \right\} \cdot R \right)$. The operator

$$U = \sum_n w_n P_{[\phi_n]}$$

is definite, because of the definiteness of all $P_{[\phi_n]}$ and $w_n \geq 0$, and its trace is equal to $\sum_n w_n = 1$ since $\operatorname{Tr} P_{[\phi_n]} = 1$. Hence it characterizes the mixture of states just described completely, with respect to its statistical properties:

$$\rho' = \operatorname{Tr}(UR).$$

We note that we shall have to pay attention to

these mixtures of states also, in addition to the individual states themselves. We turn first, however, to more general investigations.

Let us forget the whole of quantum mechanics but retain the following. Suppose a system $\mathbf{S}$[155] is given, which is characterized for the experimenter by the enumeration of all the effectively measurable quantities in it and their functional relations with one another. With each quantity we include the directions as to how it is to be measured -- and how its value is to be read or calculated from the indicator positions on the measuring instruments. If $\mathfrak{R}$ is a quantity and $f(x)$ any function, then the

---

[155]It is important to emphasize the conceptual difference between a system as such and a system in a certain state. This is an example of a system: A hydrogen atom, i.e., an electron and a proton with the known forces acting between them; it is described formally by these data: The configuration space has 6 dimensions: the coordinates are $q_1, \ldots, q_6$; the momenta are $p_1, \ldots, p_6$; the Hamiltonian function is

$$H(q_1, \ldots, q_6, p_1, \ldots, p_6) = \frac{p_1^2 + p_2^2 + p_3^2}{2m_e} + \frac{p_4^2 + p_5^2 + p_6^2}{2m_p} + \frac{e^2}{\sqrt{(q_1 - q_4)^2 + (q_2 - q_5)^2 + (q_3 - q_6)^2}}$$

A state is then determined by additional data. In classical mechanics this is done by the assigning of numerical values $q_1^0, \ldots, q_6^0, p_1^0, \ldots, p_6^0$ to the coordinates and momenta; in quantum mechanics by specifying the wave function $\phi(q_1, \ldots, q_6)$. One never needs more information than this: if both system and state are known, then the theory gives unambiguous directions for answering all questions by calculation.

quantity $f(\mathfrak{R})$ is defined as follows: To measure $f(\mathfrak{R})$, we measure $\mathfrak{R}$ and find the value a (for $\mathfrak{R}$). Then $f(\mathfrak{R})$ has the value $f(a)$. As we see, all quantities $f(\mathfrak{R})$ ($\mathfrak{R}$ fixed, $f(x)$ an arbitrary function) are measured simultaneously with $\mathfrak{R}$. This is a first example of simultaneously measurable quantities. In general, we call two (or more) quantities $\mathfrak{R}$, $\mathfrak{S}$ simultaneously measurable if there is an arrangement which measures both simultaneously in the same system -- except that their respective values are to be calculated in different ways from the readings. (In classical mechanics, as is well-known, all quantities are simultaneously measurable, but this is not the case in quantum mechanics, as we have seen in III.3.) For such quantities, and a function $f(x, y)$ of two variables, we can also define the quantity $f(\mathfrak{R}, \mathfrak{S})$. This is measured if we measure $\mathfrak{R}$, $\mathfrak{S}$ simultaneously -- if the values a, b are found for these, then the value of $f(\mathfrak{R}, \mathfrak{S})$ is $f(a, b)$. But it should be realized that it is completely meaningless to try to form $f(\mathfrak{R}, \mathfrak{S})$ if $\mathfrak{R}$, $\mathfrak{S}$ are not simultaneously measurable: there is no way of giving the corresponding measuring arrangement.

However, the investigation of the physical quantities related to a single object $\mathbf{S}$ is not the only thing which can be done -- especially if doubts exist relative to the simultaneous measurability of several quantities. In such cases it is also possible to observe great statistical ensembles which consist of many systems $\mathbf{S}_1, \ldots, \mathbf{S}_N$ (i.e., N models of $\mathbf{S}$, N large).[156] In such an ensemble

---

[156]Such ensembles, called collectives, are in general necessary for establishing probability theory as the theory of frequencies. They were introduced by R. v. Mises, who discovered their meaning for probability theory, and who built up a complete theory on this foundation (cf., for example, his book, "Wahrscheinlichkeit, Statistik and ihre Wahreheit," Berlin, 1928).

$[\mathbf{S}_1, \ldots, \mathbf{S}_N]$ we do not measure the "value" of a quantity $\mathfrak{R}$ but its distribution of values: i.e., for each interval $a' < a \leq a''$ (a', a'' given, $a' \leq a''$) the number of those among the $\mathbf{S}_1, \ldots, \mathbf{S}_N$ for which the value of $\mathfrak{R}$ lies in the interval -- dividing this number by N we obtain the probability function $w(a', a'') = w(a'') - w(a')$.[157] The essential advantages of the observation of such ensembles are these:

1. Even if the measurement of a quantity $\mathfrak{R}$ should alter the measured system $\mathbf{S}$ to an important degree (in quantum mechanics this is actually the case, and in III.4. we saw that this is necessarily so in the physics of elementary processes, since the measurement's interference with the observed system is of the same order of magnitude as the system or its observed parts), the statistical determination of the probability distribution of $\mathfrak{R}$ in the ensemble $[\mathbf{S}_1, \ldots, \mathbf{S}_N]$ will alter this ensemble arbitrarily little if N is sufficiently large.

2. Even if two (or more) quantities $\mathfrak{R}$, $\mathfrak{S}$ in a

---

[157] $w(a')$ is the probability of $a \leq a'$, i.e., it belongs to the interval $-\infty, a'$. This $w(a)$ or, as we shall call it, $w_{\mathfrak{R}}(a)$, in order to emphasize its dependence on $\mathfrak{R}$, is easily seen to have the following properties: $w_{\mathfrak{R}}(a) \to 0$ for $a \to -\infty$; $w_{\mathfrak{R}}(a) \to 1$ for $a \to +\infty$. For $a \geq a_0$, $a \to a_0$, $w_{\mathfrak{R}}(a) \to w_{\mathfrak{R}}(a_0)$; $a' \leq a''$, $w_{\mathfrak{R}}(a') \leq w_{\mathfrak{R}}(a'')$. (In quantum mechanics, if $E(\lambda)$ is the resolution of the identity belonging to R, $w_{\mathfrak{R}}(a) = \|E(a)\phi\|^2 = (E(a)\phi, \phi)$.)

If $w_{\mathfrak{R}}(a)$ is differentiable, then the ordinary "probability density" $\frac{d}{da} w_{\mathfrak{R}}(a)$ can be introduced in its place; if it is not continuous at $a = a_0$ (from the left), then the single point $a = a_0$ has the "discrete probability" $w_{\mathfrak{R}}(a_0) - w_{\mathfrak{R}}(a_0 - 0)$. But the general concept which is valid under all conditions is $w_{\mathfrak{R}}(a)$; cf. the reference in Note 156.

single system S are not simultaneously measurable, their probability distributions in a given ensemble $[\mathbf{S}_1, \ldots, \mathbf{S}_N]$ can be obtained with arbitrary accuracy if N is sufficiently large.

Indeed, with an ensemble of N elements it suffices to carry out the statistical inspections, relative to the distribution of values of the quantity $\mathfrak{R}$, not on all N elements $\mathbf{S}_1, \ldots, \mathbf{S}_N$, but on any subset of M ($\leq N$) elements, say $[\mathbf{S}_1, \ldots, \mathbf{S}_M]$ -- provided that M, N are both large, and that M is very small compared to N.[158] Then only the M/N -th part of the ensemble is affected by the changes which result from the measurement. The effect is an arbitrarily small one if M/N is chosen small enough -- which is possible for sufficiently large N, even in the case of large M, as was stated in **1**. In order to measure two (or several) quantities $\mathfrak{R}$, $\mathfrak{S}$ simultaneously, we need two sub-ensembles, say $[\mathbf{S}_1, \ldots, \mathbf{S}_M]$ and $[\mathbf{S}_{M+1}, \ldots, \mathbf{S}_{2M}]$ ($2M \leq N$), of such a type that the first is employed in obtaining the statistics of $\mathfrak{R}$, and the second in obtaining those of $\mathfrak{S}$. The two measurements therefore do not disturb each other, although they are performed in the same ensemble $[\mathbf{S}_1, \ldots, \mathbf{S}_N]$ and they can change this ensemble only by an arbitrarily small amount, if 2M/N is sufficiently small -- which is possible for sufficiently large N even in the case of large M, as was stated in **2**.

We see that the introduction of the statistical ensembles, i.e., of probability methods, is undertaken because of the possibility of affecting a single system by measurement, and by the possibility of non-simultaneous measurability of several quantities. A general theory must consider these circumstances, since their appearance in elementary processes was always to be suspected,[159] and

---

[158]This follows from the so-called law of large numbers, the theorem of Bernoulli.

[159]So, for example, it was considered to constitute a basic

since their presence has now become a certainty, as the exact discussion of the situation shows (cf. III.4.). The statistical ensembles eliminate these difficulties, and again make possible an objective description [which is independent of chance, as well as of whether one measures (in a given state) the one or the other of two not simultaneously measurable quantities].

For such ensembles, it is not surprising that a physical quantity $\mathfrak{R}$ does not have a sharp value, i.e., that its distribution function does not consist of a single value a_0,[160] but that several values or intervals of values are possible, and that a positive dispersion exists.[160] However, two different reasons for this

---

difficulty in defining the electric field, that the electrical test charge to be used cannot be smaller than an electron.

[160]The sharp value a_0 corresponds to the probability function $w_{\mathfrak{R}}(a)$ with

$$w_{\mathfrak{R}}(a) \begin{cases} = 1 , & \text{for } a \geq a_0 \\ = 0 , & \text{for } a < a_0 \end{cases}$$

In this case, and only in this, the dispersion ϵ^2 is zero. The mean value ρ and the dispersion ϵ^2 are in general calculated as follows (Stieltjes integrals!):

$$\rho = \int_{-\infty}^{+\infty} a dw_{\mathfrak{R}}(a) ,$$

$$\epsilon^2 = \int_{-\infty}^{+\infty} (a - \rho)^2 dw_{\mathfrak{R}}(a)$$

$$= \int_{-\infty}^{+\infty} a^2 dw_{\mathfrak{R}}(a) - 2\rho \int_{-\infty}^{+\infty} a dw_{\mathfrak{R}}(a) + \rho^2 = \int_{-\infty}^{+\infty} a^2 dw_{\mathfrak{R}}(a) - \rho^2$$

behavior are a priori conceivable:

I. The individual systems $\mathbf{S}_1, \ldots, \mathbf{S}_N$ of our ensemble can be in different states, so that the ensemble $[\mathbf{S}_1, \ldots, \mathbf{S}_N]$ is defined by their relative frequencies. The fact that we do not obtain sharp values for the physical quantities in this case is caused by our lack of information: we do not know in which state we are measuring, and therefore we cannot predict the results.

II. All individual systems $\mathbf{S}_1, \ldots, \mathbf{S}_N$ are in the same state, but the laws of nature are not causal. Then the cause of the dispersions is not our lack of information, but is nature itself, which has disregarded the "principle of sufficient cause."

Case **I** is generally well-known, while Case **II** is important and new. To be sure we shall be sceptical at first of the possibility of its existence, but we shall find an objective criterion which allows us to distinguish between its appearance or non-appearance. It appears at first that serious objections can be raised against its conceivability and its meaningfulness, but we believe that these objections are not valid, and that certain difficulties (for example, in quantum mechanics) permit no other way out but **II**. We therefore apply ourselves to the discussion of the conceptual difficulties of **II**.

One might object against **II** that nature cannot violate the "principle of sufficient cause," i.e., causality, at all, because this is merely a definition of identity. That is, the theorem that two identical objects $\mathbf{S}_1$, $\mathbf{S}_2$ -- i.e., two replicas of the system $\mathbf{S}$ which are in the same state -- will remain identical in all conceivable interactions is true because it is tautological.

$$= \int_{-\infty}^{+\infty} a^2 dw_{\mathfrak{R}}(a) - \left(\int_{-\infty}^{+\infty} a dw_{\mathfrak{R}}(a) \right)^2$$

(cf. with III.4., Note 130).

For if $\mathbf{S}_1$, $\mathbf{S}_2$ could react differently to the same intervention in their interaction (i.e., if they gave different values in the measurement of a quantity $\mathfrak{R}$), then we would not have called them identical. Therefore, in an ensemble [$\mathbf{S}_1, \ldots, \mathbf{S}_N$] which has dispersion relative to a quantity $\mathfrak{R}$, the individual systems $\mathbf{S}_1, \ldots, \mathbf{S}_N$ cannot (by definition) all be in the same state. (The application to quantum mechanics would be: Since one will obtain different values in the measurement of the same quantity $\mathfrak{R}$ in several systems, which all are in the state with the wave function ϕ -- if ϕ is not an eigenfunction of the operator R of $\mathfrak{R}$ [161] -- therefore these systems are not equal to one another -- i.e., the description by the wave function is not complete. Therefore other variables must exist, the "hidden parameters" mentioned in III.2. We shall soon see what difficulties this presents.) In a large statistical ensemble therefore, as long as any quantity $\mathfrak{R}$ possesses a dispersion in it, the possibility must exist of resolving it into several differently constituted parts (according to the various states of its elements). This is all the more plausible, since ordinarily a simple method of such a resolution seems to exist: namely, we can resolve it according to the various values which $\mathfrak{R}$ has in the ensemble. After a subdivision or resolution relative to all quantities $\mathfrak{R}$, $\mathfrak{S}$, $\mathfrak{T}, \ldots$ which are present has been carried out, a truly homogeneous ensemble would then be obtained. At the end of the process, these quantities would have no further dispersion in any of the sub-ensembles.

First of all, the statements contained in the last sentences are incorrect because we did not consider the fact that measurement changes the measured system. If

---

[161] This is a case of independent measurements in several systems: successive measurements on the same system would always give the same value again (cf. III.3.).

we measure $\mathfrak{R}$ (which for simplicity may have only two values a_1, a_2) on all objects and get perhaps a_1 on $\mathbf{S}'_1, \ldots, \mathbf{S}'_{N_1}$ and a_2 on $\mathbf{S}''_1, \ldots, \mathbf{S}''_{N-N_1}$, then there is no dispersion in $[\mathbf{S}'_1, \ldots, \mathbf{S}'_N]$ nor in $[\mathbf{S}''_1, \ldots, \mathbf{S}''_{N-N_1}]$ ($\mathfrak{R}$ always has the value a_1 or a_2 respectively). Still this is not merely a resolution of $[\mathbf{S}_1, \ldots, \mathbf{S}_N]$ into the two sets mentioned, because the individual systems would be changed by the $\mathfrak{R}$ measurement. It is true that by **1.** we have a method to determine distribution of values of $\mathfrak{R}$ in such a way that $[\mathbf{S}_1, \ldots, \mathbf{S}_N]$ is changed only slightly (we measure only in $\mathbf{S}_1, \ldots, \mathbf{S}_M$, M large, M/N small). This procedure, however, does lead to the desired resolution, since, for most of the $\mathbf{S}_1, \ldots, \mathbf{S}_N$ (namely $\mathbf{S}_{M+1}, \ldots, \mathbf{S}_N$), it does not make certain what value $\mathfrak{R}$ has in each one of them. And now we see that the method given above fails to produce completely homogeneous ensembles. For let us measure a second quantity $\mathfrak{S}$ (which may also have only two values b_1, b_2) in $[\mathbf{S}'_1, \ldots, \mathbf{S}'_{N_1}]$ and $[\mathbf{S}''_1, \ldots, \mathbf{S}''_{N-N_1}]$. Let the value b_1 be found in $\mathbf{S}'''_1, \ldots, \mathbf{S}'''_{N_{11}}$ and $\mathbf{S}^{V}_1, \ldots, \mathbf{S}^{V}_{N_{12}}$ and b_2 in $\mathbf{S}^{IV}_1, \ldots, \mathbf{S}^{IV}_{N_1-N_{11}}$ and $\mathbf{S}^{VI}_1, \ldots, \mathbf{S}^{VI}_{N-N_1-N_{12}}$. Then $\mathfrak{S}$ has no dispersion in $[\mathbf{S}'''_1, \ldots, \mathbf{S}'''_{N_{11}}]$, $[\mathbf{S}^{IV}_1, \ldots, \mathbf{S}^{IV}_{N_1-N_{11}}]$, $[\mathbf{S}^{V}_1, \ldots, \mathbf{S}^{V}_{N_{12}}]$, $[\mathbf{S}^{VI}_1, \ldots, {}^{VI}_{N-N_1-N_{12}}]$ (its value is constant, b_1, b_2, b_1 or b_2 respectively). But although the first two ensembles are parts of $[\mathbf{S}'_1, \ldots, \mathbf{S}'_{N_1}]$, and the latter two are parts of $[\mathbf{S}''_1, \ldots, \mathbf{S}''_{N-N_1}]$ in which $\mathfrak{R}$ did not have a dispersion, $\mathfrak{R}$ can have a dispersion in each one of them, because the $\mathfrak{S}$ measurement has changed the individual systems (of which they consist)! That is, we do not get ahead: Each step destroys the results of the preceding one,[162] and no

---

[162] One should consider, for example, what happens if we

further repetition of successive measurements can bring order into this confusion. In the atom we are at the boundary of the physical world, where each measurement is an interference of the same order of magnitude as the object measured, and therefore affects it basically. Thus the uncertainty relations are at the root of these difficulties.

Therefore we have no method which would make it always possible to resolve further the dispersing ensembles (without a change of their elements) or to penetrate to those homogeneous ensembles which no longer have dispersion. The last ones are the ensembles we are accustomed to consider to be composed of individual particles, all identical, and all determined causally. Nevertheless, we could attempt to maintain the fiction that each dispersing ensemble can be divided into two (or more) parts, different from each other and from it, without a change in its elements. That is, the division would be such that the superposition of two resolved ensembles would again produce the original ensemble. As we see, the attempt to interpret causality as an equality definition led to a question of fact which can and must be answered, and which might conceivably be answered negatively. This is the question: is it really possible to represent each ensemble $[\mathbf{S}_1, \dots, \mathbf{S}_N]$, in which there is a quantity $\mathfrak{R}$ with dispersion, by the superposition of two (or more) ensembles different from one another and from it? (More than two, say $n = 3, 4, \dots$ can be reduced to two, if we consider the first and the super-

---

substitute the not simultaneously measurable (because of the uncertainty relations) quantities q (cartesian coordinates) and p (momentum) for $\mathfrak{R}$, $\mathfrak{S}$. If q has a very small dispersion in an ensemble, then the p measurement with the accuracy (i.e., dispersion) ϵ sets up a q dispersion of at least $h/4\pi\epsilon$ (cf. III.4.), i.e., everything is ruined.

position of the $n - 1$ others.)

If $[\mathbf{S}_1, \ldots, \mathbf{S}_N]$ were the mixture (sum) of $[\mathbf{S}'_1, \ldots, \mathbf{S}'_P]$ and $[\mathbf{S}''_1, \ldots, \mathbf{S}''_Q]$, the probability function $w_{\mathfrak{R}}(a)$ (cf. Note 157) for each quantity $\mathfrak{R}$ could be expressed with the aid of the probability functions $w'_{\mathfrak{R}}(a)$, $w''_{\mathfrak{R}}(a)$ of the two sub-ensembles,

$$(\mathbf{M}_1)\; w_{\mathfrak{R}}(a) = \alpha w'_{\mathfrak{R}}(a) + \beta w''_{\mathfrak{R}}(a)\;,\; \alpha > 0,\; \beta > 0,\; \alpha + \beta = 1$$

Here $\alpha = P/N$, $\beta = Q/N$ $(N = P + Q)$ are independent of $\mathfrak{R}$ Fundamentally, this is a purely mathematical problem: If in an ensemble with the probability functions $w_{\mathfrak{R}}(a)$ there exist quantities $\mathfrak{R}$ with dispersion (which property of $w_{\mathfrak{R}}(a)$ this is, is given in Note 160), are there two other ensembles with the probability functions $w'_{\mathfrak{R}}(a)$ and $w''_{\mathfrak{R}}(a)$ respectively, such that for all $\mathfrak{R}$ $\mathbf{M}_1$. holds? This can also be formulated in a somewhat different way if we characterize an ensemble not by the probability functions $w_{\mathfrak{R}}(a)$ of the quantity $\mathfrak{R}$ but by their expectation values

$$\text{Exp}\;(\mathfrak{R}) = \int_{-\infty}^{\infty} a\,dw_{\mathfrak{R}}(a)$$

Then our question is the following: An ensemble is dispersion free if in it, for each $\mathfrak{R}$,

$$\text{Exp}\;([\mathfrak{R} - \text{Exp}\;(\mathfrak{R})]^2) = \text{Exp}\;(\mathfrak{R}^2) - [\text{Exp}\;(\mathfrak{R})]^2$$

is equal to zero (see Note 160), i.e.,

$$(\mathbf{Dis}_1.)\quad \text{Exp}\;(\mathfrak{R}^2) = [\text{Exp}\;(\mathfrak{R})]^2$$

If this is not the case, is it always possible to find two other ensembles with

$$\text{Exp}'(\mathfrak{R}),\; \text{Exp}''(\mathfrak{R})$$

$$(\text{Exp}\;(\mathfrak{R}) \not\equiv \text{Exp}'(\mathfrak{R}) \not\equiv \text{Exp}''(\mathfrak{R}))$$

such that

$$(\mathbf{M_2.})\quad \text{Exp}(\mathfrak{R}) = \alpha\,\text{Exp}'(\mathfrak{R}) + \beta\,\text{Exp}''(\mathfrak{R}),\quad \alpha > 0,\ \beta > 0,\ \alpha + \beta = 1$$

always holds (α, β -independent of $\mathfrak{R}$)? [It should be noted that for a single quantity $\mathfrak{R}$, the number Exp ($\mathfrak{R}$) is not a substitute for the function $w_{\mathfrak{R}}(a)$; on the other hand, the knowledge of all Exp ($\mathfrak{R}$) is equivalent to the knowledge of all $w_{\mathfrak{R}}(a)$. Indeed, if $f_a(x)$ is defined by

$$f_a(x) \begin{cases} = 1, & \text{for } x \leq a \\ = 0, & \text{for } x > a \end{cases}$$

then

$$w_{\mathfrak{R}}(a) = \text{Exp}\,(f_a(\mathfrak{R}))\ ].$$

To handle this question mathematically, it is preferable not to consider the ensembles $[\mathbf{S}_1, \ldots, \mathbf{S}_N]$ themselves, but rather the corresponding Exp ($\mathfrak{R}$). To each ensemble there belongs one such function which is defined for all physical quantities $\mathfrak{R}$ in $\mathbf{S}$, and which takes on real numbers as values, and which, conversely, completely characterizes the ensemble in all its statistical properties. (Cf. the discussion above on the relation between Exp ($\mathfrak{R}$) and $w_{\mathfrak{R}}(a)$.) Of course we must still find out which properties an $\mathfrak{R}$ function must possess in order that it be the Exp ($\mathfrak{R}$) of a suitable ensemble. But as soon as we have done this, we can define:

α) An $\mathfrak{R}$ function, which is an Exp ($\mathfrak{R}$), is said to be dispersion free if it satisfies the condition $\mathbf{Dis_1}$.

β) An $\mathfrak{R}$ function, which is an Exp ($\mathfrak{R}$), is said to be homogeneous or pure if, for it, $\mathbf{M_2}$. implies that

$$\text{Exp}(\mathfrak{R}) \equiv \text{Exp}'(\mathfrak{R}) \equiv \text{Exp}''(\mathfrak{R}).$$

It is conceptually plausible that each dispersion free Exp ($\mathfrak{R}$)-function should be pure, and we shall soon prove it. But our question at the moment is the converse one: Is each pure Exp ($\mathfrak{R}$)-function dispersion free?

It is evident that each Exp ($\mathfrak{R}$)-function must possess the following properties:

A. If the quantity $\mathfrak{R}$ is identically 1 (i.e., if the "directions for measurement" are: no measurement is necessary, because $\mathfrak{R}$ always has the value 1), then Exp ($\mathfrak{R}$) = 1 .

B. For each $\mathfrak{R}$ and each real number a , Exp (a $\mathfrak{R}$) = a Exp ($\mathfrak{R}$) .[163]

C. If the quantity $\mathfrak{R}$ is by nature non-negative, if, for example, it is the square of another quantity $\mathfrak{S}$,[163] then also Exp ($\mathfrak{R}$) ≥ 0 .

D. If the quantities $\mathfrak{R}, \mathfrak{S}, \ldots$ are simultaneously measurable, then Exp ($\mathfrak{R} + \mathfrak{S} + \ldots$) Exp ($\mathfrak{R}$) + Exp ($\mathfrak{S}$) + ...[163] (For non-simultaneously measurable quantities $\mathfrak{R}, \mathfrak{S}, \ldots, \mathfrak{R} + \mathfrak{S} + \ldots$ is undefined, cf. supra.)

All this follows immediately from the definitions of the quantities under consideration (i.e., the directions for their measurement), and from the definition of the expectation value as the arithmetic mean of all results of measurement in a sufficiently large statistical ensemble. Regarding **D**. it should be noted that its correctness depends on this theorem on probability: the expectation value of a sum is always the sum of the expectation values of the individual terms, independent of whether probability dependencies exist between these or not (in contrast, for example, to the probability of the product). That we have formulated it only for simultaneously measurable $\mathfrak{R}, \mathfrak{S}, \ldots$

---

[163] $a\mathfrak{R}$, $\mathfrak{S}^2$, $\mathfrak{R} + \mathfrak{S} + \ldots$ means that we may substitute the quantities $\mathfrak{R}, \mathfrak{S}, \mathfrak{R}, \mathfrak{S}, \ldots$ in the functions $f(x) = ax$, $f(x) = x^2$, and $f(x,y,\ldots) = x + y + \ldots$, respectively, in the sense of the definitions given above.

is natural, since otherwise $\mathfrak{R} + \mathfrak{S} + \dots$ is meaningless.

But the algorithm of quantum mechanics contains still another operation, which goes beyond the one just discussed: namely, the addition of two arbitrary quantities, which are not necessarily simultaneously observable. This operation depends on the fact that for two Hermitian operators, R, S, the sum R + S is also an Hermitian operator, even if the R, S do not commute, while, for example, the product RS is again Hermitian only in the event of commutativity (cf. II.5.). In each state ϕ the expectation values behave additively: $(R\phi, \phi) + (S\phi, \phi) = ((R + S)\phi, \phi)$ (cf. $\mathbf{E}_2$., III.1.). The same holds for several summands. We now incorporate this fact into our general set-up (at this point not yet specialized to quantum mechanics):

E. If $\mathfrak{R}, \mathfrak{S}, \dots$ are arbitrary quantities, then there is an additional quantity $\mathfrak{R} + \mathfrak{S} + \dots$ (which does not depend on the choice of the Exp ($\mathfrak{R}$)-function), such that $\mathrm{Exp}\,(\mathfrak{R} + \mathfrak{S} + \dots) = \mathrm{Exp}\,(\mathfrak{R}) + \mathrm{Exp}\,(\mathfrak{S}) + \dots$.

If $\mathfrak{R}, \mathfrak{S}$ are simultaneously measurable, this must be the ordinary sum (by **D.**). But in general the sum is characterized by **E.** only in an implicit way, and it shows no way to construct from the measurement directions for $\mathfrak{R}, \mathfrak{S}, \dots$ such directions for $\mathfrak{R} + \mathfrak{S} + \dots$.[164]

---

[164]For example, the energy operator of the Heisenberg theory of an electron moving in a potential field V(x, y, z),

$$H_0 = \frac{(P^x)^2 + (P^y)^2 + (P^z)^2}{2m} + V(Q^x, Q^y, Q^z)$$

(cf. for example III.6.), is a sum of two non-commuting operators

$$R = \frac{(P^x)^2 + (P^y)^2 + (P^z)^2}{2m}, \quad S = V(Q^x, Q^y, Q^z).$$

In addition, it must be remarked that we shall admit not only Exp ($\mathfrak{R}$)- functions representing expectation values, but also functions which correspond to relative values -- i.e., we allow the normalization condition **A.** to be dropped. If Exp (1) (which is ≥ 0 by **C.**) is finite and $\neq 0$, this is unimportant, since for Exp ($\mathfrak{R}$)/Exp (1) everything is as before. But Exp (1) $= \infty$ is an entirely different possibility, and it is actually for its sake that we want this extension. It is best illustrated by a simple example. The fact is, that there are cases in which it is better to operate with relative probabilities in place of the true probabilities -- specifically the cases with an infinite total relative probability [Exp (1) corresponds to the total probability]; the following is such an example: Let the observed system be a particle in one dimension, and let its statistical distribution be of such a kind that it lies with equal probability anywhere in the infinite interval. Then each finite interval on this line has the probability zero, but the equal probability of all places is not expressed in this way, but rather by the fact that two finite intervals have as their probability ratio the quotient of their lengths. Since 0/0 has no meaning, this can be expressed only if we introduce their lengths as relative probabili-

---

While the measurement of the quantity $\mathfrak{R}$ belonging to R is a momentum measurement, and that of the quantity $\mathfrak{S}$ belonging to S a coordinate measurement, we measure the quantity $\mathfrak{R} + \mathfrak{S}$ belonging to $H_0 = R + S$ in an entirely different way: for example, by the measurement of the frequency of the spectral lines emitted by this (bound) electron, since these lines determine (by reason of the Bohr frequency relation) the energy levels, i.e., the $\mathfrak{R} + \mathfrak{S}$ values. Nevertheless, under all circumstances,

$$\text{Exp}\ (\mathfrak{R} + \mathfrak{S}) = \text{Exp}\ (\mathfrak{R}) + \text{Exp}\ (\mathfrak{S}).$$

ties -- the relative total probability will then of course be ∞ .

Upon consideration of the foregoing, we arrive at the following form for our conditions (**A'.** corresponds to **C.**, **B'.** corresponds to **B.**, **D.**, **E.**):

A'. If the quantity $\mathfrak{R}$ is by nature non-negative, for example, if it is the square of another quantity $\mathfrak{S}$, then Exp ($\mathfrak{R}$) ≥ 0 .

B'. If $\mathfrak{R}, \mathfrak{S}, \ldots$ are arbitrary quantities and $a, b, \ldots$ real numbers, then Exp ($a\mathfrak{R} + b\mathfrak{S} + \ldots$) = a Exp ($\mathfrak{R}$)+b Exp ($\mathfrak{S}$) + ...

We emphasize:

1. Since we have considered the relative expectation values, the functions Exp ($\mathfrak{R}$) and c Exp ($\mathfrak{R}$) (c a constant, > 0) are not essentially different from each other.

2. Exp ($\mathfrak{R}$) $\equiv 0$ (for all) furnishes no information, and therefore this function is excluded.

3. Absolute, i.e., correctly normalized, expectation values exist if Exp (1) = 1 . Exp (1) is in any case ≥ 0 , by **A'.**, and if it is finite and $\neq 0$, then **1.** (with $c = 1/\text{Exp}\,(1)$) leads to the correctly normalized value. For Exp (1) = 0 implies ., as we shall show, and this case is therefore eliminated also; for Exp (1) = ∞ , however, an essentially non-normalized, (i.e., relative) statistik exists.

We must still return to our definitions **α**), **β**). By ., **M_2.** can be replaced by the following simpler condition:

$$\mathbf{M_3}.\quad \text{Exp}\,(\mathfrak{R}) \equiv \text{Exp}'(\mathfrak{R}) + \text{Exp}''(\mathfrak{R})$$

And in the case of **Dis_1** ., it is to be observed that the calculation there presupposes Exp (1) = 1 . For Exp (1) = ∞ the dispersion free character cannot be defined, since it means Exp $((\mathfrak{R} - \rho)^2) = 0$, where ρ is the absolute expectation value of $\mathfrak{R}$, i.e.,

Exp ($\mathfrak{R}$)/Exp (1) , which is in this case ∞/∞ and is therefore meaningless.[165] Therefore **α**), **β**) now runs as follows:

α') An $\mathfrak{R}$ function which is an Exp ($\mathfrak{R}$) is said to be dispersion free if Exp (1) $\neq 0$ and is finite, so that we can assume Exp (1) = 1 by **1**. Then **Dis$_1$**. is characteristic.

β') An $\mathfrak{R}$ function which is an Exp ($\mathfrak{R}$) is said to be homogeneous or pure if for it **M$_3$**. has

$$\text{Exp}'(\mathfrak{R}) = c' \text{ Exp}(\mathfrak{R}), \quad \text{Exp}''(\mathfrak{R}) = c'' \text{ Exp}(\mathfrak{R})$$

as a consequence (c', c'' constants, $c' + c'' = 1$ and, because of **A'**. and **1**., **2**., also $c' > 0$, $c'' > 0$) .

By reason of **A'**., **B'**. and **α'**), **β'**), we are now in the position to make a decision on the question of causality, as soon as we know the physical quantities in **S** , as well as the functional relations existing among them. In the following section this will be carried out for the relations of quantum mechanics.

As a conclusion to this section, two remarks should be added.

First, one which concerns the case Exp (1) = 0 . It follows from **B'**. that Exp (c) = 0 , therefore if a quantity $\mathfrak{R}$ is always $\geq c'$, $\leq c''$, then by **A"**., Exp ($c'' - \mathfrak{R}$) ≥ 0, Exp ($\mathfrak{R} - c'$) ≥ 0, and hence by **B'**., Exp (c') $\leq$ Exp ($\mathfrak{R}$) $\leq$ Exp (c'') , i.e., Exp ($\mathfrak{R}$) = 0 . Now let $\mathfrak{R}$ be arbitrary, $f_1(x), f_2(x), \ldots$ a sequence of bounded functions with

$$f_1(x) + f_2(x) + \cdots = x$$

(for example,

---

[165]For dispersion-free ensembles, however, there is no reason for not introducing the correct expectation values.

$$f_1(x) = \frac{\sin x}{x}, \quad f_n(x) = \frac{\sin x n}{x n} - \frac{\sin (n-1)x}{(n-1)x}$$

for $n = 2,3,\ldots$). Then Exp $(f_n(\mathfrak{R})) = 0$ for $n = 1,2,\ldots$, and therefore by **B'.**, Exp $(\mathfrak{R}) = 0$ also. Consequently, Exp $(1) = 0$ is excluded by **2.**, according to the proposition stated previously.

Second, it is remarkable that by **Dis₁.**, Exp $(\mathfrak{R}^2) = [\text{Exp}(\mathfrak{R})]^2$ is characteristic for the dispersion free case, although in this case

$$(\textbf{Dis}_2.)\quad \text{Exp}\,(f(\mathfrak{R})) = f(\text{Exp}\,(\mathfrak{R}))$$

must hold for each function $f(x)$, since Exp $(\mathfrak{R})$ is simply the value of $\mathfrak{R}$, and Exp $(f(\mathfrak{R}))$ the value of $f(\mathfrak{R})$. **Dis_1.** is a special case of **Dis_2.**: $f(x) = x^2$, but how is it that this suffices? The answer is the following: if **Dis_2.** holds for $f(x) = x^2$, it holds of itself for all $f(x)$. One would even be able to replace x^2 by any other continuous and convex function of x (i.e., one for which $f(\frac{x+y}{2}) < \frac{f(x)+f(y)}{2}$ for all $x \neq y$). We do not enter into the proof.

2. PROOF OF THE STATISTICAL FORMULAS

There corresponds to each physical quantity of a quantum mechanical system, a unique hypermaximal Hermitian operator, as we know (cf., for example, the discussion in III.5.), and it is convenient to assume that this correspondence is one-to-one -- that is, that actually each hypermaximal operator corresponds to a physical quantity. (We also made occasional use of this in III.3.) In such a case the following rules are valid (cf. **F.**, **L.** in III.5., as well as the discussion at the end of IV.1.):

I. If the quantity $\mathfrak{R}$ has the operator R, then the quantity $f(\mathfrak{R})$ has the operator $f(R)$.

II. If the quantities $\mathfrak{R}, \mathfrak{S}, \ldots$ have the operators $R, S, \ldots$, then the quantity $\mathfrak{R} + \mathfrak{S} + \ldots$ has the operator $R + S + \ldots$. (The simultaneous measurability of $\mathfrak{R}, \mathfrak{S}, \ldots$ is not assumed, cf. the discussion on this point above.)

A'., **B'.**, **α')**, **β')** and **I.**, **II.** form the mathematical basis of our analysis.

Let $\phi_1, \phi_2, \ldots$ be a complete orthonormal set. In place of each operator R let us consider the matrix $a_{\mu\nu} = (R\phi_\mu, \phi_\nu)$. We form the Hermitian operators with the respective matrices

$$e^{(n)}_{\mu\nu} = \left\{\begin{array}{l} 1 \text{, for } \mu=\nu=n \\ 0 \text{, otherwise} \end{array}\right| ,$$

$$f^{(mn)}_{\mu\nu} = \left\{\begin{array}{l} 1 \text{, for } \mu=m,\ \nu=n \\ 1 \text{, for } \mu=n,\ \nu=m \\ 0 \text{, otherwise} \end{array}\right| , \quad g^{(mn)}_{\mu\nu} = \left\{\begin{array}{l} i \text{, for } \mu=m,\ \nu=n \\ -i \text{, for } \mu=n,\ \nu=m \\ 0 \text{, otherwise} \end{array}\right|$$

these operators are

$$P_{[\phi_n]},\ P_{\left[\frac{\phi_m+\phi_n}{\sqrt{2}}\right]} - P_{\left[\frac{\phi_m-\phi_n}{\sqrt{2}}\right]},\ P_{\left[\frac{\phi_m+i\phi_n}{\sqrt{2}}\right]} - P_{\left[\frac{\phi_m-i\phi_n}{\sqrt{2}}\right]},$$

let the corresponding quantities be $\mathfrak{U}^{(n)}, \mathfrak{V}^{(mn)}, \mathfrak{W}^{(mn)}$. Evidently (because $a_{nm} = \overline{a}_{mn}$)

$$a_{\mu\nu} = \sum_n a_{nn} e^{(n)}_{\mu\nu} + \sum_{\substack{m,n \\ m<n}} \operatorname{Re} a_{mn} f^{(mn)}_{\mu\nu} + \sum_{\substack{m,n \\ m<n}} \operatorname{Im} a_{mn} g^{(mn)}_{\mu\nu},$$

therefore

$$\mathfrak{R} = \sum_n a_{nn} \mathfrak{U}^{(n)} + \sum_{\substack{m,n \\ m<n}} \operatorname{Re} a_{mn} \mathfrak{V}^{(mn)} + \sum_{\substack{m,n \\ m<n}} \operatorname{Im} a_{mn} \mathfrak{W}^{(mn)},$$

and because of **II.** and **B'.**,

$$\text{Exp}\,(\,\mathfrak{R}\,) = \sum_{n} a_{nn}\,\text{Exp}\,(\mathfrak{U}^{(n)}) + \sum_{\substack{m,n\\ m<n}} \text{Re}\; a_{mn}\,\text{Exp}\,(\;{}^{(mn)})\;.$$

$$+ \sum_{\substack{m,n\\ m<n}} \text{Im}\; a_{mn}\,\text{Exp}\,(\mathfrak{W}^{(mn)})\,.$$

Therefore, if we set

$$\mu_{nn} = \text{Exp}\,(\mathfrak{U}^{(n)})$$

$$\left.\begin{aligned} \mu_{mn} &= \tfrac{1}{2}\,\text{Exp}\,(\mathfrak{V}^{(mn)}) + \tfrac{i}{2}\,\text{Exp}\,(\mathfrak{W}^{(mn)}) \\ \mu_{nm} &= \tfrac{1}{2}\,\text{Exp}\,(\mathfrak{V}^{(mn)}) - \tfrac{i}{2}\,\text{Exp}\,(\mathfrak{W}^{(mn)}) \end{aligned}\right\}\ (m < n)$$

then

$$\text{Exp}\,(\,\mathfrak{R}\,) = \sum_{m,n} \mu_{nm} a_{mn}$$

Since $\mu_{nm} = \mu_{nm}$ we can define a Hermitian operator U by $(U\phi_m, \phi_n) = \mu_{mn}$,[166] and the right side is the $\text{Tr}\,(UR)$

---

[166]That is,

$$U\phi_m = \sum_{n} \mu_{mn}\phi_n$$

where the finiteness of $\sum_{n} |\mu_{mn}|^2$ is of course necessary. This can be established in the following way: If $\sum_{n} |x_n|^2 = 1$, then $R = P_{[\phi]}$ has the matrix $\overline{x}_\mu x_\nu$ for $\phi = \sum_{n} x_n\phi_n$, and its $\mathfrak{R}$ has the expectation value $\sum_{m,n} \mu_{nm}\overline{x}_m x_n$. Because of $P_{[\phi]} = P^2_{[\phi]}$, $1 - P_{[\phi]} = (1 - P_{[\phi]})^2$, this is ≥ 0, $\leq \text{Exp}\,(1)$, therefore ≥ 0, ≤ 1

(cf. II.11.). Therefore we obtain the formula

$$\text{(Tr.)} \qquad \text{Exp}\ (\mathfrak{R}) = \text{Tr}\ (UR)$$

U is a Hermitian operator[167] independent of R , and

---

at least for normalized Exp ($\mathfrak{R}$) . If $x_{N+1} = x_{N+2} = \ldots = 0$, this means that the N-dimensional Hermitian form

$$\sum_{m,n=1}^{N} \mu_{nm}\bar{x}_m x_n$$

has values $\geq 0, \leq 1$ for

$$\sum_{n=1}^{N} |x_n|^2 = 1$$

i.e., the eigenvalues of the matrix $\mu_{\rho\sigma}, \rho,\sigma=1,\ldots,N$ are $\geq 0, \leq 1$. Therefore the length of the vector

$$y_m = \sum_{n=1}^{N} \mu_{mn} x_n$$

is always $\leq$ than that of the vector x_m .

$$\text{For}\quad x_m = \begin{cases} 1, & \text{for } m = \bar{m} \\ 0, & \text{otherwise} \end{cases} \quad \text{then}\quad y_m = \mu_{m\bar{m}},$$

therefore,

$$\sum_{m=1}^{N} |x_m|^2 \geq \sum_{m=1}^{N} |y_m|^2, \quad 1 \geq \sum_{m=1}^{N} |\mu_{m\bar{m}}|^2 .$$

Since this holds for each N , $\sum_{n} |\mu_{\bar{m}n}|^2 \leq 1$.

[167]The whole consideration is rigorous only if all $\phi_1, \phi_2, \ldots$ belong to the domain of R . Now for each R

therefore is determined by the ensemble itself.

With regard to **II.**, **Tr.** satisfies **B'.** for each choice of U : Therefore we have only to determine what limitation **A'.** implies for U .

If $\|\phi\| = 1$ but ϕ is otherwise arbitrary, then $\mathfrak{R}^2 = \mathfrak{R}$ for the quantity $\mathfrak{R}$ belonging to P_ϕ because of $P_\phi^2 = P_\phi$ and **I.** Therefore, by **A'.**, $\mathrm{Exp}(\mathfrak{R}) \geq 0$. Consequently $\mathrm{Tr}(UP_{[\phi]}) = (U\phi, \phi) \geq 0$. If f is arbitrary, then for $f \neq 0$, ϕ can be written $f/\|f\|$ and then $(U\phi, \phi) = (Uf, f)/\|f\|^2$. Hence $(Uf, f) \geq 0$; for $f = 0$, this holds automatically. Consequently U is definite. But the definiteness of U , which thus follows from **A'.** is also sufficient for the validity of **A'.**

Indeed: **A'.** asserts that each $\mathrm{Exp}(\mathfrak{S}^2) \geq 0$ and no more; because, if $\mathfrak{R}$ is capable of only non-negative values, then for $f(x) = |x|$, $f(\mathfrak{R}) = \mathfrak{R}$ and since $(g(x))^2 = f(x)$ identically for $g(x) = \sqrt{|x|}$, so

---

we can find such a complete orthonormal set $\phi_1, \phi_2, \ldots$ (cf. II.11.), but if R does not have meaning everywhere, then this set depends upon R . Actually therefore, for each complete orthonormal set $\phi_1, \phi_2, \ldots$ we have a U dependent on this set, such that

$$\mathrm{Exp}(\mathfrak{R}) = \mathrm{Tr}(UR)$$

need by valid only for those R to whose domain $\phi_1, \phi_2, \ldots$ belong.

However, all these U are equal to one another. Because if U', U" are two such, then the above formula holds for both, provided that R has meaning everywhere, i.e., then $\mathrm{Tr}(U'R) = \mathrm{Tr}(U''R)$. For $R = P_{[\phi]}$ therefore, $(U'\phi, \phi) = (U''\phi, \phi)$, $((U' - U'')\phi, \phi) = 0$. Since this holds for all ϕ with $\|\phi\| = 1$, and therefore for all elements of the Hilbert space, $U' - U'' = 0$, and therefore $U' = U''$.

$(g(\mathfrak{R}))^2 = f(\mathfrak{R})$, $\mathfrak{R} = \mathfrak{S}^2$ with $\mathfrak{S} = g(\mathfrak{R})$.[168] Hence we must prove this only: If S is the operator of $\mathfrak{S}$ then $\mathrm{Tr}\,(US^2) \geq 0$. Now S^2 is definite

$$((S^2 f, f) = (Sf, Sf) \geq 0)$$

Therefore, if we write A, B in place of U, S^2 , the problem reduces to the proof of the following theorem: if A, B are Hermitian and definite, then $\mathrm{Tr}\,(AB) \geq 0$. But we have proved this in II.11. by use of a general theorem on definite operators (cf. Note 114).[169]

---

[168]We cannot substitute $\mathfrak{S} = \sqrt{\mathfrak{R}}$ directly, i.e., $\mathfrak{S} = h(\mathfrak{R})$, $h(x) = \sqrt{x}$, because we only considered real valued functions defined for all real x , and $\sqrt{x}$ is not such -- since it is imaginary for negative x .

[169]It is also possible to give a simple direct proof. Let $\phi_1, \phi_2, \ldots$ be a complete orthonormal set, $a_{\mu\nu} = (A\phi_\mu, \phi_\nu)$ $b_{\mu\nu} = (B\phi_\mu, \phi_\nu)$, $\mathrm{Tr}\,(AB) = \sum_{\mu,\nu} a_{\mu\nu} b_{\nu\mu}$. This is ≥ 0 if

$$\sum_{\mu,\nu=1}^{N} a_{\mu\nu} b_{\nu\mu} \geq 0 .$$

If

$$f = \sum_{\mu=1}^{N} x_\mu \phi_\mu$$

then

$$(Af, f) = \sum_{\mu,\nu=1}^{N} a_{\mu\nu} x_\mu \overline{x}_\nu \geq 0, \quad (Bf, f) = \sum_{\mu,\nu=1}^{N} b_{\mu\nu} x_\mu \overline{x}_\nu \geq 0$$

and therefore the finite matrices $a_{\mu\nu}$, $b_{\mu\nu}$ $(\mu, \nu = 1, \ldots, N)$ are also definite. Now both the definiteness and the value of

Hence we have determined the functions Exp ($\mathfrak{R}$) completely; they correspond to the definite Hermitian operator U , and the connection is given by **Tr.** We shall call U the statistical operator of the ensemble under consideration.

The points **1.**, **2.**, **3.** in IV.1. are now easy to discuss. These are the results:

1. From the standpoint of the relative probabilities and expectation values, U and cU are not essentially different from each other (c a constant, > 0).

2. U = 0 furnishes no information and is therefore to be excluded.

3. Absolute (i.e., correctly normalized) proba-

$$\sum_{\mu,\nu=1}^{N} a_{\mu\nu} b_{\nu\mu}$$

are orthogonally invariant in N-dimensional space; since $b_{\mu\nu}$ is Hermitian, it can (in the N-dimensional space) be brought into diagonal form by an orthogonal transformation. We may therefore assume it to be diagonal in the first place, i.e., $b_{\mu\nu} = 0$ for $\mu \neq \nu$. Then

$$\sum_{\mu,\nu=1}^{N} a_{\mu\nu} b_{\nu\mu} = \sum_{\mu=1}^{N} a_{\mu\mu} b_{\mu\mu}$$

Because of the definiteness of both matrices $a_{\mu\mu} \geq 0$, $b_{\mu\mu} \geq 0$ (we set

$$\left(x_\nu \begin{cases} = 1, & \text{for } \nu = \mu \\ = 0, & \text{for } \nu \neq \mu \end{cases} \right)$$

and this implies that the above sum is ≥ 0.

bilities and expectation values are obtained if $\mathrm{Tr}\, U = 1$. So long as $\mathrm{Tr}\, U$ is finite, we can normalize U by multiplication with $c = 1/\mathrm{Tr}\, U$ (according to **1.**). (Because of the definiteness of U, $\mathrm{Tr}\, U \geq 0$, but actually, $\mathrm{Tr}\, U > 0$, since it follows from $\mathrm{Tr}\, U = 0$ that $U = 0$, as was shown at the end of IV.1. in the general case; for our case, this also follows from II.11. This is the case excluded by **2.**). It is for an infinite $\mathrm{Tr}\, U$ only, that we have essentially relative probabilities and expectation values.

Finally, we must investigate **α**), **β**) from IV.1., i.e., discover the dispersion free and homogeneous ensembles among the U.

First we consider the dispersion free ensembles. We have then to assume U as correctly normalized (cf. IV.1.), and then $\mathrm{Exp}\,(\mathfrak{R}^2) = [\mathrm{Exp}\,(\mathfrak{R})]^2$ always; i.e., $\mathrm{Tr}\,(UR^2) = [\mathrm{Tr}\,(UR)]^2$. For $R = P_{[\phi]}$, $R^2 = R = P_{[\phi]}$, $\mathrm{Tr}\,(UP_{[\phi]}) = (U\phi, \phi)$, therefore $(U\phi, \phi) = (U\phi, \phi)^2$, i.e., $(U\phi, \phi) = 0$ or 1. If $\|\phi'\| = 1$, $\|\phi''\| = 1$, then we can vary ϕ continuously, so that it begins with ϕ', is ϕ'' at the end, and at all times, $\|\phi\| = 1$.[170] In this case, $(U\phi, \phi)$ also varies continuously, and since it can only be 0 or 1, it is constant -- therefore $(U\phi', \phi') = (U\phi'', \phi'')$. Consequently $(U\phi, \phi)$ always $= 0$

---

[170]This is clear for $\phi' = \phi''$. Next, let $\phi' \neq \phi''$. "Orthogonalization" of ϕ', ϕ'' (cf. II.2.) leads to a ϕ_1, with $\|\phi_1\| = 1$, which is orthogonal to ϕ', and such that ϕ'' is a linear combination of ϕ', ϕ_1. Therefore $\phi'' = a\phi' + b\phi_1$, $\|\phi''\|^2 = |a|^2 + |b|^2 = 1$. Let $|a| = \cos\theta$, $|b| = \sin\theta$. Then $a = e^{i\alpha}\cos\theta$, $b = e^{i\beta}\cos\theta$, and hence, if we now define $a^{(x)} = e^{ix\alpha}\cos(x\theta)$, $b^{(x)} = e^{ix\beta}\sin(x\theta)$, then also $|a^{(x)}|^2 + |b^{(x)}|^2 = 1$. Hence $\|\phi^{(x)}\| = 1$ for $\phi^{(x)} = a^{(x)}\phi' + b^{(x)}\phi_1$. Also, $\phi^{(x)}$ varies continuously from $\phi'(x = 0)$ to $\phi''(x = 1)$.

or always $= 1$, from which we obtain $U = 0$ or $U = 1$, respectively. $U = 0$ is excluded by **2.**, $U = 1$ cannot be normalized ($\mathrm{Tr}\, 1$ = dimension number of the space $= \infty$), and as we can also see directly, $U = 1$ is not dispersion free. Consequently there exist no dispersion free ensembles.

Let us now go on to the homogeneous case. By $\boldsymbol{\beta}$) and **Tr.**, U is homogeneous if from

$$U = V + W$$

(V, W, like U, definite and Hermitian) it follows that $V = c'U$, $W = c''U$.[171] We assert that this property holds for $U = P_{[\phi]}$ ($\|\phi\| = 1$) and only for these.

First let U have the property mentioned. Because $U \neq 0$, there is an f_0 with $Uf_0 \neq 0$, therefore $f_0 \neq 0$, and consequently $(Uf_0, f_0) > 0$ (cf. II.5., THEOREM 19.). We form the two Hermitian operators

$$Vf = \frac{(f, Uf_0)}{(Uf_0, f_0)} \cdot Uf_0 \,, \quad Wf = Uf - \frac{(f, Uf_0)}{(Uf_0, f_0)} \cdot Uf_0$$

then

$$(Vf, f) = \frac{|(f,Uf_0)|^2}{(Uf_0,f_0)} \geq 0, \quad (Wf, f) = \frac{(Uf,f)(Uf_0,f_0)-|(f,Uf_0)|^2}{(Uf_0,f_0)} \geq 0$$

(cf. II.5., THEOREM 19.), i.e., V, W are definite, and furthermore $U = V + W$. Therefore $V = c'U$ and because $Vf_0 = Uf_0 \neq 0$, $c' = 1$, i.e., $U = V$. If we now set $\phi = \frac{1}{\|Uf_0\|} \cdot Uf_0$ ($\|\phi\| = 1$), and

---

[171] Actually, because of **2.**, we should require that $W \neq 0$, $V \neq 0$. The cases $V = 0$ or $W = 0$ are, however, included with $c' = 0, c'' = 1$ and $c' = 1, c'' = 0$ respectively.

$$c = \frac{||Uf_0||^2}{(Uf_0, f_0)}$$

$(c > 0)$, then $Uf = Vf = c(f, \phi)\phi = cP_{[\phi]}f$, i.e., $U = cP_{[\phi]}$, i.e., by **1.**, U is essentially $P_{[\phi]}$.

Conversely, let $U = P_{[\phi]}$ $(||\phi|| = 1)$. If $U = V + W$, V, W definite, then it follows from $Uf = 0$ that

$$0 \leq (Vf, f) \leq (Vf, f) + (Wf, f) = (Uf, f) = 0, (Vf, f) = 0,$$

therefore $Vf = 0$ (cf. supra). But $Uf = P_{[\phi]}f = 0$ follows from $(f, \phi) = 0$, therefore $Vf = 0$ also; and therefore for each g, $(f, Vg) = (Vf, g) = 0$. That is, everything which is orthogonal to ϕ is also orthogonal to Vg, and consequently $Vg = c_g \cdot \phi$ (c_g a number dependent on g), but we use only the case $g = \phi$ i.e., $V\phi = c'\phi$. Each f has the form $(f,\phi)\cdot\phi + f'$, where f' is orthogonal to ϕ. Therefore

$$Vf = (f, \phi)\cdot V\phi + Vf' = (f, \phi)\cdot c'\phi = c'P_{[\phi]}f = c'Uf.$$

Consequently $V = c'U$, $W = U - V = (1 - c')U$, and the proof is complete.

Hence the homogeneous ensembles correspond to the $U = P_{[\phi]}$, $||\phi|| = 1$, and in fact **Tr.** then becomes the formula **E₂.** of III.1.:

$$(\mathbf{E_2}) \qquad \mathrm{Exp}\ (\mathfrak{R}) = \mathrm{Tr}\ (P_{[\phi]}R) = (R\phi, \phi)$$

It should be observed that $\mathrm{Exp}\ (1) = \mathrm{Tr}\ (P_{[\phi]}) = 1$ (because $P_{[\phi]}$ belongs to the one dimensional $[\phi]$, or by **E₂.**), i.e., the present form of $\overline{U}$ is correctly normalized. Finally, we find out when $P_{[\phi]}$ and $P_{[\psi]}$ have the same statistics, i.e., when $P_{[\phi]} = cP_{[\psi]}$ (c a constant, > 0, cf. 1.). Since $\mathrm{Tr}\ (P_{[\phi]}) = \mathrm{Tr}\ (P_{[\psi]}) = 1$, $c = 1$, therefore $P_{[\phi]} = P_{[\psi]}$, $[\phi] = [\psi]$, and therefore $\phi = a\psi$. It follows from $\|\phi\| = \|\psi\| = 1$ that $|a| = 1$ for the constant a. This is also clearly sufficient.

Putting these things together, we can therefore say: There are no ensembles which are free from dispersion. There are homogeneous ensembles, and these correspond to the $U = P_{[\phi]}$ with $\|\phi\| = 1$, and only to these. For these U, **Tr.** goes over into $\mathbf{E}_2$.; the normalization is correct and U does not change if ϕ is replaced by an $a\phi$ (a constant, $|a| = 1$), but in every other change of ϕ it changes essentially (cf. **1**.). The homogeneous ensembles therefore correspond to the states of quantum mechanics, as these were characterized earlier: The ϕ of Hilbert space, with $\|\phi\| = 1$, in which a constant factor of absolute value one is unimportant (cf. for example, II.2.), and the statistical assertions are made by $\mathbf{E}_2$.[172]

We have derived all these results from the purely qualitative conditions **A'**., **B'**., α), β), **I**., **II**.

Hence, within the limits of our conditions, the decision is made and it is against causality; because all ensembles have dispersions, even the homogeneous.

There would still be the question of "hidden parameters," brought up in III.2., to be discussed, i.e., the question as to whether the dispersions of the homogeneous ensembles characterized by the wave functions ϕ (i.e., by $\mathbf{E}_2$.) are not due to the fact that these are not

---

[172]The deductions given in the last two sections, which lead to the concept of the homogeneous ensemble, were given by the author, Gött. Nachr., 1927. The existence of the homogeneous ensembles and their relation to the general ensembles was discovered independently by H. Weyl, Z. Phys. 46 (1927) and the author (in the reference above). A special case of the more general ensembles (namely, for two coupled systems, cf. the discussion in VI.2.) was produced earlier by J. Landau, Z. Physik, 45 (1927).

the real states, but only mixtures of several states -- while for the knowledge of the actual state additional data, besides the data of the wave function ϕ, would be necessary (these would be the "hidden parameters"), which together would determine everything causally, i.e., lead to dispersion free ensembles. The statistics of the homogeneous ensemble ($U = P_{[\phi]}$, $\|\phi\| = 1$) would then have resulted from the averaging over all the actual states of which it was composed, i.e., by the averaging over that region of values of the "hidden parameters" which is involved in those states. But this is impossible for two reasons: First, because then the homogeneous ensemble in question could be represented as a mixture of two different ensembles,[173] contrary to its definition. Second, because the dispersion free ensembles, which would have to correspond to the "actual" states (i.e., which consist only of systems in their own "actual" states), do not exist. It should be noted that we need not go any further into the mechanism of the "hidden parameters," since we now know that the established results of quantum mechanics can never be re-derived with their help. In fact, we have even ascertained that it is impossible that the same physical quantities exist with the same function connections (i.e., that **I**., **II**. hold), if other variables (i.e., "hidden parameters") should exist in addition to the wave functions.

Nor would it help if there existed other, as yet

---

[173] If the "hidden parameters," the totality of which we shall denote by π, take on only discrete values $\pi_1, \pi_2, \ldots, \pi_n$ $(n > 1)$, we obtain the two ensembles whose superposition is the original one, by assuming the systems with $\pi = \pi_1$ in the one, and the systems with $\pi \neq \pi_1$ in the other. If π varies continuously over a region Π, then let Π' be a sub-region of Π, and the one ensemble would then contain the systems with π from Π', and the other those whose π does not belong to Π'.

undiscovered, physical quantities, in addition to those represented by the operators in quantum mechanics, because the relations assumed by quantum mechanics (i.e., I., II.) would have to fail already for the by now known quantities, those that we discussed above. It is therefore not, as is often assumed, a question of a re-interpretation of quantum mechanics, -- the present system of quantum mechanics would have to be objectively false, in order that another description of the elementary processes than the statistical one be possible.

The following circumstance is also worthy of mention. The indeterminacy relations have at first glance a certain similarity to the basic postulate of relativity theory. There it is maintained that it is impossible in principle to determine the simultaneity of two events occurring at points a distance r apart, more precisely than within a time interval of magnitude r/c (c is the velocity of light); while the indeterminacy relations predict that it is impossible in principle to give the position of a material point in phase space more precisely than within a region of volume $(\frac{h}{4\pi})^3$.[174] Nevertheless there

[174] The phase space is 6-dimensional, its 6 coordinates are the three cartesian coordinates of the mass particles, q_1, q_2, q_3, and the three corresponding momenta p_1, p_2, p_3. By III.4., we have for the relative dispersions $\epsilon_1, \epsilon_2, \epsilon_3, \eta_1, \eta_2, \eta_3$,

$$\epsilon_1 \eta_1 \geq \frac{h}{4\pi}, \quad \epsilon_2 \eta_2 \geq \frac{h}{4\pi}, \quad \epsilon_3 \eta_3 \geq \frac{h}{4\pi},$$

i.e.,

$$\epsilon_1 \epsilon_2 \epsilon_3 \eta_1 \eta_2 \eta_3 \geq \left(\frac{h}{4\pi}\right)^3,$$

and the position in the phase space of classical mechanics is in each case undetermined within this volume.

exists a fundamental difference. The relativity theory denies the possibility of an objective, precise, measurement of distant simultaneity, but in spite of this, it is possible, by the introduction of a Galilean frame of reference, to put a coordinate system in the world which makes a simultaneity definition possible that is in reasonable agreement with our normal concepts on this subject. An objective meaning will not be attributed to such a definition of distant simultaneity only because this coordinate system can be chosen in an infinite number of different ways, so that infinitely many different distant simultaneity definitions can be obtained, all of which are equally good. That is, behind the impossibility of the measurement we find an infinite multiplicity of possible theoretical definitions. It is otherwise in quantum mechanics, where it is in general not possible to describe a system with the wave function ϕ by points in phase space, not even if we introduce new (hypothetical, unobserved) coordinates, the "hidden parameters," -- since this would lead to dispersion free ensembles. That is, not only is the measurement impossible, but so is any reasonable theoretical definition, i.e., any definition which, although incapable of experimental proof, would also be incapable of experimental refutation. The principle of impossibility of the measurement thus arises in one case from the fact that there is an infinite number of ways in which the relevant concepts can be defined without conflicting directly with experience (or, with the general, basic assumptions of the theory) -- while in the other case no such way exists at all.

To summarize, the position of causality in modern physics can therefore be characterized as follows: In the macroscopic case there is no experiment which supports it, and none can be devised because the apparent causal order of the world in the large (i.e., for objects visible to the naked eye) has certainly no other cause than the "law of large numbers" and it is completely independent of whether the natural laws governing the elementary processes are causal or not.[175] That macroscopically identical objects exhibit identical behavior has little to do with causality: They are in fact not equal at all, since the coordinates which determine the states of their atoms almost never coincide exactly, and the macroscopic method of observation averages over these coordinates (here they are the "hidden parameters"). The number of these coordinates is, however, great (for 1 gram of matter, about 10^{25}), and therefore the above mentioned averaging process entails an extensive diminution of all dispersions, according to the well-known laws of the calculus of probability. (Naturally this is true only in the general case; in suitable special cases -- such as Brownian motion, unstable states, among others -- this apparent macroscopic causality fails.) The question of causality could be put to a true test only in the atom, in the elementary processes themselves, and here everything in the present state of our knowledge militates against it. The only formal theory existing at the present time which orders and summarizes our experiences in this area in a half-way satisfactory manner, i.e., quantum mechanics, is in compelling logical contradiction with causality. Of course it would be an exaggeration to maintain that causality has thereby been done away with: quantum mechanics has, in its present form, several serious lacunae, and it may even be that it is false, although this latter possibility is highly unlikely, in the face of its startling capacity in the qualitative explanation of general problems, and in the quantitative calculation of special ones. In spite of the fact that quantum mechanics agrees well with experiment, and that it has opened up for us a qualitatively new side of the world, one can never say of the theory that it has been proved by experience, but only

---

[175] Cf. the extremely lucid discussions of Schrödinger on this subject: Naturwiss. 17 (1929), p. 37.

that it is the best known summarization of experience. However, mindful of such precautions, we may still say that there is at present no occasion and no reason to speak of causality in nature -- because no experiment indicates its presence, since the macroscopic are unsuitable in principle, and the only known theory which is compatible with our experiences relative to elementary processes, quantum mechanics, contradicts it.

To be sure, we are dealing with an age old way of thinking of all mankind, but not with a logical necessity (this follows from the fact that it was at all possible to build a statistical theory), and anyone who enters upon the subject without preconceived notions has no reason to adhere to it. Under such circumstances, is it sensible to sacrifice a reasonable physical theory for its sake?

3. CONCLUSIONS FROM EXPERIMENTS

The last section has taught us that the most general statistical ensemble which is compatible with our qualitative basic assumptions is characterized, according to the law **Tr.**, by a definite operator U. Those particular ensembles which we had denoted as homogeneous, were characterized by $U = P_{[\phi]}$ $(\|\phi\| = 1)$ and since these are the actual states of the systems **S** (i.e., not capable of further resolution), we also call them states (specifically, $U = P_{[\phi]}$ is the state ϕ).

If U has a pure discrete spectrum, perhaps with the eigenvalues $w_1, w_2, \cdots$ and the eigenfunctions $\phi_1, \phi_2, \cdots$ (which form a complete orthonormal set), then (cf. II.8.),

$$U = \sum_n w_n P_{[\phi_n]} .$$

Because of the definiteness of U, all $w_n \geq 0$ [indeed $U\phi_n = w_n\phi_n$, therefore $(U\phi_n, \phi_n) = w_n$, and therefore this is ≥ 0 ,] and $\sum_n w_n = \sum_n (U\phi_n, \phi_n) = \text{Tr}\, U$ (cf. the beginning of IV.1. also), that is, $\sum_n w_n = 1$ if U is correctly normalized. By the remarks at the beginning of IV.1., U can be interpreted as a superposition of the states $\phi_1, \phi_2, \ldots$ with the respective relative weights $w_1, w_2, \ldots$ -- if U is correctly normalized, then these relative weights are also the absolute weights.

But a correctly normalized U, i.e., $\text{Tr}\, U = 1$, is totally continuous (by II.11., cf. in particular Note 115), and therefore has a pure discrete spectrum. The same is true if $\text{Tr}\, U$ is finite. (An infinite $\text{Tr}\, U$ can be regarded as a limiting case which we shall not go into here.) In the really interesting case therefore, the observed ensemble can be represented as a superposition of states, which we have actually chosen to be pair-wise orthogonal. We shall then call the general ensembles mixtures (in contrast to the homogeneous ensembles which are the states).

If all the eigenvalues of U are simple, i.e., if the $w_1, w_2, \ldots$ are different from each other, then, as we know, the $\phi_1, \phi_2, \ldots$ are uniquely determined except for a constant factor of absolute value 1. The corresponding states (and the $P_{[\phi_1]}, P_{[\phi_2]}, \ldots$) are then uniquely determined. Likewise, the weights $w_1, w_2, \ldots$ are uniquely determined -- except for a permutation of the sequence. In this case therefore, we can state uniquely from which (pair-wise orthogonal) states the mixture U is formed. If U has multiple eigenvalues ("degeneracies"), however, the situation is quite different. Exactly how the $\phi_1, \phi_2, \ldots$ can be chosen was discussed in II.8. This can be done in infinitely many ways, all essentially different (while the $w_1, w_2, \ldots$ are still uniquely determined). We must write down those among the $w_1, w_2, \ldots$ which are different from one another: $w', w'', \ldots$, and then form the closed linear manifold of its eigenfunctions for each weight

$w = w', w'', \ldots$ (i.e., the set of all solutions of $Uf = wf$), the $\mathfrak{M}_{w'}, \mathfrak{M}_{w''}, \ldots$. After this we proceed in the following manner. From each $\mathfrak{M}_{w'}, \mathfrak{M}_{w''}, \ldots$ choose an arbitrary orthonormal set spanning that manifold, $\chi_1', \chi_2', \ldots$; $\chi_1'', \chi_2'', \ldots$; $\ldots$ respectively. The $\chi_1', \chi_2', \ldots, \chi_1'', \chi_2'', \ldots, \ldots$ are then the $\phi_1, \phi_2, \ldots$; and the corresponding eigenvalues $w', w', \ldots, w'', w'', \ldots, \ldots$ are the $w_1, w_2, \ldots$. As soon as an $\mathfrak{M}_w$ has more than one dimension, i.e., a multiple eigenvalue exists, the corresponding $\chi_1, \chi_2, \ldots$ are no longer determined within a constant factor of absolute value one (χ_1 for example can be any normalized element of $\mathfrak{M}_w$) -- i.e., the states themselves are also multivalued.

This phenomenon can also be formulated as follows: If the states $\chi_1, \chi_2, \ldots$ are pair-wise orthogonal (i.e., $\chi_1, \chi_2, \ldots$ form an orthonormal set which may be finite or infinite), and if we mix them in such a way that all get the same weight (i.e., the relative weights $1:1:\ldots$), then the resulting mixture depends only on the closed linear manifold which is spanned by the $\chi_1, \chi_2, \ldots$. In fact,

$$U = P_{[\chi_1]} + P_{[\chi_2]} + \cdots = P_{\mathfrak{M}} .$$

If the number of the $\chi_1, \chi_2, \ldots$ is finite, say s : $\chi_1, \ldots, \chi_s$ then this U can also be considered as a mixture of all normalized elements of $\mathfrak{M}$, i.e., of all states of $\mathfrak{M}$. These are the

$$\chi = x_1\chi_1 + \cdots + x_s\chi_s , \quad |x_1|^2 + \cdots + |x_s|^2 = 1 .$$

Actually, if we set $x_1 = u_1 + iv_1, \ldots, x_s = u_s + iv_s$, and call the $2s - 1$ dimensional unit-sphere surface $u_1{}^2 + v_1{}^2 + \cdots + u_s{}^2 + v_s{}^2 = 1$ (corresponding to $|x_1|^2 + \cdots + |x_s|^2 = 1$) K, and its surface element do, then for

$$U' = \iint_K \cdots \iint P_{[\chi]} do ,$$

we have

$$(U'f, g) = \iint_K \cdots \iint (P_{[\chi]}f, g) do = \iint_K \cdots \iint (f, \chi)\overline{(g, \chi)} do$$

$$= \iint_K \cdots \iint (f, \sum_{\mu=1}^{s} (u_\mu + iv_\mu)\chi_\mu)\overline{(g, \sum_{\mu=1}^{s} (u_\mu + iv_\mu)\chi_\mu)} do$$

$$= \iint_K \cdots \iint \sum_{\mu,\nu}^{s} (f, \chi_\mu)\overline{(g, \chi_\nu)}(u_\mu - iv_\mu)(u_\nu + iv_\nu) do$$

$$= \sum_{\mu,\nu=1}^{s} (f, \chi_\mu)\overline{(g, \chi_\nu)} \iint_K \cdots \iint [(u_\mu u_\nu + v_\mu v_\nu) +$$

$$+ i(u_\mu v_\nu - u_\nu v_\mu)] do .$$

Therefore, since all $u_\mu v_\nu$, $u_\nu v_\mu$ -integrals, as well as the $u_\mu u_\nu$, $v_\mu v_\nu$ -integrals for $\mu \neq \nu$, are $= 0$ by symmetry,[176]

---

[176] $u_\mu \longrightarrow -u_\rho$ (or $u_\nu \longrightarrow -u_\nu$ or $v_\mu \longrightarrow -v_\mu$) is a symmetry operation of K, in which the former integrands change their signs, and their integrals are therefore equal to zero. $u_\mu \longrightarrow v_\mu$, $v_\mu \longrightarrow u_\mu$ or $u_\mu \longrightarrow u_\nu$, $u_\nu \longrightarrow u_\mu$ are such symmetry operations of K, in which the latter integrals are exchanged. Their integrals are therefore equal, and hence equal to $1/2\,s$ times their sum:

$$\iint_K \cdots \iint (u_1^2 + v_1^2 + \cdots + u_s^2 + v_s^2) dv =$$

$$= \iint_K \cdots \iint dv = \text{surface of } K ,$$

which may be called C.

and for $\mu = \nu$ the latter are all $= C/2s$ $(C > 0)$,[176] we have further

$$(U'f, g) = \frac{C}{s} \sum_{\mu=1}^{s} (f, \chi_\mu)\overline{(g, \chi_\mu)} = \frac{C}{s} \sum_{\mu=1}^{s} (P_{[\chi_\mu]}f, g)$$

$$= (\{\frac{C}{s} \sum_{\mu=1}^{s} P_{[\chi_\mu]}\}f, g) .$$

Consequently,

$$U' = \frac{C}{s} \sum_{\mu=1}^{s} P_{[\chi_\mu]} = \frac{C}{s} U ,$$

i.e., U' and U are not essentially different.

These results are of great significance for the nature of the quantum mechanical statistics, and we shall therefore repeat them:

1. If a mixture is made up of mutually orthogonal states with exactly equal weights, then it can no longer be determined what these states were. Or, equivalently, we can produce the same mixture from different (mutually orthogonal) components by mixing in exactly equal proportions.

2. The mixture so obtained is, if the number of components is finite, identical with the mixture of all states which are linear combinations of these components.

The simplest example of this type is the following: if we mix ϕ, ψ (orthogonal) in the proportion $1:1$, then we obtain the same result as if we mix, for example,

$$\frac{\phi + \psi}{\sqrt{2}} , \frac{\phi - \psi}{\sqrt{2}}$$

in the proportion $1:1$ or even all $x\phi + y\psi$ $(|x|^2 + |y|^2 = 1)$. If we mix two non-orthogonal ϕ, ψ (the proportion need not be $1:1$), then we are still less able to determine the composition of the final mixture since this mixture could certainly have also been obtained by mixing orthogonal states.

We postpone a further investigation into the nature of mixtures until the thermodynamical discussions in V.2. and thereafter. ——

The formula **Tr.** in IV.2. states how the expectation value of the quantity $\mathfrak{R}$ with the operator R is to be calculated in a mixture with the statistical operator U : it is $\mathrm{Tr}\,(UR)$. The probability therefore that the value a of R lies in the interval $a' < a \leq a''$ (a', a'' given, $a' \leq a''$) , is to be found as in III.1. or III.5.: If the quantity $F(\mathfrak{R})$ is formed with the function

$$F(x) \begin{cases} = 1 \text{ , for } a' < x \leq a'' \\ = 0 \text{ , otherwise} \end{cases}$$

then its expectation value is the probability mentioned. Now $F(\mathfrak{R})$ has (by I. in IV.2.) the operator $F(R)$ and if $E(\lambda)$ is the resolution of the identity belonging to R , then, as we have calculated more than once, $F(R) = E(a'') - E(a')$ and the desired probability is $w(a', a'') = \mathrm{Tr}\, U(E(a'') - E(a'))$. Consequently the probability function which describes the statistics of $\mathfrak{R}$ is $w(a) = \mathrm{Tr}\, UE(a)$ [cf. IV.1., Note 175; for states, i.e., $U = P_{[\phi]}$, we again have $w(a) = \mathrm{Tr}\, P_{[\phi]}E(a) = (E(a)\phi, \phi)$]. Naturally these probabilities are only relative if U is not correctly normalized.

The question as to when the quantity $\mathfrak{R}$ with the operator R , in the mixture with the statistical operator U , takes on the value λ^* with certainty, can be answered directly with the aid of $w(a)$: for $a < \lambda^*$ we must require $w(a) = 0$, and for $a \geq \lambda^*$, we must require $w(a) = 1$, or, if U is not correctly normalized, $w(a) = \mathrm{Exp}\,(1) = \mathrm{Tr}\,(U)$. That is, $\mathrm{Tr}\, UE(a) = 0$ for $a < \lambda^*$,

$\mathrm{Tr}\, U(1 - E(a)) = 0$ for $a \geq \lambda^*$.[177] Now for definite operators A, B, $\mathrm{Tr}\,(AB) = 0$ has $AB = 0$ as a consequence (cf. II.11.), and therefore $UE(a) = 0$ for $a < \lambda^*$, and $UE(a) = U$ for $a \geq \lambda^*$ or, what is equivalent, $E(a)U = 0$ or U respectively, since the factors must commute because of the Hermitian nature of the product. That is, for $f = Ug$,

$$E(a)f \begin{cases} = f, & \text{for } a \geq \lambda^* \\ = 0, & \text{for } a < \lambda^* \end{cases}$$

and by the discussion carried out in II.8., this means that $Rf = \lambda^* f$ i.e., $RUg = \lambda^* Ug$, identically in g. Consequently the ultimate condition is $RU = \lambda^* U$. Or, if we denote the closed linear manifold formed by all solutions h of $Rh = \lambda^* h$ by $\mathfrak{M}$: Uf always lies in $\mathfrak{M}$.

The same result could also have been obtained from the vanishing of the dispersion, i.e., the (possibly relative) expectation value of $(\mathfrak{R} - \lambda^*)^2$. ——

In III.3., we answered the following questions (let $\mathfrak{R}$, $\mathfrak{S}$, ... be physical quantities, $R, S, \ldots$ their respective operators):

1. When is $\mathfrak{R}$ absolutely exactly measurable? Answer: whenever R has only a discrete spectrum.

---

[177] If $\mathrm{Tr}\, U$ is infinite, the latter formulas, obtained by subtractions, may appear doubtful. They can, however, also be established as follows: That $\mathfrak{R}$ has the value λ^* means that $w(a', a'') = 0$, i.e., $\mathrm{Tr}\, U(E(a'') - E(a')) = 0$ for $a'' < \lambda^*$ or $a' \geq \lambda^*$. Since this trace is always ≥ 0, and since it is monotone increasing with a'' as well as monotone decreasing with a', it suffices to consider the $\lim_{a'' \to -\infty}$ for $a'' < \lambda^*$ and the $\lim_{a' \to +\infty}$ for $a' \geq \lambda^*$. That is, $\mathrm{Tr}\, U(1 - E(a')) = 0$ for $a' \geq \lambda^*$, and $\mathrm{Tr}\, UE(a'') = 0$ for $a'' < \lambda^*$.

2. When are $\mathfrak{R}$, $\mathfrak{S}$ measurable simultaneously with absolute accuracy? Answer: whenever R, S have only discrete spectra and commute.

3. When are several quantities $\mathfrak{R}, \mathfrak{S}, \ldots$ measurable simultaneously with absolute accuracy? Answer: whenever R, S have only discrete spectra and all commute.

4. When are several quantities $\mathfrak{R}, \mathfrak{S}, \ldots$ measurable simultaneously with arbitrary accuracy? Answer: whenever R, S, ... all commute.

In this case we used the following principle abstracted from the result of the Compton-Simons experiment:

(M.) If the physical quantity $\mathfrak{R}$ is measured twice in succession in a system **S**, then we get the same value each time. This is the case even though $\mathfrak{R}$ has a dispersion in the original state of **S**, and the $\mathfrak{R}$ measurement can change the state of **S**.

We had discussed the physical meaning of **M.** in detail in III.3. Further assumptions in the answering of **1.** - **4.** were: the statistical formula $\mathbf{E}_2$. of III.3. for states; the assumption **F.** of III.3., according to which F($\mathfrak{R}$) has the operator F(R) if $\mathfrak{R}$ has the operator R; the assumption according to which $\mathfrak{R} + \mathfrak{S}$ has the operator R + S if the (simultaneously measurable) quantities $\mathfrak{R}, \mathfrak{S}$ have the respective operators R, S.

Since these three assumptions are again at our disposal (the first follows from the formula **Tr.** in IV.2., the other two correspond to **I.**, **II.** in IV.2.), and **M.** should also be assumed as being correct -- because we have perceived that it is indispensable for the conceptual structure of quantum mechanics -- the proofs given in III.3. for **1.** - **4.** also hold here. The answers given are therefore again correct.

In III.5., we investigated those physical quantities which take on only two values. 0, 1. They corresponded to the properties $\mathfrak{E}$ uniquely. Indeed, if $\mathfrak{E}$ was given, then the quantity could be defined like this: it is measured by distinguishing whether $\mathfrak{E}$ is present or not, and its value is then 1 or 0 respectively. Conversely, if the quantity was given, then $\mathfrak{E}$ was this property: the quantity in question has the value 1 (i.e., not 0). From the **F.** there (i.e., **1.** in IV.2.) it followed that the corresponding operators E are actually the projections and only these. The probability therefore that $\mathfrak{E}$ be present was equal to the expectation value of the quantity defined above. In III.5., it was calculated only for states (i.e., $U = P_{[\phi]}$, $\|\phi\| = 1$), but we can determine it in general from **Tr.**: it is Tr (UE). (Relative! The absolute value is obtained only if U is correctly normalized, i.e., if $\mathrm{Tr}\, U = 1$.)

Since we have made certain of **1.** - **4.**, the statements derived from them, α) - ζ), in III.5., are likewise valid. Of course it should be observed that in the former case, α) gave information only for the states, but we have extended it here to all mixtures:

α') The property $\mathfrak{E}$ is present or not present in the mixture with the statistical operator U, with the respective probabilities

$$\mathrm{Tr}\,(UE) \quad \text{and} \quad \mathrm{Tr}\,(U(1 - E))$$

(Relative probabilities! These are absolute only if U is properly normalized, i.e., $\mathrm{Tr}\, U = 1$.)

If several quantities $\mathfrak{R}_1, \ldots, \mathfrak{R}_l$ are investigated and if $\mathfrak{R}_1, \ldots, \mathfrak{R}_l$ have the respective operators $R_1, \ldots, R_l$ to which correspond the respective resolutions of the identity $E_1(\lambda), \ldots, E_l(\lambda)$; if further, l intervals $I_1: \lambda_1' < \lambda \leq \lambda_1'', \ldots, I_l: \lambda_l' < \lambda \leq \lambda_l''$ are given, and if $E_1(I_1) = E_1(\lambda_1'') - E_1(\lambda_1'), \ldots, E_l(I_l) = E_l(\lambda_l'') - E_l(\lambda_l')$; then the projections $E_1(I_1), \ldots, E_l(I_l)$ (cf. ζ)) belong to the respective properties "$\mathfrak{R}_1$ lies in I_1," ..., "$\mathfrak{R}_l$ lies in I_l." The commutativity of the $E_1(I_1), \ldots, E_l(I_l)$ is then characteristic (cf. γ)) for these to be simultaneously decidable, and the projection for their simultaneous validity is $E = E_1(I_1) \ldots E_l(I_l)$ (cf. ϵ)). Hence the probability of the last mentioned event is Tr (UE) (cf. α')).

Let us now follow the converse path: Let us assume that we do not know the state of a system **S** , but that we have made certain measurements on **S** and know their results. In reality, it always happens this way, because we can learn something about the state of **S** only from the results of measurements. More precisely, the states are only a theoretical construction, only the results of measurement are actually available, and the problem of physics is to furnish relations between the results of past and future measurements. To be sure, this is always accomplished through the introduction of the auxiliary concept "state," but the physical theory must then tell us on the one hand how to make from past measurements inferences about the present state, and on the other hand, how to go from the present state to the results of future measurements. Up to now, we have dealt only with the latter question, and we must now apply ourselves to the former.

If anterior measurements do not suffice to determine the present state uniquely, then we may still be able to infer from those measurements, under certain circumstances, with what probabilities particular states are present. (This holds in causal theories, for example, in classical mechanics, as well as in quantum mechanics.) The proper problem is then this: Given certain results of measurements, find a mixture whose statistics are the same as those which we shall expect for a system **S** of which we know only that these measurements were carried out on it and that they had the results mentioned. Of course we must actually be more precise, and must describe what it means to say that we "know only this," and no more, about **S** -- and how this can lead to a set of statistics.

In any case, the connection with statistics must be the following: If, for many systems $\mathbf{S}'_1, \ldots, \mathbf{S}'_M$ (replicas of **S**) , these measurements give the results mentioned, then this ensemble $[\mathbf{S}'_1, \ldots, \mathbf{S}'_M]$ coincides in all its statistical properties with the mixture that corresponds to the results of the measurements. That the results of the measurements are the same for all $\mathbf{S}'_1, \ldots, \mathbf{S}'_M$ can be attributed, by **M.**, to the fact that originally a large ensemble $[\mathbf{S}_1, \ldots, \mathbf{S}_N]$ was given in which the measurements were carried out, and then those elements for which the desired results occurred were collected into a new ensemble. This is then $[\mathbf{S}'_1, \ldots, \mathbf{S}'_M]$. Of course everything depends on how $[\mathbf{S}_1, \ldots, \mathbf{S}_N]$ is chosen. This initial ensemble gives, so to speak, the a priori probabilities of the individual states of the system $\mathbf{S}$. The whole state of affairs is well-known from the general probability theory: to be able to conclude from the results of the measurements to the states, i.e., from effect to cause, i.e., to be able to calculate a posteriori probabilities, we must know the a priori probabilities. In general these can be chosen in many different ways, and accordingly our problem cannot be solved uniquely. However, we shall see that under the special conditions of quantum mechanics, a certain determination of the initial ensemble $[\mathbf{S}_1, \ldots, \mathbf{S}_N]$ (i.e., of the a priori probabilities) is particularly satisfactory.

The results are quite different if the results of measurements at our disposal suffice for us to determine the state of $\mathbf{S}$ completely. Then the answer to each question must be unique. We shall soon see how this circumstance makes itself felt.

Finally, let us mention the following. Instead of saying that several results of measurement (on $\mathbf{S}$) are known, we can also say that $\mathbf{S}$ was examined in relation to a certain property $\mathfrak{E}$ and its presence was ascertained. From α) - ζ) we know how these things are related: if, for example, the results of measurements (which can be established simultaneously) are available, according to which the values of the quantities $\mathfrak{R}_1, \ldots, \mathfrak{R}_l$ lie in the respective intervals $I_1, \ldots, I_l$, then the projection of E (with the symbols used earlier) is $E = E_1(I_1) \ldots E_l(I_l)$.

The information about $\mathbf{S}$ therefore always amounts to the presence of a certain property $\mathfrak{E}$ which is formally characterized by stating the projection E. Let us investigate the statistical operator U of the equivalent ensemble $[\mathbf{S}_1', \ldots, \mathbf{S}_M']$; as well as the statistical operator of the general initial ensemble $[\mathbf{S}_1, \ldots, \mathbf{S}_N]$, U_0 . What are the mathematical relations of E, U, U_0 ?

Because of **M.**, $\mathfrak{E}$ is certainly present in $[\mathbf{S}_1', \ldots, \mathbf{S}_M']$, i.e., the quantity corresponding to $\mathfrak{E}$ has the value 1 . This means that $EU = U$, as we saw in the beginning of this section, i.e., Uf always lies in $\mathfrak{M}$ where $\mathfrak{M}$ is the set of all f with $Ef = f$, i.e., the closed linear manifold belonging to E .

Instead of $EU = U$ we can also write $UE = U$, $U(1 - E) = 0$, i.e., $Ug = 0$ for all $g = (1 - E)f$, i.e., for all g of the closed linear manifold belonging to $1 - E$, i.e., for all g of $\mathfrak{R} - \mathfrak{M}$. Therefore Uf is equal to 0 for all f of $\mathfrak{R} - \mathfrak{M}$, and for the f of $\mathfrak{M}$ it also lies in $\mathfrak{M}$. Nothing further can be said about U in this way.

This determines U (essentially, i.e., except for a constant factor) if and only if $\mathfrak{M}$ is 0 or 1-dimensional. In fact, for $\mathfrak{M} = [0]$ we have $U = 0$ which is impossible, by IV.2., Remark 1; for $\mathfrak{M} = [\phi]$ ($\phi \neq 0$, therefore $\|\phi\|$ may be taken equal to 1) $U\phi = c\phi$, therefore for all f of $\mathfrak{M}$ (since these are equal to $a\phi$) $Uf = cf$. Hence, in general, $Uf = UEf = cEf$, $U = cE = cP_{[\phi]}$, and since $c > 0$ (because U is definite and $\neq 0$) , U is essentially $= E = P_{[\phi]}$. For $\geq$ 2-dimensional $\mathfrak{M}$ we can choose two orthonormal ϕ, ψ from $\mathfrak{M}$. Then $P_{[\phi]}$, $P_{[\psi]}$ are two essentially different U which satisfy our condition. Therefore $E = 0$ is impossible; for $E = P_{[\phi]}$ ($\|\phi\| = 1$), $U = E = P_{[\phi]}$; otherwise U is many valued.

The fact that $E = 0$ is incompatible with finding any U at all, would be disastrous, if it were possible for an $\mathbf{S}$ to possess such a property $\mathfrak{E}$. However, this is excluded by η): Such an $\mathfrak{E}$ can never be present, its probability is always 0 . A one dimensional $\mathfrak{M}$, i.e., an $E = P_{[\phi]}$ ($\|\phi\| = 1$) , determines U uniquely and fixes the state ϕ -- therefore this is that kind of measurement which determines the state of **S** completely if it turns out in the affirmative, and in fact determines it to be ϕ.[178] All other measurements are incomplete and do not succeed in determining a unique state.

In the general case we proceed as follows. If we call the quantity corresponding to $\mathfrak{E}$ also $\mathfrak{E}$, then U obtains like this: $\mathfrak{E}$ is measured on the whole ensemble $([\mathbf{S}_1, \ldots, \mathbf{S}_N])$ belonging to U_0, and all elements in which the value 1 results are collected, forming the ensemble of U $([\mathbf{S}_1', \ldots, \mathbf{S}_M'])$. The measurement of $\mathfrak{E}$ might be effected in many different ways, for example, another quantity $\mathfrak{R}$, of which $\mathfrak{E}$ is a known function: $\mathfrak{E} = F(\mathfrak{R})$, could be measured. To be more specific: Let $\phi_1, \phi_2, \ldots$ be an orthonormal set which spans $\mathfrak{M}$, and let $\psi_1, \psi_2, \ldots$ be a corresponding set for $\mathfrak{R} - \mathfrak{M}$, then $\phi_1, \phi_2, \ldots,$ $\psi_1, \psi_2, \ldots$ spans $\mathfrak{M} + (\mathfrak{R} - \mathfrak{M}) = \mathfrak{R}$, i.e., it is complete.

---

[178]That is, if $\mathfrak{E}$ is present, the state is ϕ. If it is not present, then "not $\mathfrak{E}$" is present, where $1 - E = 1 - P_{[\phi]}$, $\mathfrak{R} - \mathfrak{M} = \mathfrak{R} - [\phi]$ appear in place of $E = P_{[\phi]}$, $\mathfrak{M} = [\phi]$. This does not determine U uniquely. (E indeed corresponds to the question: "is the state ϕ ?"). A measurement which for each process determines the state uniquely is a measurement of a quantity R whose operator R has a pure discrete spectrum with simple eigenvalues, cf. III.3. After the measurement, one of the states $\phi_1, \phi_2, \ldots$ (the eigenfunctions of R) is present, i.e., the state of **S** is in general changed by the measurement. By analogy, the $\mathfrak{E}$ measurement also changes the state, since afterwards, for a positive result, $U = P_{[\phi]}$, and for a negative one, $U(1 - P_{[\phi]}) = U$, $UP_{[\phi]} = 0$ i.e., $U\phi = 0$; while previously, neither was necessarily the case. This quantum mechanical "determination" of the state thus alters it, as was to be expected.

Let $\lambda_1, \lambda_2, \ldots, \mu_1, \mu_2, \ldots$ be distinct real numbers, and define the operator R by

$$R\left(\sum_n x_n\phi_n + \sum_n y_n\psi_n\right) = \sum_n \lambda_n x_n\phi_n + \sum_n \mu_n y_n\psi_n .$$

R clearly has the pure discrete spectrum $\lambda_1, \lambda_2, \ldots, \mu_1, \mu_2, \ldots$, with the respective eigenfunctions $\phi_1, \phi_2, \ldots, \psi_1, \psi_2, \ldots$, and all the eigenvalues are simple. If $F(x)$ is any function with

$$F(\lambda_n) = 1 \ , \ F(\mu_n) = 0$$

then $F(R)$ has the eigenvalue 1 for $\phi_1, \phi_2, \ldots$ and therefore also for each f of $\mathfrak{M}$, and $F(R)$ has the eigenvalue 0 for $\psi_1, \psi_2, \ldots$, and therefore for each f of $\mathfrak{R} - \mathfrak{M}$ -- consequently $E = F(R)$. If R belongs to $\mathfrak{R}$, then $\mathfrak{E} = F(\mathfrak{R})$. The $\mathfrak{E}$ measurement can therefore be interpreted as an $\mathfrak{R}$ measurement.

In this case we can calculate how U_0, U are related. According to the $\mathfrak{R}$ measurement each system is in one of the states $\phi_1, \phi_2, \ldots, \psi_1, \psi_2, \ldots$, the particular one depending on which of the values $\lambda_1, \lambda_2, \ldots, \mu_1, \mu_2, \ldots$ was found. The respective probabilities are

$$\mathrm{Tr}\,(U_0 P_{[\phi_1]}) = (U_0\phi_1, \phi_1),\ \mathrm{Tr}\,(U_0 P_{[\phi_2]}) = (U_0\phi_2, \phi_2),\ \ldots,$$

$$\mathrm{Tr}\,(U_0 P_{[\psi_1]}) = (U_0\psi_1, \psi_1),\ \mathrm{Tr}\,(U_0 P_{[\psi_2]}) = (U_0\psi_2, \psi_2),\ \ldots,$$

(cf. the observations of III.3., the validity of which we have established). That is, these fractions of the U_0-ensemble go over into the ensembles $P_{[\phi_1]}, P_{[\phi_2]}, \ldots, P_{[\psi_1]}, P_{[\psi_2]}, \ldots$. Since $\mathfrak{E} = 1$ corresponds to $\mathfrak{R} = \lambda_1, \lambda_2, \ldots$, the U ensemble arises by the inclusion of the first group. Consequently,

$$U = \sum_n (U_0\phi_n, \phi_n)P_{[\phi_n]} .$$

Now each $P_{[\phi_n]}$ commutes with R,[179] and therefore it must also commute with U. That is, if U is not commutable with each R arising in the manner described above, then certain measurement processes (namely, those depending on the corresponding $\mathfrak{R}$) are eliminated from U_0 in the production of U. Then we know more about U than the mere fact that it was created by an $\mathfrak{E}$ measurement. But since U should represent just this state of our knowledge, we try to adhere to this condition: if there exists a U for which no measurement process of $\mathfrak{E}$ need be excluded, then we shall make use of such a U. Therefore, let us investigate whether there are such U and what they are!

As we saw, U must commute with all R produced in the manner described. From this it follows that $RU\phi_n = UR\phi_n = U(\lambda_n\phi_n) = \lambda_n U\phi_n$ i.e., $U\phi_n$ is an eigenfunction of R with the eigenvalue λ_n -- therefore $U\phi_n = a_n\phi_n$. In particular, $U\phi_1 = a\phi_1$. If any ϕ of $\mathfrak{M}$ with $\|\phi\| = 1$ is given, then we can so choose $\phi_1, \phi_2, \ldots, \psi_1, \psi_2, \ldots$ that $\phi_1 = \phi$, and therefore each such ϕ is an eigenfunction of U. All these ϕ must belong to the same eigenvalue: Indeed, if ϕ, ψ belong to different eigenvalues, then they must be orthogonal. Next,

$$\frac{\phi + \psi}{\sqrt{2}}$$

is also an eigenfunction, and since

$$\left(\frac{\phi + \psi}{\sqrt{2}}, \phi\right) = \frac{(\phi, \phi)}{\sqrt{2}} = \frac{1}{\sqrt{2}}, \quad \left(\frac{\phi + \psi}{\sqrt{2}}, \psi\right) = \frac{(\psi, \psi)}{\sqrt{2}} = \frac{1}{\sqrt{2}}$$

---

[179] For example, because

$$RP_{[\phi_n]}f = R((f, \phi_n)\cdot\phi_n) = (f, \phi_n)\cdot R\phi_n = \lambda_n(f, \phi_n)\cdot\phi_n,$$

$$P_{[\phi_n]}Rf = (Rf, \phi_n)\cdot\phi_n = (f, R\phi_n)\cdot\phi_n = \lambda_n(f, \phi_n)\cdot\phi_n.$$

it is orthogonal to neither ϕ nor ψ, and hence belongs to the same eigenvalue as both ϕ and ψ, which is impossible, since these belong to different eigenvalues. Consequently, $U\phi = a\phi$ with constant a. The restriction $\|\phi\| = 1$, to which this result is subject, may clearly be omitted. So for all f of $\mathfrak{M}$, $Uf = af$. Therefore $UEg = aEg$ always, i.e., therefore $UE = aE$, but since $U = UE$, $U = aE$. U, E are both definite and $\neq 0$. Therefore $a > 0$, and hence we can set $U = E$ without changing it essentially.

Conversely, this U fulfills the requirement for each R, i.e., for each set $\phi_1, \phi_2, \dots, \psi_1, \psi_2, \dots$, if U_o is appropriately chosen. For $U_o = 1$,

$$\sum_n (U_o\phi_n, \phi_n)P_{[\phi_n]} = \sum_n (\phi_n, \phi_n)P_{[\phi_n]} = \sum_n P_{[\phi_n]} = P_{\mathfrak{M}} = E = U.$$

Hence $E = U$ is established in the sense of the outline sketched above. Also, U_o can be determined if we assume it to be universal, i.e., independent of E and R. $U_o = 1$ then yields the desired result, and only this. Indeed

$$(U\phi_m, \phi_m) = (E\phi_m, \phi_m) = (\phi_m, \phi_m) = 1,$$

$$(U\phi_m, \phi_m) = \sum_n (U_o\phi_n, \phi_n)(P_{[\phi_n]}\phi_m, \phi_m)$$

$$= \sum_n (U_o\phi_n, \phi_n)|(\phi_n, \phi_m)|^2 = (U_o\phi_m, \phi_m).$$

Therefore $(U_o\phi_m, \phi_m) = 1$. Since each ϕ of $\mathfrak{M}$ with $\|\phi\| = 1$ can be made a ϕ_1 we have $(U_o\phi, \phi) = 1$, and from this it follows for all f of $\mathfrak{M}$ that $(U_of, f) = (f, f)$. Since $\mathfrak{M}$ is arbitrary, this holds for all f in general, and consequently $U_o = 1$.

Consider two properties $\mathfrak{E}$, $\mathfrak{F}$, not necessarily simultaneously decidable. What is the probability that a system $\mathbf{S}$, in which the property $\mathfrak{E}$ has just been found to hold, will, in an immediately following observation, also prove to possess the property $\mathfrak{F}$. By the above, this probability is $\mathrm{Tr}\,(EF) = \sum(EF)$ (E, F are the operators of $\mathfrak{E}$, $\mathfrak{F}$; the first formula holds because $U = E$, the second because $E^2 = E$, $F^2 = F$, by II.11.). Moreover, these probabilities are relative, so that $\mathfrak{E}$ should be considered fixed and $\mathfrak{F}$ variable; if $\mathrm{Tr}\,(E) = \sum(E) =$ dimension number of $\mathfrak{M}$ is finite, we can normalize it by division with this number.

In place of $\mathfrak{E}$, $\mathfrak{F}$ we can also consider physical quantities: Let $\mathfrak{R}_1, \ldots, \mathfrak{R}_j$ and $\mathfrak{S}_1, \ldots, \mathfrak{S}_l$ be the two separate sets of simultaneously measurable quantities (they need not, however, form such a set together); their respective operators $R_1, \ldots, R_j$, $S_1, \ldots, S_l$; and their resolutions of the identity $E_1(\lambda), \ldots, E_j(\lambda)$, $F_1(\lambda), \ldots, F_l(\lambda)$. Let the respective intervals be $I_1 : \lambda_1' < \lambda \leq \lambda_1'', \ldots, I_j :$ $\lambda_j' < \lambda \leq \lambda_j''$, $J_1 : \mu_1' < \lambda \leq \mu_1'', \ldots, J_l : \mu_l' < \lambda \leq \mu_l''$, and let $E_1(I_1) = E_1(\lambda_1'') - E_1(\lambda_1'), \ldots, E_j(I_j) = E_j(\lambda_j'') - E_j(\lambda_j')$, $F_1(J_1) = F_1(\mu_1'') - F_1(\mu_1'), \ldots, F_l(J_l) = F_l(\mu_l'') - F_l(\mu_l')$. The question is: $\mathfrak{R}_1, \ldots, \mathfrak{R}_j$ were measured on $\mathbf{S}$ and their values lay respectively in $I_1, \ldots, I_j$; what is the probability that the values $\mathfrak{S}_1, \ldots, \mathfrak{S}_l$, in a measurement which follows immediately, lie in $J_1, \ldots, J_l$, respectively? Clearly, we must set $E = E_1(I_1) \ldots E_j(I_j)$, $F = F_1(J_1) \ldots F_l(J_l)$. Then the desired probability is (cf. ϵ), ζ))

$$\mathrm{Tr}\,(E_1(I_1) \ldots E_j(I_j) \cdot F_1(J_1) \ldots F_j(I_j))$$

$$= \sum (E_1(I_1) \ldots E_j(I_j) \cdot F_1(I_1) \ldots F_l(J_l))$$

In conclusion, let us refer once again to the meaning of the general initial ensemble $U_0 = 1$. We obtained U from it by resolving it, in the case of the

$\mathfrak{R}$ measurement, into two parts. If we had not so resolved it, i.e., had measured $\mathfrak{R}$ on all its elements and had joined all these together again to form an ensemble, then we would again have obtained $U_o = 1$. This can easily be calculated directly, or can be proved by choosing $E = 1$. Then the $\mu_1, \mu_2, \ldots$ and the $\psi_1, \psi_2, \ldots$ are absent, and the $\lambda_1, \lambda_2, \ldots$ and $\phi_1, \phi_2, \ldots$ form a complete set. Therefore, although the measurement of $\mathfrak{R}$ changes the individual elements under certain circumstances, all these changes must exactly compensate each other, because the entire ensemble does not change. Furthermore, this property is characteristic for $U_o = 1$. Because, if for all complete orthonormal sets $\phi_1, \phi_2, \ldots$

$$U_o = \sum_{n=1}^{\infty} (U_o \phi_n, \phi_n) P_{[\phi_n]} ,$$

then U_o commutes with $P_{[\phi_1]}$, since it does with each $P_{[\phi_n]}$. That is, U_o commutes with each $P_{[\phi]}$, $\|\phi\| = 1$. Therefore

$$U_o \phi = U_o P_{[\phi]} \phi = P_{[\phi]} U_o \phi = (U_o \phi, \phi) \cdot \phi ,$$

i.e., ϕ is an eigenfunction of U_o. From this it follows that $U_o = 1$, exactly as $U = E$ was obtained earlier from the corresponding relation (with $\mathfrak{M}$, E in place of $\mathfrak{R}$, 1).

In $U_o = 1$ therefore, all possible states are in the highest possible degree of equilibrium, and no measuring action can alter this. For each complete orthonormal set $\phi_1, \phi_2, \ldots$,

$$U_o = 1 = \sum_{n=1}^{\infty} P_{[\phi_n]}$$

i.e., the superposition $1:1:\ldots$ of all states $\phi_1, \phi_2, \ldots$.

From this we learn that $U_0 = 1$ corresponds in the older quantum theory to the ordinary thermodynamic assumption of the "a priori equal probability of all simple quantum orbits." It will also play an important role in our thermodynamic considerations, to which the next sections are devoted.

CHAPTER V

GENERAL CONSIDERATIONS

1. MEASUREMENT AND REVERSIBILITY

What happens to a mixture with the statistical operator U, if a quantity $\mathfrak{R}$ with the operator R is measured in it? This operator must be thought of as measuring $\mathfrak{R}$ in each element of the ensemble and collecting the elements that have been thus treated into a new ensemble. We can answer this question -- to the extent to which it admits of an unambiguous answer.

First, let R have a pure discrete, simple spectrum, let $\phi_1, \phi_2, \ldots$ be the complete orthonormal set of eigenfunctions and $\lambda_1, \lambda_2, \ldots$ the corresponding eigenvalues (by assumption, all different from each other). After the measurement, the state of affairs is the following: In the fraction $(U\phi_n, \phi_n)$ of the original ensemble, $\mathfrak{R}$ has the value λ_n $(n = 1, 2, \ldots)$. This fraction then forms an ensemble in which $\mathfrak{R}$ has the value λ_n with certainty (**M.** in IV.3.); it is therefore in the state ϕ_n with the (correctly normalized) statistical operator $P_{[\phi_n]}$. Upon collecting these sub-ensembles, therefore, we obtain a mixture with the statistical operator

$$U' = \sum_{n=1}^{\infty} (U\phi_n, \phi_n) P_{[\phi_n]} .$$

Second, let R have just a pure discrete spectrum, and let the meaning of $\phi_1, \phi_2, \ldots$ and $\lambda_1, \lambda_2, \ldots$ be as before, except that the eigenvalues λ_n are not all simple -- i.e., among the λ_n there are coincidences. Then the measuring process of $\mathfrak{R}$ is not uniquely defined (the same was the case, for example, with $\mathfrak{E}$ in IV.3.). Indeed: Let $\mu_1, \mu_2, \ldots$ be distinct real numbers, and S the operator corresponding to the $\phi_1, \phi_2, \ldots$ and $\mu_1, \mu_2, \ldots$. Let $\mathfrak{S}$ be the corresponding quantity. If F(x) is a function with

$$F(\mu_n) = \lambda_n \qquad (n = 1, 2, \ldots)$$

then $F(S) = R$, therefore $F(\mathfrak{S}) = \mathfrak{R}$. Hence the $\mathfrak{S}$ measurement can also be regarded as an $\mathfrak{R}$ measurement. This now changes U into the U' given above, and U' is independent of the (entirely arbitrary) $\mu_1, \mu_2, \ldots$, but not of the $\phi_1, \phi_2, \ldots$. Yet the $\phi_1, \phi_2, \ldots$ are not uniquely determined, because of the multiplicity of the eigenvalues of R . In IV.2., we stated (following II.8.), what can be said regarding the $\phi_1, \phi_2, \ldots$: Let $\lambda', \lambda'', \ldots$ be the different eigenvalues among the $\lambda_1, \lambda_2, \ldots$, let $\mathfrak{M}_{\lambda'}, \mathfrak{M}_{\lambda''}, \ldots$ be the sets of the f with $Rf = \lambda' f$, $Rf = \lambda'' f$, ... respectively. Finally, let $\chi_1', \chi_2', \ldots$; $\chi_1'', \chi_2'', \ldots$; ... , respectively be arbitrary orthonormal sets which span $\mathfrak{M}_{\lambda'}, \mathfrak{M}_{\lambda''}, \ldots$. Then $\chi_1', \chi_2', \ldots, \chi_1'', \chi_2'', \ldots, \ldots$ is the most general $\phi_1, \phi_2, \ldots$ set. Hence U' may be depending upon the choice of $\mathfrak{S}$, i.e., depending upon the actual measuring arrangement, any expression

$$U' = \sum_n (U\chi_n', \chi_n')P_{[\chi_n']} + \sum_n (U\chi_n'', \chi_n'')P_{[\chi_n'']} + \cdots .$$

This expression, however, is unambiguous only in special cases.

We determine this special case. Each individual term must be unambiguous. That is, for each eigenvalue λ , if $\mathfrak{M}_\lambda$ is the set of the f with $Rf = \lambda f$, the sum

$$\sum_n (U\chi_n, \chi_n) P_{[\chi_n]}$$

must have the same value for every choice of the orthonormal set $\chi_1, \chi_2, \cdots$ spanning the manifold $\mathfrak{M}_\lambda$. If we call this sum V, then verbatim repetition of the observations in IV.3. (in which the U_0, U, $\mathfrak{M}$ there are to be replaced by U, V, $\mathfrak{M}_\lambda$) shows that we must have $V = c_\lambda P_{\mathfrak{M}}$ (c_λ constant, > 0), and that this is equivalent to the validity of $(Uf, f) = c_\lambda(f, f)$ for all f of $\mathfrak{M}_\lambda$. Since these f are the same as the $P_{\mathfrak{M}_\lambda} g$ for all g, we require: $(UP_{\mathfrak{M}_\lambda} g, P_{\mathfrak{M}_\lambda} g) = c_\lambda (P_{\mathfrak{M}_\lambda} g, P_{\mathfrak{M}_\lambda} g)$, i.e., $(P_{\mathfrak{M}_\lambda} UP_{\mathfrak{M}_\lambda} g, g) = c_\lambda (P_{\mathfrak{M}_\lambda} g, g)$, i.e.,

$$P_{\mathfrak{M}_\lambda} UP_{\mathfrak{M}_\lambda} = c_\lambda P_{\mathfrak{M}_\lambda}$$

for all eigenvalues λ of R. But if this condition, clearly restricting U sharply, is not satisfied, then different arrangements of measurement for $\mathfrak{R}$ can actually transform U into different U'. (Nevertheless, we shall succeed in V.4. in making some statements about the result of a general $\mathfrak{R}$ measurement, on a thermodynamical basis.

Third, let R have no pure discrete spectrum. Then by III.3. (or IV.3., criterion 1.), it is not measurable with absolute precision, and $\mathfrak{R}$ measurements of limited precision (as we discussed in the case referred to) are equivalent to measurements of quantities with pure discrete spectra.

Another type of intervention in material systems, in contrast to the discontinuous, non-causal and instantaneously acting experiments or measurements, is given by the time dependent Schrödinger differential equation. This describes how the system changes continuously and causally in the course of time, if its total energy is known. For states ϕ , these equations are

$$(\mathbf{T}_1.) \qquad \frac{\partial}{\partial t}\phi_t = -\frac{2\pi i}{h} H \phi_t$$

where H is the energy operator.

For the statistical operator of the state ϕ_t , $U_t = P_{[\phi_t]}$, this means:

$$\left(\frac{\partial}{\partial t} U_t\right)f = \frac{\partial}{\partial t}(U_t f) = \frac{\partial}{\partial t}((f, \phi_t)\cdot\phi_t) = (f, \frac{\partial}{\partial t}\phi_t)\cdot\phi_t + (f, \phi_t)\cdot\frac{\partial}{\partial t}\phi_t$$

$$= -(f, \frac{2\pi i}{h} H\phi_t)\cdot\phi_t - (f, \phi_t)\frac{2\pi i}{h} H \phi_t$$

$$= \frac{2\pi i}{h}((Hf, \phi_t)\cdot\phi_t - (f, \phi_t)\cdot H\phi_t) = \frac{2\pi i}{h}(U_t H - H U_t)f ,$$

that is:

$$(\mathbf{T}_2.) \qquad \frac{\partial}{\partial t} U_t = \frac{2\pi i}{h}(U_t H - H U_t) .$$

Now if U_t is not a state, but a mixture of several states, say $P_{[\phi_t^{(1)}]}, P_{[\phi_t^{(2)}]}, \ldots$ with the respective weights $w_1, w_2, \ldots$, then it must be changed in such a way as results from the changes of the individual $P_{[\phi_t^{(1)}]}, P_{[\phi_t^{(2)}]}, \ldots$. By the addition of the corresponding equations $\mathbf{T}_2.$, we recognize that $\mathbf{T}_2.$ holds for this U_t also. Now since all U are such mixtures, or limiting cases of such (for example, each U with finite $\mathrm{Tr}\, U$ is such a mixture), we can claim the general validity of $\mathbf{T}_2.$

In $\mathbf{T}_2.$, moreover, H may also depend on t , just as in the Schrödinger differential equation $\mathbf{T}_1.$ If that is not the case, then we can even given explicit solutions: For $\mathbf{T}_1.$, as we already know,

$$(\mathbf{T}_1'.) \qquad \phi_t = e^{-\frac{2\pi i}{h} t \cdot H} \phi_0 \quad ,$$

and for $\mathbf{T}_2$.,

$$(\mathbf{T}_2'.) \qquad U_t = e^{-\frac{2\pi i}{h} t \cdot H} U_0 e^{\frac{2\pi i}{h} t H} \; .$$

(It is easily verified that these are solutions, and also that they follow from each other. It is clear also that there is only one solution with a fixed initial ϕ_0 or U_0 respectively: the differential equations $\mathbf{T}_1$., $\mathbf{T}_2$. are of first order in t.)

We therefore have two fundamentally different types of interventions which can occur in a system $\mathbf{S}$ or in an ensemble $[\mathbf{S}_1, \ldots, \mathbf{S}_N]$. First, the arbitrary changes by measurements which are given by the formula

$$(\mathbf{1}.) \quad U \longrightarrow U' = \sum_{n=1}^{\infty} (U\phi_n, \phi_n) P_{[\phi_n]}$$

($\phi_1, \phi_2, \ldots$ a complete orthonormal set, cf. supra). Second, the automatic changes which occur with passage of time. These are given by the formula

$$(\mathbf{2}.) \quad U \longrightarrow U_t = e^{-\frac{2\pi i}{h} \cdot tH} U e^{\frac{2\pi i}{h} tH}$$

(H is the energy operator, t the time; H is independent of t). If H depends on t, then we may divide the time interval under consideration into small time intervals in each one of which H does not change -- or changes only very slightly, and apply **2**. to these individual intervals. Superposition then gives the final result.

We must now analyze in more detail these two types of intervention, their nature, and their relation one to another.

First of all, it is noteworthy that the time

dependence of H is included in **2.** (in the manner described there), so that one should expect that **2.** would suffice to describe the intervention caused by a measurement: Indeed, a physical intervention can be nothing else than the temporary insertion of a certain energy coupling into the observed system, i.e., the introduction of an appropriate time dependency of H (prescribed by the observer). Why then do we need the special process **1.** for the measurement? The reason is this: In the measurement we cannot observe the system **S** by itself, but must rather investigate the system **S + M**, in order to obtain (numerically) its interaction with the measuring apparatus **M**. The theory of the measurement is a statement concerning **S + M**, and should describe how the state of **S** is related to certain properties of the state of **M** (namely, the positions of a certain pointer, since the observer reads these). Moreover, it is rather arbitrary whether or not one includes the observer in **M**, and replaces the relation between the **S** state and the pointer positions in **M** by the relations of this state and the chemical changes in the observer's eye or even in his brain (i.e., to that which he has "seen" or "perceived"). We shall investigate this more precisely in VI.1. In any case, therefore, the application of **2.** is of importance only for **S + M**. Of course, we must show that this gives the same result for **S** as the direct application of **1.** on **S**. If this is successful, then we have achieved a unified way of looking at the physical world on a quantum mechanical basis. We postpone the discussion of this question until VI.3.

Second, it is to be noted, with regard to **1.**, that we have repeatedly shown that a measurement in the sense of **1.** must be instantaneous, i.e., must be carried through in so short a time that the change of U given by **2.** is not yet noticeable. (If we wanted to correct this by calculating the changed U_t by **2.**, we would still gain nothing, because to apply any U_t, we must first know t, the moment of measurement, exactly, i.e., the time duration of the measurement must be short.) This is now questionable in principle, because it is well-known that there is a quantity which, in classical mechanics, is canonically conjugate with the time: the energy.[180] Therefore it is to be expected that for the canonically conjugate pair time-energy, there must exist indeterminacy relations similar to those of the pair cartesian coordinate-momentum.[181] Note that the special relativity theory shows that a far reaching analogy must exist: the three space coordinates and time form a "four vector" as do the three momentum coordinates and the energy. Such an indeterminacy relation would mean that it is not possible to carry out a very precise measurement of the energy in a very short time. In fact, one would expect for the error of measurement (in the energy) and the time duration τ a relation of the form

$$\epsilon\tau \sim h \ .$$

A physical discussion, similar to that carried out in III.4. for p, q, actually leads to this result.[181] Without going into details, we shall consider the case of a light quantum. Its energy uncertainty ϵ is, because of the Bohr frequency condition, h times the frequency uncertainty: $h\Delta\nu$. But, as we discussed in Note 137, $\Delta\nu$ is at best the reciprocal of the time duration, $1/\tau$, i.e., $\epsilon \gtrsim h/\tau$ -- and in order that the monochromatic nature of the light quantum be established in the entire time interval τ, the measurement must extend over this entire time interval. The case of the light quantum is characteristic,

---

[180] Any textbook of classical (Hamiltonian) mechanics gives an account of these connections.

[181] The uncertainty relations for the pair time-energy have been discussed frequently. Cf. the comprehensive treatment of Heisenberg, Die Physikalischen Prinzipien der Quantentheorie, II.2.d., Leipzig, 1930.

since the atomic energy levels, as a rule, are determined from the frequency of the corresponding spectral lines. Since the energy behaves in such fashion, a relation between the precision of measurement for other quantities $\mathfrak{R}$ and the duration of the measurement is also possible. Then how can our assumption of instantaneous measurements be justified?

First of all we must admit that this objection points at an essential weakness which is, in fact, the chief weakness of quantum mechanics: its non-relativistic character, which distinguishes the time t from the three space coordinates x, y, z, and presupposes an objective simultaneity concept. In fact, while all other quantities (especially those x, y, z closely connected with t by the Lorentz transformation) are represented by operators, there corresponds to the time an ordinary number-parameter t, just as in classical mechanics. Or: a system consisting of 2 particles has a wave function which depends on its $2 \times 3 = 6$ space coordinates, and only upon one time t, although, because of the Lorentz transformation, two times would be desirable. It may be connected with this non-relativistic character of quantum mechanics that we can ignore the natural law of minimum duration of the measurements. This might be a clarification, but not a happy one!

A more detailed investigation of the problem, however, shows that the situation is really not so bad as this. For what we really need is not that the change of t be small, but only that it have little effect in the calculation of the probabilities $(U\phi_n, \phi_n)$, and therefore in the formation of

$$U' = \sum_{n=1}^{\infty} (U\phi_n, \phi_n) P_{[\phi_n]}$$

whether we start out from U itself or from a

$$U_t = e^{-\frac{2\pi i}{h} tH} U e^{\frac{2\pi i}{h} tH} .$$

Because of

$$(U_t\phi_n, \phi_n) = \left(e^{-\frac{2\pi i}{h}tH} U e^{\frac{2\pi i}{h}tH}\phi_n, \phi_n\right)$$
$$= \left(\overline{U}e^{\frac{2\pi i}{h}tH}\phi_n, e^{\frac{2\pi i}{h}tH}\phi_n\right),$$

this can be accomplished by so changing H by an appropriate perturbation energy that

$$e^{\frac{2\pi i}{h}tH}\phi_n$$

differs from ϕ_n only by a constant factor of absolute value 1 . That is, the state ϕ_n should be essentially constant under the influence of **2**., i.e., a stationary state; or equivalently $H\phi_n$ must be equal to a real constant times ϕ_n , i.e., ϕ_n an eigenfunction of H. At first glance, such a change of the energy operator H, which makes the eigenfunctions of R stationary, and therefore eigenfunctions of H (i.e., R, H commutative) may seem implausible. But this is not really the case, and one can even see that the typical arrangements of measurement aim at exactly this sort of effect on H.

In fact, each measurement results in the emission of a light quantum or a mass particle, with a certain energy, in a certain direction. It is then by these characteristics, i.e., by its momentum, that the particle expresses the result of the measurement or, a mass point (for example, a pointer on a scale) comes to rest, and its cartesian coordinates give the result of the measurement. In the case of light quanta, using the terminology of III.6., the desired measurement is thus equivalent to the statement as to which $M_n = 1$ (the rest being $= 0$) , i.e., to the enumeration of all $M_1, M_2, \ldots$ values. For a moving (departing) mass point, the statement of its three momentum components P^x, P^y, P^z is the corresponding equivalent; for a mass point at rest (the index point), the statement of its three cartesian coordinates x, y, z, or, using their operators, of the Q^x, Q^y, Q^z. But the measurement is completed only if the light quantum or mass point is actually borne "away," i.e., only when the light quantum is not in danger of absorption; or when the mass point may no longer be deflected by potential energies; or, if the mass point is actually at rest, in which case a large mass is necessary.[182] (This latter is certainly necessary because of the uncertainty relations, since the velocity must be near 0, and therefore its dispersion must be small, although its product with the mass -- the momentum -- has a large dispersion, because of the small dispersion of the coordinates. Ordinarily, the pointers are macroscopic objects, i.e., enormous.) Now the energy operator H, so far as it concerns the light quantum, is (III.6, page 270)

$$\sum_{n=1}^{\infty} h\rho_n \cdot M_n$$

$$+ \sum_{p=1}^{\infty} \sum_{n=1}^{\infty} w_{kj}^{n} \left(\sqrt{M_n + 1} \cdot \left(\begin{matrix} k \\ M_n \end{matrix} \rightleftarrows \begin{matrix} j \\ M_n + 1 \end{matrix} \right) + \sqrt{M_n} \cdot \begin{matrix} k \\ M_n \end{matrix} \rightleftarrows \begin{matrix} j \\ M_n - 1 \end{matrix} \right)$$

while for both mass point examples, H is given by

$$\frac{(P^x)^2 + (P^y)^2 + (P^z)^2}{2m} + V(Q^x, Q^y, Q^z)$$

---

[182] All other details of the measuring arrangement aim only at the connection of the quantity $\mathfrak{R}$, which is actually of interest, or of its operator R, with the M_n or the P^x, P^y, P^z or the Q^x, Q^y, Q^z, respectively, that have

(m the mass, V the potential energy). Our criteria say: the w^n_{kj} should vanish, or V should be constant, or m should be very large. But this actually produces the effect that the P^x, P^y, P^z and the Q^x, Q^y, Q^z respectively commute with the H given above.

In conclusion, it should be mentioned that the making stationary of the really interesting states (here the $\phi_1, \phi_2, \ldots$) plays a role elsewhere, too, in theoretical physics. The assumptions on the possibility of the interruption of chemical reactions (i.e., their "poisoning"), which are often unavoidable in physical-chemical "ideal experiments," are of this nature.[183]

The two interventions **1.** and **2.** are fundamentally different from one another. That both are formally unique, i.e., causal, is unimportant; indeed, since we are working in terms of the statistical properties of mixtures, it is not surprising that each change, even if it is statistical, effects a causal change of the probabilities and the expectation values. Indeed, it is precisely for this reason, that one introduces statistical ensembles and probabilities! On the other hand, it is important that **2.** does not increase the statistical uncertainty existing in U, but that **1.** does: **2.** transforms states into states

$$\left(P_{[\phi]} \quad \text{into} \quad P_{[e^{-\frac{2\pi i}{h} tH} \phi]} \right)$$

while **1.** can transform states into mixtures. In this sense, therefore, the development of a state according to **1.** is statistical, while according to **2.** it is causal.

---

been mentioned. Of course, this is the most important practical aspect of the measuring technique.

[183] Cf. e.g., Nernst, Theoretische Chemie, Stuttgart (numerous editions since 1893), Book IV, Discussion of the thermodynamic proof of the "mass action law."

Furthermore, for fixed H and t , **2.** is simply a unitary transformation of all U : $U_t = AUA^{-1}$, $A = e^{-\frac{2\pi i}{h} tH}$ is unitary. That is, $Uf = g$ implies that $U_t(Af) = Ag$, so that U_t results from U by the unitary transformation A of Hilbert space, that is, by an isomorphism which leaves all our basic geometric concepts invariant (cf. the principles set down in I.4.). Therefore it is reversible: it suffices to replace A by A^{-1} -- and this is possible, since A, A^{-1} can be regarded as entirely arbitrary unitary operators because of the far reaching freedom in the choice of H, t . Just as in classical mechanics therefore, **2.** does not reproduce one of the most important and striking properties of the real world, namely its irreversibility, the fundamental difference between the time directions, "future" and "past."

1. behaves in a fundamentally different fashion: the transition

$$U \longrightarrow U' = \sum_{n=1}^{\infty} (U\phi_n, \phi_n) P_{[\phi_n]}$$

is certainly not prima facie reversible. We shall soon see that it is in general irreversible, in the sense that it is not possible in general to come back from a given U' to its U by repeated applications of any processes ., **2.**

Therefore, we have reached a point at which it is desirable to utilize the thermodynamical method of analysis, because it alone makes it possible for us to understand correctly the difference between **1.** and **2.**, into which reversibility questions obviously enter.

2. THERMODYNAMICAL CONSIDERATIONS

We shall investigate the thermodynamics of quantum mechanical ensembles according to two different points of view. First, let us assume the validity of both fundamental laws of thermodynamics, i.e., the impossibility of perpetual motion of the first and second kind (energy law and entropy law),[184] and calculate the entropy for each ensemble from this. In this case, normal methods of the phenomenological thermodynamics are applied, and quantum mechanics plays a role only insofar as our thermodynamical observations relate to such objects whose behavior is regulated by the laws of quantum mechanics (our ensembles, as well as their statistical operators U) -- but the correctness of both laws will be assumed and not proved. Afterwards we shall prove the validity of these fundamental laws in quantum mechanics. Since the energy law holds in any case, only the entropy law has to be considered. That is, we shall show that the interventions **1.**, **2.** never decrease the entropy, as calculated by the first method. This order may seem somewhat unnatural, but it is based on the fact that it is by the phenomenological discussion that we obtain that overall view of the problem which is required for considerations of the second kind.

We therefore begin with the phenomenological consideration, which will also permit us to solve a well-known paradox of classical thermodynamics. First we must emphasize that the unusual character of our "ideal experiments," i.e., their practical infeasibility, does not impair their demonstrative power: In the sense of phenomenological thermodynamics, each conceivable process constitutes valid evidence, provided that it does not conflict with the two fundamental laws of thermodynamics.

---

[184]The phenomenological system of thermodynamics built upon this foundation can be found in numerous texts. For example, Planck, Treatise on Thermodynamics, London, 1927. For the following, the statistical aspect of these laws is of chief importance. This is analyzed in the following treatises: Einstein, Verh. d. dtsch. physik, Ges. 12 (1914); Szilard, Z. Physik 32 (1925).

Our purpose is to determine the entropy of an ensemble $[\mathbf{s}_1, \ldots, \mathbf{s}_N]$ with the statistical operator U, where U is assumed to be correctly normalized, i.e., Tr U = 1. In the terminology of classical statistical mechanics, we are dealing with a Gibbs ensemble: i.e., the application of statistics and thermodynamics will be made not on the (interacting) components of a single, very complicated mechanical system with many (only imperfectly known) degrees of freedom[185] -- but on an ensemble of very many (identical) mechanical systems, each of which may have an arbitrarily large number of degrees of freedom, and each of which is entirely separated from the others, and does not interact with any of them.[186] As a consequence of the complete separation of the systems $\mathbf{s}_1, \ldots, \mathbf{s}_N$, and of the fact that we shall apply to them the ordinary methods of enumeration of the calculus of probability, it is evident that ordinary statistics be used, and that the Bose-Einstein and Fermi-Dirac statistics, which differ from those and which are applicable to certain ensembles of indistinguishable and interacting particles (namely, for light quanta or electrons and protons, cf. III.6., in particular, Note 147), do not enter into the problem.

---

[185]This is the Maxwell-Boltzmann method of statistical mechanics (cf. the review in the article of P. and T. Ehrenfest in Enzykl. d. Math. Wiss., Vol. II.4. D., Leipzig, 1907). In the gas theory for example, the "very complicated" system is the gas which consists of many (interacting) molecules, and the molecules are investigated statistically.

[186]This is the Gibbs method (cf. the reference in Note 185). Here the individual system is the entire gas, and many replicas of the same system (i.e., of the same gas) are considered simultaneously, and their properties are evaluated statistically.

The method introduced by Einstein for the thermodynamical treatment of such ensembles $[\mathbf{S}_1, \ldots, \mathbf{S}_N]$ is the following:[187] Each system $\mathbf{S}_1, \ldots, \mathbf{S}_N$ is confined in a box $\mathbf{K}_1, \ldots, \mathbf{K}_N$, whose walls are impenetrable to all transmission effects -- which is possible for this system because of the lack of interaction. Furthermore, each box must have a very large mass, so that the possible state (and hence energy and mass) changes of the $\mathbf{S}_1, \ldots, \mathbf{S}_N$ affects its mass only slightly. Also, their velocities in the ideal experiments which are to be carried out are thereby kept so small that the calculations may be performed non-relativistically. We then enclose these boxes into a very large box $\overline{\mathbf{K}}$ (i.e., the volume $\mathcal{V}$ of $\overline{\mathbf{K}}$ should be much larger than the sum of the volumes of the $\mathbf{K}_1, \ldots, \mathbf{K}_N$). For simplicity, no force field will be present in $\overline{\mathbf{K}}$ (in particular, it should be free from all gravitational fields, and so large that the masses of the $\mathbf{K}_1, \ldots, \mathbf{K}_N$ have no relevant effects either. We can therefore regard the $\mathbf{K}_1, \ldots, \mathbf{K}_N$ (which contain $\mathbf{S}_1, \ldots, \mathbf{S}_N$ respectively) as the molecules of a gas which is enclosed in the large container $\overline{\mathbf{K}}$. If we now bring $\overline{\mathbf{K}}$ into contact with a very large heat reservoir of temperature $\mathbf{T}$, then the walls of $\overline{\mathbf{K}}$ also take on this temperature, and its (true) molecules assume the corresponding Brownian motion. Therefore they will contribute momentum to the adjacent $\mathbf{K}_1, \ldots, \mathbf{K}_N$, so that these engage in motion, and transfer momentum to the other $\mathbf{K}_1, \ldots, \mathbf{K}_N$. Soon all $\mathbf{K}_1, \ldots, \mathbf{K}_N$ will be in motion and will be exchanging momentum (on the wall of $\overline{\mathbf{K}}$) with the (true) molecules of the wall, and with each other (in the interior of $\overline{\mathbf{K}}$) by collision processes. The stationary equilibrium state of motion is then obtained if the $\mathbf{K}_1, \ldots, \mathbf{K}_n$ have taken on that velocity distribution which is in equilibrium with the Brownian motion of the wall

[187] See the reference in Note 184. This was further developed by L. Szilard.

molecules (of temperature $\mathbf{T}$) -- i.e., the Maxwellian velocity distribution of a gas of temperature $\mathbf{T}$, the "molecules" of which are the $\mathbf{K}_1, \ldots, \mathbf{K}_N$.[188] We can then say: the $[\mathbf{s}_1, \ldots, \mathbf{s}_N]$-gas has taken on the temperature $\mathbf{T}$. For brevity, we shall call the ensemble $[\mathbf{s}_1, \ldots, \mathbf{s}_N]$ with the statistical operator U the U-ensemble, and the $[\mathbf{s}_1, \ldots, \mathbf{s}_N]$-gas the U-gas.

The reason that we concern ourselves with such a gas is that we must determine the entropy difference of the U-ensemble and the V-ensemble (U, V definite operators with $\mathrm{Tr}\, U = 1$, $\mathrm{Tr}\, V = 1$, and with the corresponding ensembles $[\mathbf{s}_1, \ldots, \mathbf{s}_N]$ and $[\mathbf{s}'_1, \ldots, \mathbf{s}'_N]$). The determination requires by definition a reversible transformation of the former ensemble into the latter,[189] and this is best accomplished by the aid of the U- and V-gases. That is, we maintain that the entropy difference of the U- and V-ensembles is exactly the same as that of the U- and V-gases -- if both are observed at the same temperature $\mathbf{T}$, but are otherwise arbitrary. If $\mathbf{T}$ is very near 0, then this is obviously the case with arbitrary precision; because the difference between the U-ensemble and the V-gas vanishes at the temperature 0, since the $\mathbf{K}_1, \ldots, \mathbf{K}_N$ of the latter have then no motion of their own, and the

---

[188]The kinetic theory of gases, as is well-known, describes in this way that process in which the walls communicate their temperature to the gas enclosed by them. Cf. the references in Notes 184 and 185.

[189]In this transformation, if the heat quantities $\mathbf{Q}_1, \ldots, \mathbf{Q}_i$ are required at the respective temperatures $\mathbf{T}_1, \ldots, \mathbf{T}_i$, then the entropy difference is equal to

$$\frac{\mathbf{Q}_1}{\mathbf{T}_1} + \frac{\mathbf{Q}_2}{\mathbf{T}_2} + \cdots + \frac{\mathbf{Q}_i}{\mathbf{T}_i} .$$

Cf. the reference in Note 184.

presence of the $K_1, \ldots, K_N$, $\bar{K}$, when they are at rest is thermodynamically unimportant (and likewise for V). Therefore we shall have accomplished our aim if we can show that for a given change of T, the entropy of the U-gas changes just as much as the entropy of the V-gas. The entropy change of a gas which is heated from T_1 to T_2 depends only upon its caloric equation of state, or more precisely, upon its specific heat.[190] Naturally, the gas must not be assumed to be an ideal gas here if, as in our case, T_1 must be chosen near 0.[191] On the other hand, it is certain that both gases (U and V) have the same equation of state and the same specific heats because, by kinetic theory, the boxes $K_1, \ldots, K_N$ dominate and cover completely the systems $S_1, \ldots, S_N$ and $S'_1, \ldots, S'_N$ which are enclosed in them. In this heating process therefore, the difference of U and V is not noticeable, and the two entropy differences coincide, as was maintained. In the following therefore, we shall compare only the U- and V-gases with each other, and we shall choose the temperature T so high that these can be regarded as ideal gases.[192] In this way, we control its kinetic behavior

---

[190] If $c(T)$ is the specific heat at the temperature T of the gas quantum under discussion, then in the temperature interval T, $T + dT$ it takes on the quantity of heat $c(T)dT$. By Note 185, the entropy difference is then

$$\int_{T_1}^{T_2} \frac{c(T)dT}{T}$$

[191] For an ideal gas, $c(T)$ is constant; for very small T, this certainly fails. Cf. for example, the reference in Note 6.

[192] In addition to this, it is required that the volume V of $\bar{K}$ be large in comparison to the total volume of the

completely, and we can apply ourselves to the real problem: to transform U-gas reversibly into V-gas. In this case, in contrast to the processes used so far, we shall also have to consider the $\mathbf{s}_1, \ldots, \mathbf{s}_N$ found in the interior of the $\mathbf{K}_1, \ldots, \mathbf{K}_N$ i.e., we shall have to "open" the boxes $\mathbf{K}_1, \ldots, \mathbf{K}_N$.

Next, we show that all states $U = P_{[\phi]}$ have the same entropy, i.e., that the reversible transformation of the $P_{[\phi]}$ ensemble into the $P_{[\psi]}$ ensemble is accomplished without the absorption or liberation of heat energy (mechanical energy must naturally be consumed or produced if the expectation value of the energy in $P_{[\phi]}$ is different from that in $P_{[\psi]}$), cf. Note 189. In fact, we shall not even have to refer to the gases just considered. This transformation succeeds even at the temperature 0, i.e., with the ensembles themselves. It should be mentioned, furthermore, that as soon as this is proved, we shall be able to and shall so normalize the entropies of the U ensembles that all states have the entropy 0.

Moreover, the transformation of $P_{[\phi]}$ into $P_{[\psi]}$ described above does not need to be reversible: Because if it is not so, then the entropy difference must be $\geq$ the expression given in Note 189 (cf. reference in Note 185), therefore ≥ 0. Permutation of $P_{[\phi]}$, $P_{[\psi]}$ shows that this value must also be ≤ 0. Therefore the value is $= 0$.

$\mathbf{K}_1, \ldots, \mathbf{K}_N$; furthermore that the "energy per degree of freedom" $\kappa \mathbf{T}$ (κ = Boltzmann's constant) be large in comparison to

$$h^2 / \mu \mathcal{V}^{2/3}$$

(h = Planck's constant, μ = mass of the individual molecule; this quantity is of the dimensions of energy). Cf. for example, Fermi, Z. Physik, 36 (1926).

The simplest process would be to refer to the time dependent Schrödinger differential equation, i.e., our process **2**., in which an energy operator H and a numerical value of t must be found such that the unitary operator

$$e^{-\frac{2\pi i}{h} t H}$$

transforms ϕ into ψ. Then, in t seconds, $P_{[\phi]}$ would change spontaneously into $P_{[\psi]}$. The process is also reversible, and no mention has been made of the heat (cf. V.1.). However, we prefer to avoid assumptions regarding the possible forms of the energy operators H and to apply the process **1**. alone, i.e., measuring interventions. The simplest such measurement would be to measure the quantity $\mathfrak{R}$ in the ensemble $P_{[\phi]}$, whose operator R has a pure discrete spectrum with simple eigenvalues $\lambda_1, \lambda_2, \ldots$, and in which ψ occurs among the eigenfunctions $\psi_1, \psi_2, \ldots$, say $\psi_1 = \psi$. This measurement transforms ϕ into a mixture of the states $\psi_1, \psi_2, \ldots$, and there $\psi_1 = \psi$ will be present along with the other states ψ_n. However, this procedure is unsuitable, because $\psi_1 = \psi$ occurs only with the probability $|(\phi, \psi)|^2$, while the portion $1 - |(\phi, \psi)|^2$ goes over into other states. In fact, the latter portion is the entire result for orthogonal ϕ, ψ. A different experiment however will accomplish our purpose. By repetition of a great number of different measurements, we shall change $P_{[\phi]}$ into such an ensemble, which differs from $P_{[\psi]}$ by an arbitrarily small amount. That all these operators are (or at least, can be) irreversible is unimportant, as we discussed above.

We assume ϕ, ψ orthogonal, since we could otherwise choose a χ ($\|\chi\| = 1$) orthogonal to both, and could go from ϕ to χ, and then from χ to ψ. Now let $k = 1, 2, \ldots$ be a number which is at our disposal, and set $\psi^{(\nu)} = \cos \frac{\pi\nu}{2k} \cdot \phi + \sin \frac{\pi\nu}{2k} \cdot \psi$ ($\nu = 0, 1, \ldots, k$). Clearly, $\psi^{(0)} = \phi$, $\psi^{(k)} = \psi$, and $\|\psi^{(\nu)}\| = 1$. We extend each $\psi^{(\nu)}$ $(\nu = 1,2,\ldots,k)$ to a complete orthonormal set $\psi_1^{(\nu)},\psi_2^{(\nu)},\ldots$ with $\psi_1^{(\nu)} = \psi^{(\nu)}$. Let $R^{(\nu)}$ be an operator with a pure discrete spectrum and different eigenvalues, say $\lambda_1^{(\nu)},\lambda_2^{(\nu)},\ldots$, whose eigenfunctions are the $\psi_1^{(\nu)},\psi_2^{(\nu)},\ldots$, and $\mathfrak{R}^{(\nu)}$ the corresponding quantity. We observe further that

$$(\psi^{(\nu-1)}, \psi^{(\nu)}) = \cos\frac{\pi(\nu - 1)}{2k}\cos\frac{\pi\nu}{2k} + \sin\frac{\pi(\nu - 1)}{2k}\sin\frac{\pi\nu}{2k}$$

$$= \cos\left(\frac{\pi\nu}{2k} - \frac{\pi(\nu - 1)}{2k}\right) = \cos\frac{\pi}{2k} .$$

In the ensemble with $U^{(0)} = P_{[\phi^{(0)}]} = P_{[\phi]}$ we now measure the quantity $\mathfrak{R}^{(1)}$, in which case $U^{(1)}$ results. We then measure the quantity $\mathfrak{R}^{(2)}$ on $U^{(1)}$, when $U^{(2)}$ results, etc. We finally measure the quantity $\mathfrak{R}^{(k)}$ on $U^{(k-1)}$ whence $U^{(k)}$ results. That $U^{(k)}$, for sufficiently large k, lies arbitrarily close to $P_{[\psi^{(k)}]} = P_{[\psi]}$ can easily be established. If we measure $\mathfrak{R}^{(\nu)}$ on $\psi^{(\nu-1)}$, then the fraction $|(\psi^{(\nu-1)}, \psi^{(\nu)})|^2 = (\cos\frac{\pi}{2k})^2$ goes over into $\psi^{(\nu)}$, and in the successive measurements of $\mathfrak{R}^{(1)}, \mathfrak{R}^{(2)},\ldots, \mathfrak{R}^{(k)}$ therefore, at least the fraction $(\cos\frac{\pi}{2k})^2$ will go over from $\psi^{(0)} = \phi$ over $\psi^{(1)},\psi^{(2)},\ldots,\psi^{(k-1)}$ into $\psi = \psi^{(k)}$. And since $(\cos\frac{\pi}{2k})^{2k} \longrightarrow 1$ as $k \longrightarrow \infty$, ψ results as nearly exclusively as one may wish, if k is sufficiently large. The exact proof runs as follows. Since the process **1.** does not change the trace, and since $\operatorname{Tr} U^{(0)} = \operatorname{Tr} P_{[\phi]} = 1$, therefore $\operatorname{Tr} U^{(1)} = \operatorname{Tr} U^{(2)} = \ldots = \operatorname{Tr} U^{(k)} = 1$. On the other hand,

$$(U^{(\nu)}f, f) = \sum_n (U^{(\nu-1)}\psi_n^{(\nu)}, \psi_n^{(\nu)})(P_{[\psi_n^{(\nu)}]}f, f)$$

$$= \sum_n (U^{(\nu-1)}\psi_n^{(\nu)}, \psi_n^{(\nu)})|(\psi_n^{(\nu)}, f)|^2 .$$

Therefore, for $\nu = 1,\ldots,k-1$ and $f = \psi_1^{(\nu+1)} = \psi^{(\nu+1)}$, and for $\nu = k$ and $f = \psi_1^{(k)} = \psi^{(k)} = \psi$, we have:

$$(U^{(\nu)}\psi^{(\nu+1)},\psi^{(\nu+1)}) \geq (U^{(\nu-1)}\psi^{(\nu)},\ \psi^{(\nu)})|(\psi^{(\nu)},\ \psi^{(\nu+1)})|^2$$

$$= (\cos\tfrac{\pi}{2k})^2\cdot(U^{(\nu-1)}\psi^{(\nu)},\ \psi^{(\nu)}) ,$$

$$(U^{(k)}\psi^{(k)},\ \psi^{(k)}) = (U^{(k-1)}\psi^{(k)},\ \psi^{(k)})$$

together with

$$(\overline{U}^{(o)}\psi^{(1)},\ \psi^{(1)}) = (P_{[\psi^{(o)}]}\psi^{(1)},\ \psi^{(1)}) = |(\psi^{(o)},\ \psi^{(1)})|^2$$

$$= (\cos\tfrac{\pi}{2k})^2 ,$$

this gives

$$(U^{(k)}\psi,\ \psi) \geq (\cos\tfrac{\pi}{2k})^{2k} .$$

Since $\mathrm{Tr}\, U^{(k)} = 1$ and $(\cos\frac{\pi}{2k})^{2k} \longrightarrow 1$ as $k \longrightarrow \infty$, we can apply the result obtained in II.11.: $U^{(k)}$ converges to $P_{[\psi]}$. Hence our aim is accomplished.

How far may we use one of the main instruments of "ideal experiments" of phenomenological thermodynamics, namely the so-called semipermeable walls, when dealing with quantum mechanical systems?

In phenomenological thermodynamics, this theorem holds: If I and II are two different states of the same system **S**, then it is permissible to assume the existence of a wall which is completely permeable for I and not permeable for II[193] -- this is, so to speak, the thermodynamical definition of difference, and therefore of

---

[193]Cf. for example, the reference in Note 184.

equality also, for two systems. How far is such an assumption permissible in quantum mechanics?

We first show that if $\phi_1, \phi_2, \ldots, \psi_1, \psi_2, \ldots$ is an orthonormal set, then there is a semi-permeable wall which lets the system $\mathbf{S}$ in each of the states $\phi_1, \phi_2, \ldots$ pass through unhindered, and which reflects unchanged the system in each of the states $\psi_1, \psi_2, \ldots$. Systems which are in other states may, on the other hand, be changed by collision with the wall.

The system $\phi_1, \phi_2, \ldots, \psi_1, \psi_2, \ldots$ can be assumed to be complete, since otherwise it could be made so by additional $\chi_1, \chi_2, \ldots$ which one could then add to the $\phi_1, \phi_2, \ldots$. We now choose an operator R with a pure discrete spectrum, and only simple eigenvalues $\lambda_1, \lambda_2, \ldots, \mu_1, \mu_2, \ldots$ whose eigenfunctions are $\phi_1, \phi_2, \ldots, \psi_1, \psi_2, \ldots$ respectively. In fact, let the $\lambda_n < 0$ and the $\mu_n > 0$. Let the quantity $\mathfrak{R}$ belong to R. We construct many windows in the wall, each of which is defined as follows: each "molecule" $\mathbf{K}_1, \ldots, \mathbf{K}_N$ of our gas (we are again considering U-gases at the temperature $\mathbf{T} > 0$) is detained there, opened, the quantity $\mathfrak{R}$ measured on the system $\mathbf{S}_1$ or $\mathbf{S}_2$ or $\ldots$ $\mathbf{S}_N$ contained in it. Then the box is closed again, and according to whether the measured value of $\mathfrak{R}$ is < 0 or > 0, the box, together with its contents, penetrates the window or is reflected, with unchanged momentum. That this contrivance satisfies the desired end is clear -- it remains only to discuss what changes remain in it after such collisions, and how closely it is related to the so-called "Maxwell's demon" of thermodynamics.[194]

In the first place, it must be said that since the measurement (under certain circumstances) changes the

---

[194] Cf. the reference in Note 185. The reader will find a detailed discussion of the difficulties connected with the concept of "Maxwell's demon" in L. Szilard, Z. Physik, <u>53</u> (1929).

state of **S**, and perhaps its energy expectation value also, this difference in the mechanical energy must be added or absorbed by the measurement action, in the sense of the first law of thermodynamics (for example, by installing a spring which can be extended or compressed, or something similar). Since it is a case of a purely automatically functioning measuring mechanism, and since only mechanical (not heat!) energies are transformed, certainly no entropy changes occur, and at present, only this is of importance to us. (If **S** is in one of the states $\phi_1, \phi_2, \ldots, \psi_1, \psi_2, \ldots$, then the $\mathfrak{R}$ measurement does not, in general, change **S**, and no compensating changes remain in the measuring apparatus.)

The second point is more doubtful. Our arrangement is rather similar to "Maxwell's demon," i.e., to a semi-permeable wall which transmits molecules coming from the right and reflects those coming from the left. If we insert such a wall in the midst of a container filled with a gas, then all the gas is soon on the left hand side -- i.e., the volume is halved without entropy consumption. This means an uncompensated entropy increase of the gas, and therefore, by the second law of thermodynamics, such a wall cannot exist. Nevertheless, our semi-permeable wall is essentially different from this thermodynamically unacceptable one; because reference is made with it only to the internal properties of the "molecules" $\mathbf{K}_1, \ldots, \mathbf{K}_N$ (i.e., the state of $\mathbf{S}_1$ or ... or $\mathbf{S}_N$ enclosed therein), and not to the exterior (i.e., whether it comes from the right or left, or something similar). This, however, is the decisive circumstance. A thorough going analysis of this question is made possible by the researches of L. Szilard, which clarified the nature of the semi-permeable wall, "Maxwell's demon," and the general role of the "intervention of an intelligent being in thermodynamical systems." We cannot go any further into these things here, especially since the reader can find a treatment of this in the references to Note 194.

In particular, the above treatment shows that two states ϕ, ψ of the system **S** can be certainly divided by a semi-permeable wall if they are orthogonal. We now want to prove the converse: if ϕ, ψ are not orthogonal, then the assumption of such a semi-permeable wall contradicts the second law of thermodynamics. That is, the necessary and sufficient condition for the separability by semi-permeable walls is $(\phi, \psi) = 0$, and not, as in classical theory, $\phi \neq \psi$ (we write ϕ, ψ instead of the I, II used above). This clarifies an old paradox of the classical form of thermodynamics, namely, the uncomfortable discontinuity in the operations with semi-permeable walls: states whose differences are arbitrarily small are always 100% separable, the absolutely equal states are in general not separable! We now have a continuous transition: It will be seen that 100% separability exists only for $(\phi, \psi) = 0$ and for increasing (ϕ, ψ) it becomes steadily worse. Finally, at maximum (ϕ, ψ), i.e., $|(\phi, \psi)| = 1$ (here $||\phi|| = ||\psi|| = 1$, and therefore it follows from $|(\phi, \psi)| = 1$ that $\phi = c\psi$, c constant, $|c| = 1$), the states ϕ, ψ are identical, and the separation is completely impossible.

In order to carry out these considerations, we must anticipate the end result of this section, the value of the entropy of the U-ensemble. Naturally we shall not use this result in its derivation.

Let us then assume that there is a semi-permeable wall separating ϕ and ψ. We shall then prove $(\phi, \psi) = 0$. We consider a $\frac{1}{2}(P_{[\phi]} + P_{[\psi]})$ gas (i.e., of $N/2$ systems in the state ϕ and $N/2$ systems in the state ψ, the trace of this operator is 1), and choose $\mathcal{V}$ (i.e., $\overline{\mathbf{K}}$), and $\mathbf{T}$ so that the gas is ideal. Let $\overline{\mathbf{K}}$ have the longitudinal cross section shown in Fig. 3: 1 2 3 4 1. We insert a semi-permeable wall at one end aa, and then move it halfway, up to the center bb. The temperature of the gas is kept fixed by contact with a large heat reservoir **W** of temperature **T** at the other end 2 3.

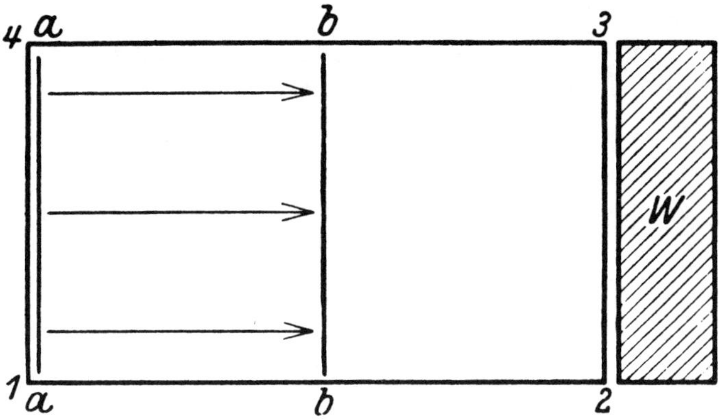

In this process, nothing happens to the ϕ molecules, but the ψ molecules are pushed into the right half of $\overline{\mathbf{K}}$ (between bb and 2 3). That is, the $\frac{1}{2}(P_{[\phi]} + P_{[\psi]})$ gas is a 1:1 mixture of a $P_{[\phi]}$ gas and a $P_{[\psi]}$ gas. Nothing happens to the former, but the latter is isothermally compressed to one half its original volume. From the equation of state of the ideal gas, it follows that in this process the mechanical work $\frac{N}{2}\,\kappa\,\mathbf{T}\ln 2$ is performed ($N/2$ is the number of the molecules of the $P_{[\psi]}$ gas, κ is Boltzmann's constant),[195] and since the energy of the gas is not changed (because of the isothermy),[196] this quantity of energy is taken over by the heat reservoir $\mathbf{W}$. The entropy change of the reservoir is then

---

[195]If an ideal gas consists of M molecules, then its pressure is $p = \frac{M\kappa\mathbf{T}}{\mathcal{V}}$. In the compression from the volume $\mathcal{V}_1$ to the volume $\mathcal{V}_2$ therefore, the mechanical work

$$\int_{\mathcal{V}_1}^{\mathcal{V}_2} p\,d\mathcal{V} = M\kappa\mathbf{T}\int_{\mathcal{V}_1}^{\mathcal{V}_2}\frac{d\mathcal{V}}{\mathcal{V}} = M\kappa\mathbf{T}\ln\frac{\mathcal{V}_2}{\mathcal{V}_1}$$

is done. In our case, $M = N/2$, $\mathcal{V}_1 = \mathcal{V}/2$, $\mathcal{V}_2 = \mathcal{V}$.

[196]The energy of an ideal gas, as is well known, depends only on its temperature.

$\mathbf{Q}/\mathbf{T} = N\kappa\cdot\frac{1}{2}\ln 2$ (see Note 186).

After this process, the half of the original gas is present to the left of bb , i.e., N/4 molecules. To the right of bb on the other hand, there is the half of the original $P_{[\phi]}$ gas, i.e., N/4 molecules, and the entire $P_{[\psi]}$ gas, i.e., N/2 molecules -- therefore a total of 3N/4 molecules of a $\frac{1}{3}P_{[\phi]} + \frac{2}{3}P_{[\psi]}$ gas. We compress or expand these gases to the volumes $\mathcal{V}/4$ and $3\mathcal{V}/4$ respectively, and mechanical work is again taken from or given to the heat reservoir $\mathbf{W}$: this amounts to $\frac{N}{4}\kappa\mathbf{T}\ln 2$ and $\frac{3N}{4}\kappa\mathbf{T}\ln\frac{3}{2}$ respectively (see Note 195), and the entropy increase of the reservoir is then $N\kappa\cdot\frac{1}{4}\ln 2$ and $-N\kappa\cdot\frac{3}{4}\ln\frac{3}{2}$ respectively. Altogether:

$$N\kappa\cdot\left(\frac{1}{2}\ln 2 + \frac{1}{4}\ln 2 - \frac{3}{4}\ln\frac{3}{2}\right) = N\kappa\cdot\frac{3}{4}\ln\frac{4}{3}\ .$$

Finally, we have a $P_{[\phi]}$ and a $P_{[\psi]}$ gas of N/4 and 3N/4 molecules respectively, with the respective volumes $\mathcal{V}/4$ and $3\mathcal{V}/4$. Originally there was a $\frac{1}{2}P_{[\phi]} + \frac{1}{2}P_{[\psi]}$ gas of N molecules in the volume $\mathcal{V}$ i.e., if we will, two $\frac{1}{2}P_{[\phi]} + \frac{1}{2}P_{[\psi]}$ gases with N/4 and 3N/4 molecules respectively, in the volumes $\mathcal{V}$ and $3\mathcal{V}/4$ respectively. The change effected by the entire process is then this: N/4 molecules in volume $\mathcal{V}/4$ changed from a $\frac{1}{2}P_{[\phi]} + \frac{1}{2}P_{[\psi]}$ gas into a $P_{[\phi]}$ gas, 3N/4 molecules in the volume $3\mathcal{V}/4$ changed from a $\frac{1}{2}P_{[\phi]} + \frac{1}{2}P_{[\psi]}$ gas into a $\frac{1}{3}P_{[\phi]} + \frac{2}{3}P_{[\psi]}$ gas, and the entropy of $\mathbf{W}$ increased by $N\kappa\cdot\frac{3}{4}\ln\frac{4}{3}$. Since the process was reversible, the entire entropy increase must be zero, i.e., the two gas-entropy changes must entirely compensate the change of entropy of $\mathbf{W}$. We must therefore find the entropy changes of the gases.

As we shall see, a U-gas of N molecules has the entropy $-M\kappa\cdot\mathrm{Tr}\,(U\ln U)$ if that of the $P_{[\chi]}$ -gas of equal volume and temperature is taken as zero (see above). If therefore U has a pure discrete spectrum with the eigenvalues $w_1, w_2, \cdots$, then this is

$$- M\kappa \cdot \sum_{n=1}^{\infty} w_n \ln w_n$$

(therefore $x \ln x$ is to be set equal to 0 for $x = 0$). As may easily be calculated, $P_{[\phi]}$, $\frac{1}{2} P_{[\phi]} + \frac{1}{2} P_{[\psi]}$, $\frac{1}{3} P_{[\phi]} + \frac{2}{3} P_{[\psi]}$ have the respective eigenvalues $1, 0$ and $\frac{1 + \alpha}{2}, \frac{1 - \alpha}{2}, 0$ and

$$\frac{3 + \sqrt{1 + 8\alpha^2}}{6}, \frac{3 - \sqrt{1 - 8\alpha^2}}{6}, 0$$

($\alpha = |(\phi, \psi)|$, therefore $\geq 0, \leq 1$), in which the multiplicity of the zero is always infinite, but in which the others are simple.[197] Therefore the entropy of the

---

[197]We determine the eigenvalues of $aP_{[\phi]} + bP_{[\psi]}$. The requirement is

$$(aP_{[\phi]} + bP_{[\psi]})f = \lambda f \ .$$

Since the left side is a linear combination of the ϕ, ψ, the right side is also, therefore also f, is too if $\lambda \neq 0$. $\lambda = 0$ is certainly an infinitely multiple eigenvalue, since each f orthogonal to ϕ, ψ belongs to it. It therefore suffices to consider $\lambda \neq 0$ and $f = x\phi + y\psi$ (let ϕ, ψ be linearly independent, otherwise, $\phi = c\psi$, $|c| = 1$, and the two states are identical).

The above equation then becomes

$$a(x + y(\psi, \phi))\cdot\phi + b(x(\phi, \psi) + y)\cdot\psi = \lambda x\cdot\phi + \lambda y\cdot\psi \ ,$$

i.e.,

$$a\cdot x + \overline{a(\phi, \psi)}\cdot y = \lambda\cdot x, \quad b(\phi, \psi)\cdot x + b\cdot y = \lambda\cdot y \ .$$

gas has increased by

$$-\frac{N}{4}\kappa\cdot 0 - \frac{3N}{4}\kappa\cdot\left(\frac{3+\sqrt{1+8\alpha^2}}{2}\ln\frac{3+\sqrt{1+8\alpha^2}}{6} + \frac{3-\sqrt{1+8\alpha^2}}{2}\ln\frac{3-\sqrt{1+8\alpha^2}}{6}\right.$$

$$+ N\kappa\cdot\left(\frac{1+\alpha}{2}\ln\frac{1+\alpha}{2} + \frac{1-\alpha}{2}\ln\frac{1-\alpha}{2}\right).$$

This should equal 0 when the entropy increase $N\kappa\cdot\frac{3}{4}\ln\frac{4}{3}$ of **W** is added to it. If we divide by $N\kappa/4$ then we have

$$-\frac{3+\sqrt{1+8\alpha^2}}{2}\ln\frac{3+\sqrt{1+8\alpha^2}}{6} - \frac{3-\sqrt{1+8\alpha^2}}{2}\ln\frac{3-\sqrt{1+8\alpha^2}}{6}$$

$$+ 2(1+\alpha)\ln\frac{1+\alpha}{2} + 2(1-\alpha)\ln\frac{1-\alpha}{2} + 3\ln\frac{4}{3} = 0$$

Also $0 \leq \alpha \leq 1$.

Now it can easily be seen that the left side increases monotonically as α varies from 0 to 1,[198]

The determinant of these equations must vanish:

$$\begin{vmatrix} a-\lambda, & a\overline{(\phi,\psi)} \\ b(\phi,\psi), & b-\lambda \end{vmatrix} = 0,\quad (a-\lambda)(b-\lambda) - ab|(\phi,\psi)|^2 = 0,$$

$$\lambda^2 - (a+b)\lambda + ab(1-\alpha^2) = 0,$$

$$\lambda = \frac{a+b\pm\sqrt{(a+b)^2 - 4ab(1-\alpha^2)}}{2} = \frac{a+b\pm\sqrt{(a-b)^2+4\alpha^2 ab}}{2}.$$

If we put $a = 1$, $b = 0$ or $a = 1/2$, $b = 1/2$ or $a = 1/3$, $b = 2/3$ respectively, then the formulas of the text are obtained.

[198]Since $(x \ln x)' = \ln x + 1$, therefore

and in fact from 0 to $3 \ln \frac{4}{3}$; therefore α must be zero (for $\alpha \neq 0$ the inverse process to that described

---

$$\left(\frac{1+y}{2} \ln \frac{1+y}{2} + \frac{1-y}{2} \ln \frac{1-y}{2}\right)' = \frac{1}{2}\left(\ln \frac{1+y}{2} + 1\right) - \frac{1}{2}\left(\ln \frac{1-y}{2} + 1\right)$$

$$= \frac{1}{2} \ln \frac{1+y}{1-y}$$

and the derivative of our expression is

$$- 3 \cdot \frac{1}{2} \ln \frac{3 + \sqrt{1+8\alpha^2}}{3 - \sqrt{1+8\alpha^2}} \cdot \frac{1}{3} \frac{8\alpha}{\sqrt{1+8\alpha^2}} + 4 \cdot \frac{1}{2} \ln \frac{1+\alpha}{1-\alpha}$$

$$= 2 \left(\ln \frac{1+\alpha}{1-\alpha} - \frac{2\alpha}{\sqrt{1+8\alpha^2}} \ln \frac{3 + \sqrt{1+8\alpha^2}}{3 - \sqrt{1+8\alpha^2}} \right) .$$

That this is > 0 means that

$$\ln \frac{1+\alpha}{1-\alpha} > \frac{2\alpha}{\sqrt{1+8\alpha^2}} \ln \frac{3 + \sqrt{1+8\alpha^2}}{3 - \sqrt{1+8\alpha^2}} ,$$

i.e.,

$$\frac{1}{2\alpha} \ln \frac{1+\alpha}{1-\alpha} > \frac{2}{3} \cdot \frac{1}{2\beta} \ln \frac{1+\beta}{1-\beta} , \quad \beta = \frac{\sqrt{1+8\alpha^2}}{3} .$$

We shall prove this with $8/9$ in place of $2/3$. Since $1 - \beta^2 = \frac{8}{9}(1 - \alpha^2)$ and $\alpha < \beta$ (which follows from the former, since $\alpha < 1$) , this means that

$$\frac{1-\alpha^2}{2\alpha} \ln \frac{1+\alpha}{1-\alpha} > \frac{1-\beta^2}{2\beta} \ln \frac{1+\beta}{1-\beta}$$

and this is proved if $\frac{1-x^2}{2x} \ln \frac{1+x}{1-x}$ is shown to be monotonically decreasing in $0 < x < 1$. This last property, however, follows, for example, from the power series expansion:

would be entropy decreasing, contrary to the second law). Therefore $(\phi, \psi) = 0$ has been proved.——

After these preparations, we can go on to determine the entropy of a U-gas of N molecules in the volume $\mathcal{V}$ and at temperature $\mathbf{T}$ -- i.e., more precisely, its entropy excess with respect to a $P_{[\phi]}$ gas under the same conditions. By our earlier remarks and in the sense of the normalization given above, this is the entropy of a U-ensemble of N individual systems. Let $\mathrm{Tr}\, U = 1$, as was done above.

The U, as we know, has a pure discrete spectrum $w_1, w_2, \dots$ with $w_1 \geq 0, w_2 \geq 0, \dots, w_1 + w_2 + \dots = 1$. Let the corresponding eigenfunctions be $\phi_1, \phi_2, \dots$. Then

$$U = \sum_{n=1}^{\infty} w_n P_{[\phi_n]}$$

(cf. IV.3.). Consequently, our U-gas is composed of a mixture of $P_{[\phi_1]}, P_{[\phi_2]}, \dots$ gases of $w_1 N, w_2 N, \dots$ molecules respectively, all in the volume $\mathcal{V}$. Let $\mathbf{T}$, $\mathcal{V}$ again be such that all these gases are ideal, and let $\overline{\mathbf{K}}$ be of rectangular cross section. Now we will apply the following reversible interventions in order to separate the $\phi_1, \phi_2, \dots$ molecules from each other (cf. Fig. 4.). We add an equally large rectangular box $\overline{\mathbf{K}}'$ (1 2 5 6 1) on to $\overline{\mathbf{K}}$ (2 3 4 5 2), and replace the common wall 2 5 by two walls lying next to each other. Let the one (2 5) be fixed and semi-permeable -- transparent for ϕ_1, but opaque for $\phi_2, \phi_3, \dots$; let the other wall (bb) be movable, but an ordinary, absolutely impenetrable wall. In addition, we insert another semi-permeable wall at dd,

---

$$\frac{1-x^2}{2x} \ln \frac{1+x}{1-x} = (1 - x^2)(1 + \frac{x^2}{3} + \frac{x^4}{5} + \dots)$$

$$= 1 - (1 - \tfrac{1}{3})x^2 - (\tfrac{1}{3} - \tfrac{1}{5})x^4 - \dots .$$

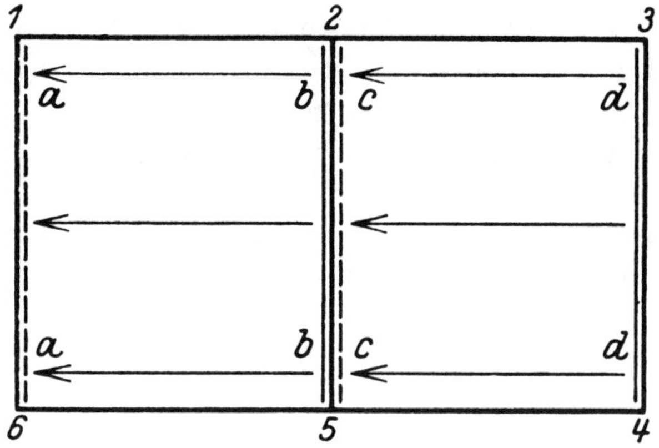

close to 3 4, which is transparent for $\phi_2, \phi_3, \ldots$ and opaque for ϕ_1. We then push bb and dd, the distance between them being kept constant, to aa and cc respectively (i.e., close to 1 6 and 2 5 respectively). By this means, the $\phi_2, \phi_3, \ldots$ are not affected, but the ϕ_1 are forced to remain between the moving walls bb, dd. Since the distance between these walls is a constant, no work is done (against the gas pressure), and no heat development takes place. Finally, we replace the walls 2 5, cc by a rigid, absolutely impenetrable wall 2 5, and remove aa -- in this way the boxes $\overline{\mathbf{K}}, \overline{\mathbf{K}}'$ are restored. There is, however, this change. All ϕ_1 molecules are in $\mathbf{K}'$, i.e., we have transferred all these from $\overline{\mathbf{K}}$ into the same sized box $\overline{\mathbf{K}}'$, reversibly and without any work being done, without any evolution of heat or tenperature change.[199]

Similarly, we "tap off" the $\phi_2, \phi_3, \ldots$ molecules into the equal boxes $\overline{\mathbf{K}}'', \overline{\mathbf{K}}''', \ldots$, and have finally, $P_{[\phi_1]}, P_{[\phi_2]}, \ldots$ gases, consisting of $w_1N, w_2N, \ldots$ mole-

---

[199]Cf., for example, the reference in Note 184 for this artifice which is characteristic of the phenomenological thermodynamical method.

cules, respectively, each in the volume $\mathcal{V}$. We now compress these isothermally to the volumes $w_1\mathcal{V}, w_2\mathcal{V}, \ldots$ respectively. We must therefore add the quantities of heat $w_1 N\kappa T \ln w_1, w_2 N\kappa T \ln w_2, \ldots$, respectively, as compensation, from a large heat reservoir (of temperature T, so that the process may be reversible; the quantities of heat are all less than zero), since the amounts of work done in compressing the individual gases are the negatives of these values (cf. Note 191). Therefore, the entropy increase for this process amounts to

$$\sum_{n=1}^{\infty} w_n N\kappa \cdot \ln w_n \ .$$

Finally, we transform the $P_{[\phi_1]}, P_{[\phi_2]}, \ldots$ gases all into a $P_{[\phi]}$ gas (reversibly, cf. above, ϕ an arbitrarily chosen state). We have then only $P_{[\phi]}$ gases of $w_1 N, w_2 N, \ldots$ molecules respectively, in the volumes $w_1\mathcal{V}, w_2\mathcal{V}, \ldots$. Since all of these are identical and of equal density $(N/\mathcal{V})$, we can mix them, and this is also reversible. We then obtain a $P_{[\phi]}$ gas of N molecules in the volume $\mathcal{V}$ (since

$$\sum_{n=1}^{\infty} w_n = 1 \ ) \ .$$

Consequently, we have carried out the desired reversible process. The entropy has increased by

$$N\kappa \sum_{n=1}^{\infty} w_n \ln w_n$$

and since it is zero in the final state, it was

$$- N\kappa \sum_{n=1}^{\infty} w_n \ln w_n$$

in the initial state.

Since U has the eigenfunctions $\phi_1, \phi_2, \ldots$ with the eigenvalues $w_1, w_2, \ldots$, U ln U has the same eigenfunctions, but the eigenvalues $w_1 \ln w_1, w_2 \ln w_2, \ldots$. Consequently,

$$\operatorname{Tr}(U \ln U) = \sum_{n=1}^{\infty} w_n \ln w_n \ .$$

It may be observed that $w_n \geq 0, \leq 1$, therefore $w_n \ln w_n \leq 0$, and in fact equals zero only for $w_n = 0, 1$. Note that for $w_n = 0$, $w_n \ln w_n$ is to be taken equal to zero -- this follows from the circumstance that in our above considerations, the vanishing w_n are not considered at all. The same conclusion may also be obtained from continuity considerations.

We have then determined the entropy of a U-ensemble, consisting of N individual systems, to be $-N\kappa \operatorname{Tr}(U \ln U)$. The previous discussion on $w_n \ln w_n$ shows that it is always ≥ 0, and in order that it be 0, all w_n must be zero or 1. Since $\operatorname{Tr} U = 1$, exactly one $w_n = 1$, while the others $= 0$, therefore $U = P_{[\phi]}$. That is, the states have an entropy $= 0$, and the other mixtures have entropies > 0.

3. REVERSIBILITY AND EQUILIBRIUM PROBLEMS

We can now prove the irreversibility of the measurement process as asserted in V.1. For example, if U is a state, $U = P_{[\phi]}$, then in the measurement of a quantity $\mathfrak{R}$ whose operator R has the eigenfunctions $\phi_1, \phi_2, \ldots$, it goes over into the ensemble

$$U' = \sum_{n=1}^{\infty} (P_{[\phi]}\phi_n, \phi_n) \cdot P_{[\phi_n]} = \sum_{n=1}^{\infty} |(\phi, \phi_n)|^2 P_{[\phi_n]}$$

and if U' is not a state, then an entropy increase has

occurred (the entropy of U was 0 , that of U' is > 0) , so that the process is irreversible. If U' , too, is to be a state, it must be a $P_{[\phi_n]}$, and since the ϕ_n are its eigenfunctions, this means that all $|(\phi, \phi_n)|^2 = 0$ except one (that one $= 1$) i.e., ϕ is orthogonal to all ϕ_n, $n \neq \bar{n}$ -- but then $\phi = c\phi_{\bar{n}}$, where $|c| = 1$, and therefore $P_{[\phi]} = P_{[\phi_{\bar{n}}]}$, U = U' . Therefore, each measurement on a state is irreversible, unless the eigenvalue of the measured quantity (i.e., this quantity in the given state) has a sharp value, in which case the measurement does not change the state at all. As we see, the noncausal behavior is thus unambiguously related to a certain concomitant thermodynamical phenomena.

We shall now discuss in complete generality when the process **1.**,

$$U \longrightarrow U' = \sum_{n=1}^{\infty} (U\phi_n, \phi_n) \cdot P_{[\phi_n]}$$

increases the entropy.

U has the entropy $-N\kappa \operatorname{Tr}(U \ln U)$. If $w_1, w_2, \ldots$ are its eigenvalues and $\psi_1, \psi_2, \ldots$ its eigenfunctions then this is equal to

$$-N\kappa \sum_{n=1}^{\infty} w_n \ln w_n = -N\kappa \sum_{n=1}^{\infty} (U\psi_n, \psi_n) \ln (U\psi_n, \psi_n) .$$

U' has the eigenvalues $(U\phi_1, \phi_1), (U\phi_2, \phi_2), \ldots,$ and therefore its entropy is

$$-N\kappa \sum_{n=1}^{\infty} (U\phi_n, \phi_n) \ln (U\phi_n, \phi_n) .$$

Consequently the entropy of U is $\geq$ that of U' depending on whether

$$* \quad \sum_{n=1}^{\infty} (U\psi_n, \psi_n) \ln (U\psi_n, \psi_n) \lesseqgtr \sum_{n=1}^{\infty} (U\phi_n, \phi_n) \ln (U\phi_n, \phi_n) .$$

We next show that in *, $\geq$ holds in any case, i.e., that the process $U \rightarrow U'$ is not entropy-diminishing -- this is indeed clear thermodynamically, but it is of importance for our subsequent purposes to have a purely mathematical proof of this fact. We proceed in such a way that U, and with it $\psi_1, \psi_2, \ldots$, are fixed, while the $\phi_1, \phi_2, \ldots$ run through all complete orthonormal sets.

Next, for reasons of continuity, we may limit ourselves to such sets $\phi_1, \phi_2, \ldots$ in which only a finite number of ϕ_n are different from the corresponding ψ_n. Then, for example, let $\phi_n = \psi_n$ for $n > M$. Then the ϕ_n, $n \leq M$ are linear combinations of the ψ_n, $n \leq M$, and conversely -- therefore,

$$\phi_m = \sum_{n=1}^{M} x_{mn}\psi_n \qquad (m = 1, \ldots, M),$$

and the M dimensional matrix $\{x_{mn}\}$ is obviously unitary. We obtain $(U\psi_m, \psi_m) = w_m$ and, as can easily be calculated,

$$(U\phi_m, \phi_m) = \sum_{n=1}^{N} w_n |x_{mn}|^2 \quad (m = 1, \ldots, M)$$

so that

$$\sum_{m=1}^{M} w_m \ln w_m \geq \sum_{m=1}^{M} \left(\sum_{n=1}^{M} w_n |x_{mn}|^2 \right) \ln \left(\sum_{n=1}^{M} w_n |x_{mn}|^2 \right)$$

is to be proved. Since the right side is a continuous function of the M^2 bounded variables x_{mn}, it has a maximum, and it also assumes its maximum value ($\{x_{mn}\}$ unitary); since the left side is its value for

$$x_{mn} \begin{cases} = 1 & \text{for } m = n \\ = 0 & \text{for } m \neq n \end{cases}$$

we must show: the maximum just mentioned occurs at this x_{mn}-complex.

Therefore, let x^{o}_{mn} $(m, n = 1,\dots,M)$ be a set of values for which the maximum occurs. If we multiply the matrix $\{x^{o}_{mn}\}$ by the unitary matrix

$$\left\{\begin{array}{cccc:c} \alpha, & \beta, & 0, & & 0 \\ -\bar{\beta}, & \bar{\alpha}, & 0, & & 0 \\ 0, & 0, & 1, & & 0 \\ \cdots & \cdots & \cdots & \cdots & \cdots \\ 0 & 0 & 0 & & 1 \end{array}\right\}, \quad |\alpha|^2 + |\beta|^2 = 1,$$

then we obtain a unitary matrix $\{x'_{mn}\}$, and therefore an acceptable x_{mn}-complex. Now, let $\alpha = \sqrt{1 - \epsilon^2}$, $\beta = \theta\epsilon$ (ϵ real, $|\theta| = 1$). ϵ will be small, and in the following we shall carry in our calculations the $1, \epsilon, \epsilon^2$ terms only, and neglect the $\epsilon^3, \epsilon^4, \dots$ terms. Then $\alpha \approx 1 - \frac{1}{2}\epsilon^2$, and in the new matrix $\{x'_{mn}\}$,

$$x'_{1n} \approx (1 - \tfrac{1}{2}\epsilon^2)x^{o}_{1n} + \theta\epsilon x^{o}_{2n},$$

$$x'_{2n} \approx -\bar{\theta}\epsilon x^{o}_{1n} + (1 - \tfrac{1}{2}\epsilon^2)x^{o}_{2n},$$

$$x'_{mn} = x^{o}_{mn} \quad (m \geq 3),$$

therefore

$$\sum_{n=1}^{M} w_n|x'_{1n}|^2 \approx \sum_{n=1}^{M} w_n|x^{o}_{1n}|^2 + \sum_{n=1}^{M} 2w_n\,\Re(\bar{\theta}x^{o}_{1n}\bar{x}^{o}_{2n})\cdot\epsilon$$
$$+ \sum_{n=1}^{M} w_n(-|x^{o}_{1n}|^2 + |x^{o}_{2n}|^2)\cdot\epsilon^2,$$

$$\sum_{n=1}^{M} w_n|x'_{2n}|^2 \approx \sum_{n=1}^{M} w_n|x^o_{2n}|^2 - \sum_{n=1}^{M} 2w_n\,\Re(\overline{\theta}x^o_{1n}\overline{x}^o_{2n})\cdot\epsilon - \sum_{n=1}^{M} w_n(-|x^o_{1n}|^2 + |x^o_{2n}|^2)\cdot\epsilon^2 ,$$

$$\sum_{n=1}^{M} w_n|x'_{mn}|^2 = \sum_{n=1}^{M} w_n|x^o_{mn}|^2 \qquad (m \geq 3)\ .$$

If we substitute these expressions in $f(x) = x \ln x$, in which

$$f'(x) = \ln x + 1\ ,\ f''(x) = \frac{1}{x}$$

and add the resulting expressions together, then

$$\sum_{m=1}^{M}\left(\sum_{n=1}^{M} w_n|x'_{mn}|^2\right)\ln\left(\sum_{n=1}^{M} w_n|x'_{mn}|^2\right)$$

$$\approx \sum_{m=1}^{M}\left(\sum_{n=1}^{M} w_n|x^o_{mn}|^2\right)\ln\left(\sum_{n=1}^{M} w_n|x^o_{mn}|^2\right)$$

$$\left(\ln\left(\sum_{n=1}^{M} w_n|x^o_{1n}|^2\right) - \ln\left(\sum_{n=1}^{M} w_n|x^o_{2n}|^2\right)\right)\cdot\sum_{n=1}^{M} 2w_n\,\Re(\overline{\theta}x^o_{1n}\overline{x}^o_{2n})\cdot\epsilon$$

$$\left[-\left(\ln\left(\sum_{n=1}^{M} w_n|x^o_{1n}|^2\right) - \ln\left(\sum_{n=1}^{M} w_n|x^o_{2n}|^2\right)\right)\left(\sum_{n=1}^{M} w_n|x^o_{1n}|^2\right) - \left(\sum_{n=1}^{M} w_n|x^o_{2n}|^2\right)\right.$$

$$\left.\frac{1}{2}\left(\frac{1}{\sum_{n=1}^{M} w_n|x^o_{1n}|^2} + \frac{1}{\sum_{n=1}^{M} w_n|x^o_{2n}|^2}\right)\left(\sum_{n=1}^{M} 2w_n\Re(\overline{\theta}x^o_{1n}\overline{x}^o_{2n})\right)^2\right]\cdot\epsilon^2\ .$$

In order that the first term on the right be the maximum value, the ϵ coefficient must be $= 0$, and the ϵ^2 coefficient ≤ 0. The former has two factors,

$$\ln\left(\sum_{n=1}^{M} w_n |x_{1n}^0|^2\right) - \ln\left(\sum_{n=1}^{M} w_n |x_{2n}^0|^2\right)$$

and

$$\sum_{n=1}^{M} 2w_n \,\Re(\bar{\theta} x_{1n}^0 \bar{x}_{2n}^0) \ .$$

If the first is zero, then the first term in the ϵ^2 coefficient $= 0$ (this is always ≤ 0), so that the second term, which is clearly ≥ 0 always, must vanish in order that the entire coefficient be ≤ 0. This means that

$$\sum_{n=1}^{M} 2w_n \,\Re(\bar{\theta} x_{1n}^0 \bar{x}_{2n}^0) = 0 \ .$$

Therefore, the second factor of the ϵ coefficient is $= 0$ in any case, which can also be written

$$2\,\Re\left(\bar{\theta} \sum_{n=1}^{M} w_n x_{1n}^0 \bar{x}_{2n}^0\right) \ .$$

Since this goes over into the absolute value of the

$$\sum_{n=1}^{M}$$

for appropriate θ, this must disappear:

$$\sum_{n=1}^{M} w_n x_{1n}^0 \bar{x}_{2n}^0 = 0 \ .$$

Since we can replace 1, 2 by any two different

$k, j = 1,\ldots,M$, we have

$$\sum_{n=1}^{M} w_n x^o_{kn} \bar{x}^o_{jn} \quad \text{for} \quad k \neq j \quad .$$

That is, the unitary coordinate transformation with the matrix $\{x^o_{mn}\}$ brings the diagonal matrix with the elements $w_1,\ldots,w_n$ again into diagonal form. Since the diagonal elements are the multipliers (or eigenvalues) of the matrix, they are not changed by the coordinate transformation, and are at most permuted. Before the transformation they were the w_m $(m = 1,\ldots,M)$, afterwards, they are the

$$\sum_{n=1}^{M} w_n |x^o_{mn}|^2$$

$(m = 1,\ldots,N)$. The sums

$$\sum_{n=1}^{M} w_n \ln w_n \, , \quad \sum_{m=1}^{M} \left(\sum_{n=1}^{M} w_n |x^o_{mn}|^2 \right) \ln \left(\sum_{n=1}^{M} w_n |x^o_{mn}|^2 \right)$$

then have the same values. Hence there is at any rate a maximum at

$$x_{mn} \begin{cases} = 1 & \text{for} \quad m = n \\ = 0 & \text{for} \quad m \neq n \end{cases} \Bigg|$$

too, as was asserted.

Let us determine when the equality holds in * . If it does hold, then

$$\sum_{n=1}^{\infty} (U\chi_n, \chi_n) \ln (U\chi_n, \chi_n)$$

takes on its maximum value not only for $\chi_n = \psi_n$ $(n = 1,2,\ldots)$ (these are the eigenfunctions of U , cf.

above), but also for $x_n = \phi_n$ $(n = 1, 2, \ldots)$ $(x_1, x_2, \ldots$ running through all complete orthonormal sets). This holds in particular if only the first M among the ϕ_n are transformed (i.e., $x_n = \phi_n$ for $n > M$) and hence, of course, transformed unitarily among each other. Let $\mu_{mn} = (U\phi_m, \phi_n)$ $(m, n = 1, \ldots, M)$, let $v_1, \ldots, v_N$ be the eigenvalues of the finite (and at the same time Hermitian and definite) matrix $\{\mu_{mn}\}$, and $\{\alpha_{mn}\}$ $(m, n = 1, \ldots, M)$ the matrix that transforms $\{\mu_{mn}\}$ to the diagonal form. This transforms the $\phi_1, \ldots, \phi_M$ into $\omega_1, \ldots, \omega_M$,

$$\phi_m = \sum_{n=1}^{M} \alpha_{mn}\omega_n$$

$(m = 1, \ldots, M)$, and then

$$U\omega_n = v_n\omega_n \text{, therefore } (U\omega_m, \omega_n) = \left\{ \begin{array}{ll} v_n & \text{, for } m = n \\ 0 & \text{, for } m \neq n \end{array} \right| .$$

For

$$\xi_m = \sum_{n=1}^{M} x_{mn}\omega_n$$

$(m = 1, \ldots, M$, let $\{x_{mn}\}$ also be unitary),

$$(U\xi_k, \xi_j) = \sum_{n=1}^{M} v_n x_{kn}\overline{x}_{jn} \quad .$$

Because of the assumption on the $\phi_1, \ldots, \phi_M$,

$$\sum_{n=1}^{M} \left(\sum_{n=1}^{M} v_n |x_{mn}|^2 \right) \ln \left(\sum_{n=1}^{M} v_n |x_{mn}|^2 \right)$$

takes on its maximum for $x_{mn} = \alpha_{mn}$. According to our previous proof, it follows from this that

$$\sum_{n=1}^{M} v_n \alpha_{kn} \overline{\alpha}_{jn} = 0$$

for $k \neq j$, i.e., $(U\phi_k, \phi_j) = 0$ for $k \neq j$, $k, j = 1, \ldots, M$.

This must hold for all M , therefore $U\phi_k$ is orthogonal to all ϕ_j , $k \neq j$ -- hence it is equal to $w'_k \phi_k$ (w'_k a constant). Consequently, the $\phi_1, \phi_2, \ldots$ are the eigenfunctions of U . The corresponding eigenvalues are $w'_1, w'_2, \ldots$ (and therefore a permutation of the $w_1, w_2, \ldots$) . But under these circumstances,

$$U' = \sum_{n=1}^{\infty} (U\phi_n, \phi_n) \cdot P_{[\phi_n]} = \sum_{n=1}^{\infty} w'_n \cdot P_{[\phi_n]} = U \quad .$$

We have therefore found.

The process **1.**,

$$U \longrightarrow U' = \sum_{n=1}^{\infty} (U\phi_n, \phi_n) \cdot P_{[\phi_n]}$$

($\phi_1, \phi_2, \ldots$ are the eigenfunctions of the operator R of the measured quantity $\mathfrak{R}$) , never diminishes the entropy. It actually increases it, unless all $\phi_1, \phi_2, \ldots$ are eigenfunctions of U , in which case $U = U'$.

In the case mentioned moreover, U commutes with R , and this is actually characteristic for it (because it is equivalent to the existence of the common eigenfunctions $\phi_1, \phi_2, \ldots$, cf. II.10.).

Hence the process **1.** is irreversible in all cases in which it effects a change at all.

The reversibility question should now be treated for the processes **1.**, **2.**, independently of phenomenological thermodynamics, as was announced as the second point of the program in V.2.. The mathematical method with which this can be accomplished we already know: if the second law of thermodynamics holds, the entropy must be equal to

$-N\kappa \operatorname{Tr}(U \ln U)$, and this may not decrease in any process **1.**, **2.** We must then treat $-N\kappa \operatorname{Tr}(U \ln U)$ merely as a calculated quantity, independently of its meaning as entropy, and find out what it does in **1.**, **2.**[200]

In **2.**, we obtain

$$U_t = e^{-\frac{2\pi i}{h} tH} U e^{\frac{2\pi i}{h} tH}$$

from U, i.e., if we designate the unitary operator

$$e^{-\frac{2\pi i}{h} tH}$$

by A, $U \rightarrow U_t = AUA^{-1}$. Since $f \rightarrow Af$, because of the unitary nature of A, is an isomorphic mapping of Hilbert space on itself, which transforms each operator P into APA^{-1}, therefore always $F(APA^{-1}) = AF(P)A^{-1}$. Consequently $U_t \ln U_t = A \cdot U \ln U \cdot A^{-1}$. Hence $\operatorname{Tr}(U_t \ln U_t) = \operatorname{Tr}(U \ln U)$, i.e., our quantity $-N\kappa \operatorname{Tr}(U \ln U)$ is constant in **2.** We have already ascertained what happens in **1.**, and in fact, without reference to the second law of thermodynamics. If U changes (i.e., $U \neq U'$), then $-N\kappa \operatorname{Tr}(U \ln U)$ increases, while for unchanged U (i.e., $U = U'$; or $\psi_1, \psi_2, \cdots$ eigenfunctions of U; or U, R commutative), it naturally remains unchanged. In an intervention composed of several process **1.** and **2.** (in arbitrary number and order) $-N\kappa \operatorname{Tr}(U \ln U)$ remains unchanged if each process **1.** is ineffective (i.e., causes no change), but in all other cases it increases.

Therefore, if only interventions **1.**, **2.** are taken into consideration, then each process **1.**, which effects a change at all, is irreversible.

It is worth noting, there are also other, simpler

---

[200]Naturally, we could neglect the factor $N\kappa$ and consider $-\operatorname{Tr}(U \ln U)$. Or, preserving the proportionality with the number of elements N, $-N \operatorname{Tr}(U \ln U)$.

expressions than $-\operatorname{Tr}(U \ln U)$ which do not decrease in **1.**, and are constant in **2.**: for example, the largest eigenvalue of U. Indeed: For **2.**, it is invariant, as are all eigenvalues of U -- while in **1.**, the eigenvalues $w_1, w_2, \ldots$ of U go over into the eigenvalues of U':

$$\sum_{n=1}^{\infty} w_n |x_{1n}|^2, \sum_{n=1}^{\infty} w_n |x_{2n}|^2, \ldots$$

(cf. the earlier considerations of this section), and since, by the unitary nature of the matrix $\{x_{mn}\}$,

$$\sum_{n=1}^{\infty} |x_{1n}|^2 = 1, \sum_{n=1}^{\infty} |x_{2n}|^2 = 1, \ldots$$

all these numbers are $\leq$ than the largest w_n. (A maximum w_n exists, since all $w_n \geq 0$, and since

$$\sum_{n=1}^{\infty} w_n = 1$$

$w_n \longrightarrow 0$.) Now since it is possible so to change U that

$$-\operatorname{Tr}(U \ln U) = -\sum_{n=1}^{\infty} w_n \ln w_n$$

remains invariant, but that the largest w_n decreases, we see that these are changes which are possible according to phenomenological thermodynamics -- therefore they are actually possible of execution with our gas processes -- but which can never be brought about by successive applications of **1.**, **2.** alone. This proves that our introduction of gas processes was indeed necessary.

Instead of $-\operatorname{Tr}(U \ln U)$ we can also consider $\operatorname{Tr}(F(U))$ for appropriate functions $F(x)$. That this increases in **1.** for $U \neq U'$ (for $U = U'$, as well as in **2.**, it is of course invariant), can also be proved, as was done for $F(x) = -x \ln x$, if the special properties of this function, which we used above are also present in $F(x)$. These are: $F''(x) < 0$, and the monotonic decrease of $F'(x)$; but the latter follows from the former. Therefore, for our non-thermodynamical irreversibility considerations, we can use each $\mathrm{Tr}\, F(U)$, if $F(x)$ is a function that is convex from above, i.e., if $F''(x) < 0$ (in $0 \leq x \leq 1$ since all eigenvalues of U lie in that interval).

Finally, it should be shown that the mixing of two ensembles U, V (say in the ratio $\alpha:\beta$; $\alpha > 0, \beta > 0, \alpha + \beta = 1$) is also not entropy-diminishing, i.e.,

$$- \mathrm{Tr}\, ((\alpha U + \beta V) \ln (\alpha U + BV))$$

$$\geq - \alpha\, \mathrm{Tr}\, (U \ln U) - \beta\, \mathrm{Tr}\, (V \ln V)$$

This also holds for each convex $F(x)$ in place of $- x \ln x$. The proof is left to the reader. ——

We shall now investigate the stationary equilibrium superposition, i.e., the mixture of maximum entropy, when the energy is given. The latter is, of course, to be understood to mean that the expectation value of the energy is prescribed -- only this interpretation is admissible, in view of the method indicated in Note 184 for the thermodynamical investigation of statistical ensembles. Consequently, only such mixtures will be allowed, for the U of which $\mathrm{Tr}\, U = 1$, $\mathrm{Tr}\, (U\mathsf{H}) = \mathbf{E}$, where H is the energy operator and $\mathbf{E}$ the prescribed energy expectation value. Under these auxiliary conditions, $- N\kappa\, \mathrm{Tr}\, (U \ln U)$ is to be made a maximum. We also make the simplifying assumption that H has a pure discrete spectrum; say the eigenvalues $W_1, W_2, \ldots$ and the eigenfunctions $\phi_1, \phi_2, \ldots$ (there may also be multiple values among these).

Let $\mathfrak{R}$ be a quantity whose operator R has the eigenfunctions (of H) $\phi_1, \phi_2, \ldots$, but only distinct eigenvalues. The measurement of $\mathfrak{R}$ transforms U, by **2.**, into

$$U' = \sum_{n=1}^{\infty} (U\phi_n, \phi_n)P_{[\phi_n]} \quad ,$$

and therefore $-N\kappa \operatorname{Tr}(U \ln U)$ increases, unless $U = U'$. Also, $\operatorname{Tr}(U)$, $\operatorname{Tr}(U H)$ do not change -- the latter because the ϕ_n are eigenfunctions of H, and therefore $(H\phi_m, \phi_n)$ vanishes for $m \neq n$:

$$\operatorname{Tr}(U'H) = \sum_{n=1}^{\infty} (U\phi_n, \phi_n) \operatorname{Tr}(P_{[\phi_n]} H)$$

$$= \sum_{n=1}^{\infty} (U\phi_n, \phi_n)(H\phi_n, \phi_n)$$

$$= \sum_{m,n=1}^{\infty} (U\phi_m, \phi_n)(H\phi_n, \phi_m) = \operatorname{Tr}(U H) \quad .$$

This must also be true because of the commutativity of R, H (i.e., simultaneous measurability of $\mathfrak{R}$ and energy). Consequently, the desired maximum is the same if we limit ourselves to the U', i.e., to statistical operators with eigenfunctions $\phi_1, \phi_2, \ldots$, and, furthermore it is assumed only among these.

Therefore

$$U = \sum_{n=1}^{\infty} w_n P_{[\phi_n]} \quad ,$$

and since U, UH, $U \ln U$ all have the eigenfunctions ϕ_n, but the respective eigenvalues w_n, $W_n w_n$, $w_n \ln w_n$, it suffices to make

$$-N\kappa \sum_{n=1}^{\infty} w_n \ln w_n$$

a maximum, with the auxiliary conditions

$$\sum_{n=1}^{\infty} w_n = 1 , \sum_{n=1}^{\infty} W_n w_n = \mathbf{E} .$$

But this is exactly the same problem as that which is obtained for the corresponding equilibrium problem of the ordinary gas theory,[201] and is solved in the same way. According to the well-known rules of extremum calculation, for the set of maximizing $w_1, w_2, \cdots$

$$\frac{\partial}{\partial w_n}\left(\sum_{m=1}^{\infty} w_m \ln w_m\right) + \alpha \frac{\partial}{\partial w_n}\left(\sum_{m=1}^{\infty} w_m\right) + \beta \frac{\partial}{\partial w_m}\left(\sum_{m=1}^{\infty} W_m w_m\right) = 0$$

must hold, in which α, β are suitable constants, and $n = 1, 2, \cdots$, that is,

$$(\ln w_n + 1) + \alpha + \beta W_n = 0 , \quad w_n = e^{-1-\alpha-\beta W_n} = a e^{-\beta W_n}$$

where the constant $a = e^{-1-\alpha}$ is introduced in place of α. From

$$\sum_{n=1}^{\infty} w_n = 1$$

it follows that

$$a = \frac{1}{\sum_{n=1}^{\infty} e^{-\beta W_n}} ,$$

and therefore

[201] Cf., for example, Planck, Theorie der Wärmestrahlung, Leipzig, 1913.

$$w_n = \frac{e^{-\beta W_n}}{\sum_{m=1}^{\infty} e^{-\beta W_m}} ,$$

and because of $\sum_{n=1}^{\infty} W_n w_n = \mathbf{E}$

$$\frac{\sum_{n=1}^{\infty} W_n e^{-\beta W_n}}{\sum_{n=1}^{\infty} e^{-\beta W_n}} = \mathbf{E}$$

which determines β. If, as is customary, we introduce the "partition function,"

$$z(\beta) = \sum_{n=1}^{\infty} e^{-\beta W_n} = \mathrm{Tr}\,(e^{-\beta H})$$

(cf. Notes 183, 184 for this and the following), then

$$z'(\beta) = -\sum_{n=1}^{\infty} W_n e^{-\beta W_n} = -\,\mathrm{Tr}\,(H e^{-\beta H})$$

and therefore the condition for β is

$$-\frac{z'(\beta)}{z(\beta)} = \mathbf{E} .$$

(We are making the assumption here that

$$\sum_{n=1}^{\infty} e^{-\beta W_n} \text{ and } \sum_{n=1}^{\infty} W_n e^{-\beta W_n}$$

converge for all $\beta > 0$, i.e., that $W_n \longrightarrow \infty$ for $n \longrightarrow \infty$, and in fact, with sufficient rapidity. For example, $W_n/\ln n \longrightarrow \infty$ suffices.) We then obtain the following expression for U itself:

$$U = \sum_{n=1}^{\infty} ae^{-\beta W_n} P_{[\phi_n]} = ae^{-\beta H} = \frac{e^{-\beta H}}{\mathrm{Tr}\,(e^{-\beta H})} = \frac{e^{-\beta H}}{z(\beta)} .$$

The properties of the equilibrium ensemble U (which is determined by the enumeration of the values of $\mathbf{E}$ or of β, and which therefore depends on a parameter, as it must) can now be determined with the method customary in gas theory.

The entropy of our ensemble is

$$\mathbf{S} = -N\kappa\,\mathrm{Tr}\,(U \ln U) = -N\kappa\,\mathrm{Tr}\left(\frac{e^{-\beta H}}{z(\beta)} \ln \frac{e^{-\beta H}}{z(\beta)}\right)$$

$$= -\frac{N\kappa}{z(\beta)}\,\mathrm{Tr}\,(e^{-\beta H}(-\beta H - \ln z(\beta)))$$

$$= \frac{\beta N\kappa}{z(\beta)}\,\mathrm{Tr}\,(He^{-\beta H}) + \frac{\ln z(\beta) N\kappa}{z(\beta)}\,\mathrm{Tr}\,(e^{-\beta H})$$

$$= N\kappa\left[-\frac{\beta z'(\beta)}{z(\beta)} + \ln z(\beta)\right] ,$$

and the total energy

$$N\mathbf{E} = -N\,\frac{z'(\beta)}{z(\beta)}$$

(this, and not $\mathbf{E}$ itself, is to be considered in conjunction with $\mathbf{S}$). Thus U, $\mathbf{S}$, $N\mathbf{E}$ are expressed by β. Instead of inverting the last relationship, i.e., expressing β by $\mathbf{E}$, it is more practical to determine the temperature $\mathbf{T}$ of the equilibrium mixture, and to reduce everything to this. This is done as follows: Our equilibrium mixture is brought into contact with a heat reservoir of temperature $\mathbf{T}'$, and the energy $Nd\mathbf{E}$ is transferred to it from that reservoir. The two laws of thermodynamics imply, then, that the total energy must remain unchanged, and that the entropy must not decrease. Consequently, the heat reservoir loses the energy $Nd\mathbf{E}$, and therefore its entropy increase is $-Nd\mathbf{E}/\mathbf{T}'$, and we must now have

$$d\,\mathbf{S} - \frac{Nd\mathbf{E}}{\mathbf{T}'} = \left(\frac{d\mathbf{S}}{Nd\mathbf{E}} - \frac{1}{\mathbf{T}'}\right)Nd\mathbf{E} \geq 0 \quad .$$

On the other hand, $Nd\mathbf{E} \gtrless 0$ must hold according to whether $\mathbf{T}' \gtrless \mathbf{T}$, because the colder body absorbs energy from the warmer -- consequently $\mathbf{T}' \gtrless \mathbf{T}$ implies $\frac{d\mathbf{S}}{Nd\mathbf{E}} - \frac{1}{\mathbf{T}'} \gtrless 0$ i.e.,

$$\mathbf{T}' \gtrless \frac{Nd\mathbf{E}}{d\mathbf{S}} = \frac{N\frac{d\mathbf{E}}{d\beta}}{\frac{d\mathbf{S}}{d\beta}} \quad .$$

Hence

$$\mathbf{T} = \frac{N\frac{d\mathbf{E}}{d\beta}}{\frac{d\mathbf{S}}{d\beta}} = -\frac{1}{\kappa}\,\frac{\left(\frac{z'(\beta)}{z(\beta)}\right)'}{\left(\ln z(\beta) - \beta\,\frac{z'(\beta)}{z(\beta)}\right)'} = -\frac{1}{\kappa}\,\frac{\left(\frac{z'(\beta)}{z(\beta)}\right)'}{-\beta\left(\frac{z'(\beta)}{z(\beta)}\right)'} = \frac{1}{\kappa\beta} \; ,$$

i.e.,

$$\beta = \frac{1}{\kappa\mathbf{T}} \quad .$$

Therefore U, $\mathbf{S}$, $N\mathbf{E}$ are now all expressed as functions of the temperature.

The analogy of the expressions obtained above for the entropy, equilibrium ensemble etc., with the corresponding results of the classical thermodynamical theory is striking. First, the entropy $-N\kappa\,\mathrm{Tr}\,(U \ln U)$.

$$U = \sum_{n=1}^{\infty} w_n P_{[\phi_n]}$$

is a mixture of the ensembles $P_{[\phi_1]}, P_{[\phi_2]}, \ldots$ with the relative weights $w_1, w_2, \ldots$, i.e., Nw_1 ϕ_1-systems, Nw_2 ϕ_2-systems, ... The Boltzmann entropy of this ensemble is obtained with the aid of the "thermodynamical probability" $N!/(Nw_1)!(Nw_2)!\ldots$. It is its κ fold logarithm.[201] Since N is large, we may approximate the factorial by the Stirling formula, $x! \approx \sqrt{2\pi x}\, e^{-x} x^x$ and then $\kappa \ln \frac{N!}{(Nw_1)!(Nw_2)!\ldots}$ becomes essentially

$$- N\kappa \sum_{n=1}^{\infty} w_n \ln w_n$$

-- and this is exactly $- N\kappa \operatorname{Tr}(U \ln U)$.

Furthermore, if we had the equilibrium ensemble

$$U = e^{-\frac{H}{\kappa T}}$$

(we neglect the normalization factor $\frac{1}{Z(\beta)}$), this is equal to

$$\sum_{n=1}^{\infty} e^{-\frac{W_n}{\kappa T}} P_{[\phi_n]} ,$$

therefore a mixture of the states $P_{[\phi_1]}, P_{[\phi_2]}, \ldots$, i.e., of the stationary states with the energies $W_1, W_2, \ldots$, and with the respective (relative) weights

$$e^{-\frac{W_1}{\kappa T}}, e^{-\frac{W_2}{\kappa T}}, \ldots \quad .$$

If an energy value is multiple, say $W_{n_1} = \ldots = W_{n_\nu} = W$, then $P_{[\phi_{n_1}]} + \ldots + P_{[\phi_{n_\nu}]}$ appears in the equilibrium ensemble with the weight

$$e^{-\frac{W}{\kappa T}} ,$$

i.e., the correctly normalized mixture

$$\frac{1}{\nu}\left(P_{[\phi_{n_1}]} + \cdots + P_{[\phi_{n_\nu}]}\right)$$

(cf. the beginning of IV.3.) appears with the weight

$$\nu e^{-\frac{W}{\kappa T}} .$$

But the classical "canonical" ensemble is defined in exactly the same way (aside from the appearance of the specifically quantum mechanical form

$$\frac{1}{\nu}\left(P_{[\phi_{n_1}]} + \cdots + P_{[\phi_{n_\nu}]}\right)):$$

this is known as Boltzmann's Theorem.[201]

For $T \longrightarrow 0$, the weights

$$e^{-\frac{W_n}{\kappa T}}$$

approach 1, therefore our U tends to

$$\sum_{n=1}^{\infty} P_{[\phi_n]} = 1 .$$

Consequently, $U \approx 1$ is the absolute equilibrium state, if no energy limitations apply -- a result that we had already obtained in IV.3. We see that the "a priori equal probability of the quantum orbits" (i.e., of the simple, non-degenerate ones -- in general the multiplicity of the eigenvalues is the "a priori" weight, cf. discussion above) follows automatically from this theory.

It remains to ascertain how much can be said non-thermodynamically about the equilibrium ensemble U of given energy -- i.e., only from the fact that U is stationary (does not change in the course of time, process **2.**), and that it remains unchanged in all measurements which do not affect the energy (i.e., in measurements of quantities that are measurable simultaneously with the energy, process **1.** with commutative R, H , i.e., $\phi_1, \phi_2, \ldots$ eigenfunctions of H).

Because of the differential equation $\frac{\partial}{\partial t} U = \frac{2\pi i}{h}(U H - H U)$ the former means only that H U commute. The latter means that if $\phi_1, \phi_2, \ldots$ are usable, as a complete eigenfunction set of H , then $U = U'$, i.e., $\phi_1, \phi_2, \ldots$ are also eigenfunctions of U . Let the corresponding H-eigenvalues be $W_1, W_2, \ldots$, those of U , $w_1, w_2, \ldots$. If $W_j = W_k$, then we can replace ϕ_j, ϕ_k by

$$\frac{\phi_j + \phi_k}{\sqrt{2}}, \quad \frac{\phi_j - \phi_k}{\sqrt{2}}$$

for H , and therefore these are also eigenfunctions of U , from which it follows that $w_j = w_k$. Therefore, a function $F(x)$ with $F(W_n) = w_n$ $(n = 1, 2, \ldots)$ can be constructed, and $F(H) = U$. It is clear that this is sufficient, and also that it implies the commutativity of H and U .

Hence there results $U = F(H)$, but a determination of $F(x)$ (it is, as we know $F(x) = \frac{1}{Z(\beta)} e^{-\beta x}$, $\beta = \frac{1}{\kappa T}$) is not accomplished. From $\mathrm{Tr}\, U = 1$, $\mathrm{Tr}\,(U H) = \mathbf{E}$, it follows that

$$\sum_{n=1}^{\infty} F(W_n) = 1 \ , \ \sum_{n=1}^{\infty} W_n F(W_n) = \mathbf{E}$$

but with this, all that this method can furnish us is exhausted.

4. THE MACROSCOPIC MEASUREMENT

Although our entropy expression, as we saw, is completely analogous to the classical entropy, it is still surprising that it is invariant in the normal evolution in time of the system (process **2.**), and only increases with measurements (process **1.**) -- in the classical theory (where the measurements in general played no role) it increased as a rule even with the ordinary mechanical evolution in time of the system. It is therefore necessary to clear up this apparently paradoxical situation.

The normal classical thermodynamical consideration runs as follows: One could take a container of volume $\mathcal{V}$, in which M molecules of a gas (for simplicity, an ideal gas) of temperature T are present in the right half (volume $\mathcal{V}/2$, separated by a partition from the other half). If we were to expand this gas isothermally and reversibly to the volume by driving back the partition with the gas pressure, utilizing the mechanical work that this performs, and by keeping the gas temperature constant by means of a large heat reservoir of temperature T), then the entropy outside (in the reservoir) would decrease by $M\kappa \ln 2$ (cf. Note 195), and therefore the gas entropy could increase by the same amount. On the other hand, if we simply remove the partition, the gas diffuses into the free left half, the volume increases to $\mathcal{V}$ -- i.e., the entropy increases by $M\kappa \ln 2$ without the corresponding compensation taking place. The process is consequently irreversible, for the entropy has increased in the course of the simple mechanical evolution in time of the system (namely, in diffusion). Why does our theory give nothing similar?

This situation is best clarified if we set $M = 1$. Thermodynamics is still valid for such a one-molecule gas, and it is true that its entropy increases by $\kappa \ln 2$ if its volume is doubled. Nevertheless, this difference is $\kappa \ln 2$ actually only so long as one knows no more about the molecule than that it is found in the volume $\mathcal{V}/2$ or $\mathcal{V}$, respectively. For example, if the molecule is in the volume $\mathcal{V}$, but it is known whether it is in the right side or left side of the middle of the container, then it suffices to insert a partition in the middle and allow this to be pushed (isothermally and reversibly) by the molecule to the left or right end of the container. In this case, the mechanical work $\kappa T \ln 2$ is performed, i.e., this energy is taken from the heat reservoir. Consequently, at the end of the process, the molecule is again in the volume $\mathcal{V}$, but we no longer know whether it is on the left or right of the middle. Hence there is a compensating entropy decrease of $\kappa \ln 2$ (in the reservoir). That is, we have exchanged our knowledge for the entropy decrease $\kappa \ln 2$.[202] Or, the entropy is the same in the volume $\mathcal{V}$ as in the volume $\mathcal{V}/2$, provided that we know in the first mentioned case, in which half of the container the molecule is to be found. Therefore, if we knew all the properties of the molecule before diffusion (position and momentum), we could calculate for each moment after the diffusion whether it is on the right or left side, i.e., the entropy has not decreased. If, however, the only information at our disposal was the macroscopic one that the volume was initially $\mathcal{V}/2$, then the entropy does increase upon diffusion.

For a classical observer, who knows all coordinates and momenta, the entropy is therefore constant, and is in fact 0, since the Boltzmann "thermodynamical probability" is 1 (cf. the reference in Note 201); just

---

[202] L. Szilard has (see reference in Note 194) shown that one cannot get this "knowledge" without a compensating entropy increase $\kappa \ln 2$. In general, $\kappa \ln 2$ is the "thermodynamic value" of the knowledge, which consists of an alternative of two cases. All attempts to carry out the process described above without the knowledge of the half of the container in which the molecule is located, can be proved to be invalid, although they may occasionally lead to very complicated automatic mechanisms.

as in our theory for states, $U = P_{[\phi]}$, since these again correspond to the highest possible state of knowledge of the observer relative to the system.

The time variations of the entropy are then based on the fact that the observer does not know everything, that he cannot find out (measure) everything which is measurable in principle. His senses allow him to perceive only the so-called macroscopic quantities. But this clarification of the apparent contradiction mentioned at the outset imposes on us the obligation of investigating the precise analog of the classical macroscopic entropy for the quantum mechanical ensemble, i.e., the entropy as seen by an observer who cannot measure all quantities, but only a few special quantities, namely, the macroscopic ones, and even these, under certain circumstances, with only limited accuracy.

In III.3., we learned that all measurements with limited accuracy can be replaced by absolutely accurate measurements of other quantities which are functions of these, and which have discrete spectra. If now $\mathfrak{R}$ is such a quantity, and R is its operator, if $\lambda^{(1)}, \lambda^{(2)}, \ldots$ are the distinct eigenvalues, then the measurement of $\mathfrak{R}$ is equivalent to the answering of the following questions: "Is $\mathfrak{R} = \lambda^{(1)}$?" "Is $\mathfrak{R} = \lambda^{(2)}$?", ... In fact, we can also say directly: Assume that $\mathfrak{S}$, with the operator S, is to be measured with limited accuracy -- say one wishes to determine within which interval $c_{n-1} < \lambda \leq c_n$ $(\ldots c_{-2} < c_{-1} < c_0 < c_1 < c_2 < \ldots)$ it lies. This is then a case of answering all these questions "Does $\mathfrak{S}$ lie in $c_{n-1} < \lambda \leq c_n$?", $n = 0, \pm 1, \pm 2, \ldots$.

Such questions now correspond, by III.5., to projections E whose quantities $\mathfrak{E}$ (which have only the two values $0, 1$) are actually to be measured. In our examples, the $\mathfrak{E}$ are the functions $F_n(\mathfrak{R})$, $n = 1, 2, \ldots$, in which

$$F_n(\lambda) \begin{cases} = 1, & \text{for } \lambda = \lambda^{(n)} \\ = 0, & \text{otherwise} \end{cases} \Bigg|$$

or the functions $G_n(\mathfrak{S})$, $n = 0, \pm 1, \pm 2, \ldots$, in which

$$G_n(\lambda) \begin{cases} = 1, & \text{for } c_{n-1} < \lambda \leq c_n \\ = 0, & \text{otherwise} \end{cases}$$

-- and the corresponding E are the $F_n(R)$ and $G_n(S)$ respectively. Therefore, instead of giving the macroscopically measurable quantities $\mathfrak{S}$ (together with the (macroscopic) measurement precision obtainable, we may equivalently give the questions $\mathfrak{E}$ which are answered by macroscopic measurements, or their projections E (cf. III.5.). This can be viewed as the characterization of a macroscopic observer. The specification of his E . (Thus, classically, one might characterize him by stating that he can measure the temperature and the pressure in each cm^3 of the gas volume [perhaps with certain limitations of precision], but nothing else).[203]

Now it is a fundamental fact with macroscopic measurements that everything which is measurable at all, is also simultaneously measurable, i.e., that all questions which can be answered separately can also be answered simultaneously, i.e., that all the E commute. The reason that the non-simultaneous measurability of quantum mechanical quantities has made such a paradoxical impression is just that this concept is so alien to the macroscopic method of observation. Because of the fundamental importance of this point, it is best to discuss it somewhat more in detail.

Let us consider the method by which two non-simultaneously measurable quantities [e.g., the coordinate q and the momentum p (cf. III.4.)] can be measured simultaneously with limited precision. Let the mean errors

---

[203]This characterization of the macroscopic observer is due to E. Wigner.

be ϵ, η respectively (according to the uncertainty principle, $\epsilon\eta \sim h$). The discussion in III.4. showed that with such precision requirements simultaneous measurement is indeed possible: the q (position) measurement is performed with light wave lengths which are not too short, the p (momentum) measurement is performed with light wave trains which are not too long. If everything is properly arranged, then the actual measurements consist in detecting two light quanta in some way, e.g., by photographing: one (in the q measurement) is the light quantum scattered by the Compton effect, the other (in the p-measurement by means of the Doppler effect) is reflected, changed in frequency and then, in the determination of this frequency, is deflected by an optical device (prism, diffraction grating). At the end of the experiment therefore, there are two light quanta or two photographic plates, and from the directions of the light quanta, or the blackened places on the plates, we must calculate q and p. But we must emphasize here that nothing prevents us from determining (with arbitrary precision) the two directions mentioned, or the blackened places, because these are obviously simultaneously measurable quantities (they are momenta or coordinates of two different objects). However, excessive precision at this point is not of much help for the measurement of q and p. As was shown in III.4., the connection of these quantities with q and p is such that the uncertainties ϵ, η remain for q and p (even if the above quantities are measured with greater precision), and the apparatus cannot be arranged so that $\epsilon\eta \ll h$.

Therefore, if we introduce the two directions mentioned, or the blackened places themselves as physical quantities (with operators Q', P'), then we see that Q', P' are commutative, but the operators Q, P belonging to q, p can be expressed by means of them with no higher precision than ϵ, η respectively. Let the quantities belonging to Q', P' be q', p'. The interpretation that the actually macroscopically measurable quantities are not the q, p themselves but the q', p' is a very plausible one (indeed the q', p' are in fact measured), and it is in accord with our postulate of the simultaneous measurability of all macroscopic quantities.

It is reasonable to attribute to this result a general significance, and to view it as disclosing a characteristic of the macroscopic method of observation. According to this, the macroscopic procedure consists of the replacing of all possible operators $A, B, C, \ldots$, which as a rule do not commute with each other, by other operators $A', B', C', \ldots$ (of which these are functions to within a certain approximation) which do commute with each other. Since we can just as well denote these functions of $A', B', C', \ldots$ themselves by $A', B', C', \ldots$, we may also say this: $A', B', C', \ldots$ are approximations of the $A, B, C, \ldots$, but commute exactly with one another. If the respective numbers $\epsilon_A, \epsilon_B, \epsilon_C, \cdots$ give a measure for the magnitudes of the operators $A' - A$, $B' - B$, $C' - C, \ldots$, then we see that $\epsilon_A \epsilon_B$ will be of the order of magnitude of $AB - BA$ (that is, $\neq 0$, generally), etc. -- this gives the limit of the approximations which can be achieved. It is, of course, advisable, in enumerating the $A, B, C, \ldots$ to restrict oneself to those operators whose physical quantities are inaccessible to macroscopic observation, at least within a reasonable approximation.

These wholly qualitative developments remain an empty program so long as we cannot show that they require only things which are mathematically practicable. Therefore, for the characteristic case Q, P, we shall discuss further the question of the existence of the above Q', P' on a mathematical basis. For this purpose, let ϵ, η be two positive numbers with $\epsilon\eta = \frac{h}{4\pi}$. We seek two commuting Q', P' such that $Q' - Q$, $P' - P$ (in a sense still to be defined more precisely), have the orders of magnitude ϵ, η respectively.

We do this with quantities q', p' which are measurable with perfect precision, i.e., Q', P' have pure discrete spectra; since they commute, there is a complete orthonormal set consisting of the eigenfunctions common to both, $\phi_1, \phi_2, \ldots$ (cf. II.10.). Let the corresponding eigenvalues of Q', P' be $a_1, a_2, \ldots$ and $b_1, b_2, \ldots$ respectively. Then

$$Q' = \sum_{n=1}^{\infty} a_n P_{[\phi_n]} \ , \ P' = \sum_{n=1}^{\infty} b_n P_{[\phi_n]} \ .$$

Arrange their measurement in such a manner, that it creates one of the states $\phi_1, \phi_2, \ldots$ -- measure a quantity $\mathfrak{R}$ whose operator R has the eigenfunctions $\phi_1, \phi_2, \ldots$ and distinct eigenvalues $c_1, c_2, \ldots$, and then Q', P' are functions of R. That this measurement implies a measurement of Q and P in approximate fashion is clearly implied by this: In the state ϕ_n the values of Q, P are expressed approximately by the respective values of Q', P', i.e., a_n, b_n. That is, their dispersions about these values are small. These dispersions are the expectation values of the quantities $(q - a_n)^2$, $(p - b_n)^2$, i.e.,

$$((Q - a_n 1)^2 \phi_n, \phi_n) = \|(Q - a_n 1)\phi_n\|^2 = \|Q\phi_n - a_n\phi_n\|^2 \ ,$$

$$((P - b_n 1)^2 \phi_n, \phi_n) = \|(P - b_n 1)\phi_n\|^2 = \|P\phi_n - b_n\phi_n\|^2 \ .$$

They are the measures for the squares of the differences of Q' and Q, P' and P respectively, i.e., they must be approximately ϵ^2 and η^2 respectively. We therefore require

$$\|Q\phi_n - a_n\phi_n\| \lesssim \epsilon \ , \ \|P\phi_n - b_n\phi_n\| \lesssim \eta \ .$$

Instead of speaking of Q', P', it is then more appropriate only to seek a complete orthonormal set $\phi_1, \phi_2, \ldots$

for which, for suitable choice of $a_1, a_2, \ldots$ and $b_1, b_2, \ldots$, the above estimates hold.

Individual ϕ (with $\|\phi\| = 1$), for which (for suitable a, b)

$$\|Q\phi - a\phi\| = \epsilon, \quad \|P\phi - b\phi\| = \eta$$

are known from III.4.:

$$\phi_{\rho,\sigma,\gamma} = \phi_{\rho,\sigma,\gamma}(q) = \left(\frac{2\gamma}{h}\right)^{\frac{1}{4}} e^{-\frac{\pi\gamma}{h}(q - \sigma)^2 + \frac{2\pi\rho}{h} iq} .$$

Hence, because of $\epsilon\eta = \frac{h}{4\pi}$ we have again

$$\epsilon = \sqrt{\frac{h\gamma}{4\pi}}, \quad \eta = \sqrt{\frac{h}{4\pi\gamma}}$$

(i.e., $\gamma = \epsilon/\eta$), and we choose $a = \sigma$, $b = \rho$. We now must construct a complete orthonormal set with the help of these $\phi_{\rho,\sigma,\gamma}$. Since σ is the Q- and ρ the P-expectation value, it is plausible that ρ, σ should each run through a set of numbers independently of each other, and in fact, in such a way that the ρ-set has approximately the density ϵ and the σ-set approximately the density η. It proves practical to choose the units

$$2\sqrt{\pi}\cdot\epsilon = \sqrt{h\gamma} \quad \text{and} \quad 2\sqrt{\pi}\cdot\eta = \sqrt{\frac{h}{\gamma}} ,$$

i.e.,

$$\rho = \sqrt{h\gamma}\,\mu, \quad \sigma = \sqrt{\frac{h}{\gamma}}\,\nu$$

($\mu, \nu = 0, \pm 1, \pm 2, \ldots$). The

$$\psi_{\mu,\nu} = \phi_{\sqrt{h\gamma}\,\mu,\ \sqrt{\frac{h}{\gamma}}\,\nu,\ \gamma}$$

($\mu, \nu = 0, \pm 1, \pm 2, \ldots$) ought then to correspond to the

ϕ_n $(n = 1,2,\ldots)$. It is obviously irrelevant that we have two indices μ, ν in place of the one n.

However, these $\psi_{\mu,\nu}$ are not yet orthogonal. (They are normalized, however, and they satisfy

$$||Q\psi_{\mu,\nu} - \sqrt{h\gamma}\,\mu\psi_{\mu,\nu}|| = \epsilon, \quad ||P\psi_{\mu,\nu} - \sqrt{\frac{h}{\gamma}}\,\nu\psi_{\mu,\nu}|| = \eta\,.)$$

If we now orthogonalize them by the E. Schmidt process (in order, cf. II.2., proof of THEOREM 8.), then we can prove the completeness of the resulting normalized orthogonal set $\psi'_{\mu,\nu}$ without any particular difficulties, and can also establish the estimates

$$||Q\psi'_{\mu,\nu} - \sqrt{h\gamma}\,\mu\psi'_{\mu,\nu}|| \leq C\epsilon, \quad ||P\psi'_{\mu,\nu} - \sqrt{\frac{h}{\gamma}}\,\nu\psi'_{\mu,\nu}|| \leq C\eta$$

with certain fixed C. A value $C \sim 60$ has been obtained in this way, and it could probably be reduced. The proof of this fact leads to rather tedious calculations, which require no new concepts, and we shall omit them. The factors $C \sim 60$ are not important, since $\epsilon\eta = h/4\pi$ measured in macroscopic (CGS) units is exceedingly small (c. 10^{-28}).

Summing up, we can then say that it is justified to assume the commutativity of all macroscopic operators, and in particular the commutativity of the macroscopic projections E introduced above.

The E correspond to all macroscopically answerable questions $\mathfrak{E}$, i.e., to all discriminations of alternatives in the system investigated, that can be carried out macroscopically. They are all commutative. We can conclude from II.5., that $1 - E$ belongs to them along with E, and that EF, $E + F - EF$, $E - EF$ belong along with E, F. It is reasonable to assume that there are only a finite number of them: $E_1, \ldots, E_n$. We introduce the notation $E^{(+)} = E$, $E^{(-)} = 1 - E$ and consider all 2^n products $E_1^{(s_1)} \cdots E_n^{(s_n)}$ $(s_1, \ldots, s_n = \pm)$. Any two different ones among these have the product zero: For if $E_1^{(s_1)} \ldots E_n^{(s_n)}$ and $E_1^{(t_1)} \ldots E_n^{(t_n)}$ are two such, and $s_\nu \neq t_\nu$, then there appear in their product the factors $E_\nu^{(s_\nu)}, E_\nu^{(t_\nu)}$ i.e., $E_\nu^{(+)} = E$ and $E_\nu^{(-)} = 1 - E$, whose product is zero. Each E_ν is the sum of several such products: Indeed,

$$E_\nu = \sum_{s_1,\ldots,s_{\nu-1},s_{\nu+1},\ldots,s_n=\pm} E_1^{(s_1)} \ldots E_{\nu-1}^{(s_{\nu-1})} \cdot E_\nu^{(+)} E_{\nu+1}^{(s_{\nu+1})} \ldots E_n^{(s_n)} .$$

Among these products consider the ones which are different from zero. Call them $E_1', \ldots, E_m'$. (Evidently $m \leq 2^n$, but actually even $m \leq n - 1$, since these must occur among the $E_1, \ldots, E_n$ and be $\neq 0$). Now clearly: $E_\mu' \neq 0$; $E_\mu' E_\nu' = 0$ for $\mu \neq \nu$; each E_μ is the sum of several E_ν': (From the latter it also follows that $n = 2^m$.) It should be noted that $E_\mu + E_\nu = E_\rho'$ can never occur, unless $E_\mu = 0, E_\nu = E_\rho'$ or $E_\mu = E_\rho', E_\nu = 0$. Otherwise, E_μ, E_ν would be sums of several E_π', and therefore E_ρ' the sum of ≥ 2 terms E_π' (possibly with repetitions). By II.4., THEOREMS 15, 16., these would all differ from one another, since their number is ≥ 2 and all are $\neq 0$, they also differ from E_ρ' -- therefore their product with E_ρ' would be zero. Hence the product of their sum with E_ρ' would also be zero, but this contradicts the assertion that the sum is $= E_\rho'$.

The properties $\mathfrak{E}_1', \ldots, \mathfrak{E}_m'$ corresponding to the $E_1', \ldots, E_m'$ are then macroscopic properties of the following type: None is absurd. Every two are mutually exclusive. Each macroscopic property obtains by disjunction of several of them. None of them can be resolved by disjunction into two sharper macroscopic properties. $\mathfrak{E}_1', \ldots, \mathfrak{E}_m'$ therefore represent the furthest that we can go in macroscopic discrimination, for they are macroscopically indecomposable.

In the following, we shall not require that their number be finite, but only that there exist macroscopically indecomposable properties $\mathfrak{E}_1', \mathfrak{E}_2, \ldots$. Let their projections be $E_1', E_2, \ldots$, all again different from zero, mutually orthogonal, and each macroscopic E the sum of several of them.

Therefore 1 is also a sum of several of them. If an E_ν' did not occur in this sum, it would be orthogonal to each term and hence to the sum, that is to 1 : $E_\nu' = E_\nu' \cdot 1 = 0$, which is impossible. Therefore $E_1' + E_2' + \cdots = 1$. We drop the prime notation: $\mathfrak{E}_1, \mathfrak{E}_2, \ldots$ and $E_1, E_2, \ldots$. The closed linear manifolds belonging to these will be called $\mathfrak{M}_1, \mathfrak{M}_2, \ldots$, and their dimension numbers $s_1, s_2, \ldots$.

If all the $s_n = 1$, i.e., all $\mathfrak{M}_n$ one dimensional, then $\mathfrak{M}_n = [\phi_n]$, $E_n = P_{[\phi_n]}$ and because $E_1 + E_2 + \cdots = 1$, the $\phi_1, \phi_2, \ldots$ would form a complete orthonormal set. This would mean that macroscopic measurements would themselves make a complete determination of the state of the observed system possible. Since this is ordinarily not the case, we have in general $s_n > 1$, and in fact, $s_n \gg 1$.

In addition, it should be observed that the E_n , which are the elementary building blocks of the macroscopic description of the world, correspond in a certain sense to the ordinary cell division of phase space in the classical theory. We have already seen that they can reproduce the behavior of non-commutative operators in an approximate fashion, in particular, that of Q, P , which are so important for phase space.

Now, what entropy does the mixture U have for a macroscopic observer whose indecomposable projections are $E_1, E_2, \ldots$? Or, more precisely, how much entropy can such an observer maximally obtain by transforming U into V -- i.e., what entropy decrease (under suitable conditions, naturally this decrease may be $\gtrless 0$) can he produce, under the most favorable circumstances, in external objects as compensation for the transition $U \longrightarrow V$?

First, it must be emphasized that he cannot distinguish between each two ensembles U, U' , if both give the same expectation value to E_n for each $n = 1, 2, \ldots$, that is, if $\mathrm{Tr}(UE_n) = \mathrm{Tr}(U'E_n)$ $(n = 1, 2, \ldots)$. After some time, of course, the discrimination may become possible, since U, U' change according to **2.**, and

$$\mathrm{Tr}(AUA^{-1}E_n) = \mathrm{Tr}(AU'A^{-1}E_n), \quad A = e^{-\frac{2\pi i}{h} tH}$$

must no longer hold.[204] But we considered only measurements which are carried out immediately. Under the above conditions we may therefore regard U, U' as indistinguishable. Furthermore, the observer can also use only such semi-permeable walls which transmit the ϕ of some E_n and reflect the remainder unchanged. This possibility suffices, as can be seen without difficulty. By means of the method of V.2., to transform a

$$U = \sum_{n=1}^{\infty} x_n E_n$$

---

[204] If E_n commutes with H , and therefore with A , the equality still holds because

$$\mathrm{Tr}(A.UA^{-1}E_n) = \mathrm{Tr}(UA^{-1}E_n.A) = \mathrm{Tr}(UA^{-1}AE_n) = \mathrm{Tr}(UE_n) .$$

But all E_n , i.e., all macroscopically observable quantities, are in no way all commutative with H . Indeed, many such quantities, for example, the center of gravity of a gas in diffusion, change appreciably with t , i.e., $\mathrm{Tr}(UE_n)$ is not constant. Since all macroscopic quantities do commute, H is never a macroscopic quantity, i.e., the

into a

$$V' = \sum_{n=1}^{\infty} y_n E_n$$

reversibly, so that the entropy difference is still $\kappa \operatorname{Tr}(U' \ln U') - \kappa \operatorname{Tr}(V' \ln V')$, i.e., the entropy of U' equals $-\kappa \operatorname{Tr}(U' \ln U')$. To be sure, in order that such U' with $\operatorname{Tr} U' = 1$ exist in general, the $\operatorname{Tr} E_n$, i.e., the numbers s_n, must be finite. We therefore assume that all s_n are finite. U' has the s_1-fold eigenvalue x_1, the s_2-fold eigenvalue $x_2, \ldots$. Therefore $-U' \ln U'$ has the s_1-fold eigenvalue $-x_1 \ln x_1$, the s_2-fold eigenvalue $-x_2 \ln x_2, \ldots$. Consequently $\operatorname{Tr} U' = 1$ implies

$$\sum_{n=1}^{\infty} s_n x_n = 1$$

and the entropy is equal to

$$-\kappa \sum_{n=1}^{\infty} s_n x_n \ln x_n \quad .$$

Because of

$$U'E_m = \sum_{n=1}^{\infty} x_n E_n E_m = x_m E_m, \quad \operatorname{Tr}(U'E_m) = x_m \operatorname{Tr} E_m = s_m x_m \, ,$$

$x_m = \dfrac{\operatorname{Tr}(U'E_m)}{s_m}$, therefore the entropy is equal to

$$-\kappa \sum_{n=1}^{\infty} \operatorname{Tr}(U'E_n) \ln \frac{\operatorname{Tr}(U'E_n)}{s_n} \quad .$$

For arbitrary U $(\operatorname{Tr} U = 1)$, the entropy must

---

energy is not measured macroscopically with complete precision. This is plausible without additional comment.

also be equal to

$$- \kappa \sum_{n=1}^{\infty} \mathrm{Tr}\,(UE_n) \ln \frac{\mathrm{Tr}\,(UE_n)}{s_n}$$

because, if we set

$$x_n = \frac{\mathrm{Tr}\,(UE_n)}{s_n}, \quad U' = \sum_{n=1}^{\infty} x_n E_n$$

then $\mathrm{Tr}\,(UE_n) = \mathrm{Tr}\,(U'E_n)$, and since U, U' are indistinguishable, they have the same entropy.

We must also mention the fact that this entropy always exceeds the customary entropy:

$$- \kappa \sum_{n=1}^{\infty} \mathrm{Tr}\,(UE_n) \ln \frac{\mathrm{Tr}\,(UE_n)}{s_n} \geq - \kappa\, \mathrm{Tr}\,(U \ln U)$$

and that the equality holds only for

$$U = \sum_{n=1}^{\infty} x_n E_n \ .$$

By the results of V.3., this is certainly the case if

$$U' = \sum_{n=1}^{\infty} \frac{\mathrm{Tr}\,(UE_n)}{s_n} E_n$$

can be obtained from U by several (not necessarily macroscopic) applications of the process 1. -- because on the left we have $- \kappa\, \mathrm{Tr}\,(U' \ln U')$, and

$$U = \sum_{n=1}^{\infty} x_n E_n$$

means the same as $U = U'$. We take an orthonormal set $\phi_1^{(n)}, \ldots, \phi_{s_n}^{(n)}$ which spans the closed linear manifold $\mathfrak{M}_n$ belonging to E_n. Because of

$$\sum_{n=1}^{\infty} E_n = 1 ,$$

all $\phi_\nu^{(n)}$ $(n = 1,2,\dots;\ \nu = 1,\dots,s_n)$ form a complete orthonormal set. Let R be an operator belonging to these eigenfunctions (with only distinct eigenvalues) and $\mathfrak{R}$ its physical quantity. In the measurement of $\mathfrak{R}$, we get from U (by **1.**)

$$U'' = \sum_{n=1}^{\infty} \sum_{\nu=1}^{s_n} (U\phi_\nu^{(n)}, \phi_\nu^{(n)}).P_{[\phi_\nu^{(n)}]} \ .$$

Then, if we set

$$\psi_\mu^{(n)} = \frac{1}{\sqrt{s_n}} \sum_{\nu=1}^{s_n} e^{\frac{2\pi i}{s_n}\mu\nu} \phi_\nu^{(n)} \qquad (\mu = 1,\dots,s_n),$$

the $\psi_1^{(n)},\dots,\psi_{s_n}^{(n)}$ form an orthonormal set which spans the same closed linear manifold as the $\phi_1^{(n)},\dots,\phi_{s_n}^{(n)}$: $\mathfrak{M}_n$. Therefore the $\psi_\nu^{(n)}$ $(n = 1,2,\dots;\ \nu = 1,2,\dots,s_n)$ also form a complete orthonormal set, and we form an operator S with these eigenfunctions, and the corresponding physical quantity $\mathfrak{S}$. We must note the validity of the following formulas:

$$\left(P_{[\phi_\nu^{(n)}]}\psi_\mu^{(m)}, \psi_\mu^{(m)}\right) \begin{cases} = 0 & \text{for } m \neq n \\ = \frac{1}{s_n} & \text{for } m = n \end{cases} ,$$

$$\sum_{\nu=1}^{s_n} P_{[\phi_\nu^{(n)}]} = \sum_{\nu=1}^{s_n} P_{[\psi_\nu^{(n)}]} = E_n \ .$$

In the measurement of $\mathfrak{S}$, therefore, U'' becomes (by **1.**)

$$\sum_{m=1}^{\infty} \sum_{\mu=1}^{s_m} (U''\psi_\mu^{(m)}, \psi_\mu^{(m)}) P_{[\psi_\mu^{(m)}]}$$

$$= \sum_{m=1}^{\infty} \sum_{\mu=1}^{s_m} \left[\sum_{n=1}^{\infty} \sum_{\nu=1}^{s_n} (U\phi_\nu^{(n)}, \phi_\nu^{(n)})(P_{[\phi_\nu^{(n)}]}\psi_\mu^{(m)}, \psi_\mu^{(m)}) \right] P_{[\psi_\mu^{(m)}]}$$

$$= \sum_{m=1}^{\infty} \sum_{\mu=1}^{s_m} \left[\sum_{\nu=1}^{s_m} \frac{(U\phi_\nu^{(m)}, \phi_\nu^{(m)})}{s_m} \right] P_{[\psi_\mu^{(m)}]} = \sum_{m=1}^{\infty} \sum_{\nu=1}^{s_m} \frac{\mathrm{Tr}\,(UE_m)}{s_m} P_{[\psi_\mu^{(m)}]}$$

$$= \sum_{m=1}^{\infty} \frac{\mathrm{Tr}\,(UE_m)}{s_m} E_m = U' \quad .$$

Consequently, two processes **1.** suffice to transform U into U' -- and this is all we needed for the proof.

This entropy for states $(U = P_{[\phi]}, \mathrm{Tr}\,(UE_m) = (E_n\phi, \phi) = \|E_n\phi\|^2)$,

$$-\kappa \sum_{n=1}^{\infty} \|E_n\phi\|^2 \ln \frac{\|E_n\phi\|^2}{s_n}$$

is no longer subject to the inconveniences of the "macroscopic" entropy: In general, it is not constant in time (i.e., in process **2.**), and not $= 0$ for all states $U = P_{[\phi]}$. In fact: that the $\mathrm{Tr}\,(UE_n)$, from which our entropy is formed, are not time constant in general, was discussed in Note 204. It is easy to determine when the state $U = P_{[\phi]}$ has the entropy 0: Since

$$\frac{\|E_n\phi\|^2}{s_n} \geq 0, \leq 1$$

all summands

$$\|E_n\phi\|^2 \ln \frac{\|E_n\phi\|^2}{s_n}$$

in the entropy expression are ≤ 0 . All these must therefore be $= 0$. That is,

$$\frac{||E_n\phi||^2}{s_n} = 0,\ 1\ .$$

The former means that $E_n\phi = 0$, the latter that $||E_n\phi|| = \sqrt{s_n}$, but since

$$||E_n\phi|| \leq 1,\ s_n \geq 1$$

this implies $s_n = 1,\ ||E_n\phi|| = ||\phi||$; i.e., $E_n\phi = \phi$; or: $s_n = 1$, ϕ in $\mathfrak{M}_n$. The latter can certainly not hold for two different n , but also, it cannot hold at all because then $E_n\phi = 0$ would always be true, and therefore $\phi = 0$ since

$$\sum_{n=1}^{\infty} E_n = 1\ .$$

Hence, for exactly one n , ϕ is in $\mathfrak{M}_n$, and then $s_n = 1$. Since we determined that in general all $s_n \gg 1$, this is impossible. That is, our entropy is always > 0 .

Since the macroscopic entropy is time variable, the next question to be answered is this: does it behave like the entropy of the phenomenological thermodynamics in the real world, i.e., does it increase predominantly? This question is answered affirmatively in classical mechanical theory by the so-called Boltzmann H-theorem. In that, however, certain statistical assumptions, the so-called "disorder assumptions" must be made.[205] In quantum

---

[205]For the classical H-theorem, see Boltzmann, Vorlesungen über Gastheorie, Leipzig, 1896, as well as the extremely instructive discussion by P. and T. Ehrenfest in the article cited in Note 185. The "disorder assumptions" which can take the place (in quantum mechanics) of those of Boltzmann

mechanics, it was possible for the author to prove the corresponding theorem without such assumptions.[206] Since the detailed discussion of this subject, as well as of the ergodic theorem closely connected with it (cf. the reference in Note 206, where this theorem is also proved) would go beyond the scope of this volume, we cannot report on these investigations. The reader who is interested in this problem can refer to the treatments in the references.

---

have been formulated by W. Pauli (Sommerfeld-Festschrift, 1928), and the H-theorem is proved there with their help. More recently, the author also succeeded in proving the classical-mechanical ergodic theorem, cf. Proc. Nat. Ac., Jan. and March, 1932, as well as the improved treatment of G. D. Birkhoff, Proc. Nat. Ac., Dec. 1931 and March, 1932.

[206] Z. Physik, 57 (1929).

CHAPTER VI

THE MEASURING PROCESS

1. FORMULATION OF THE PROBLEM

In the discussions so far, we have treated the relation of quantum mechanics to the various causal and statistical methods of describing nature. In the course of this we found a peculiar dual nature of the quantum mechanical procedure which could not be satisfactorily explained. Namely, we found that on the one hand, a state ϕ is transformed into the state ϕ' under the action of an energy operator H in the time interval $0 \leq \tau \leq t$:

$$\frac{\partial}{\partial t} \phi_\tau = - \frac{2\pi i}{h} H \phi_\tau \qquad (0 \leq \tau \leq t),$$

so if we write $\phi_0 = \phi$, $\phi_t = \phi'$, then

$$\phi' = e^{-\frac{2\pi i}{h} tH} \phi$$

which is purely causal. A mixture U is correspondingly transformed into

$$U' = e^{-\frac{2\pi i}{h} tH} U e^{\frac{2\pi i}{h} tH} .$$

Therefore, as a consequence of the causal change of ϕ

into ϕ', the states $U = P_{[\phi]}$ go over into the states $U' = P_{[\phi']}$ (process **2.** in V.1.). On the other hand, the state ϕ -- which may measure a quantity with a pure discrete spectrum, distinct eigenvalues and eigenfunctions $\phi_1, \phi_2, \ldots$ -- undergoes in a measurement a non-causal change in which each of the states $\phi_1, \phi_2, \ldots$ can result, and in fact does result with the respective probabilities $|(\phi, \phi_1)|^2, |(\phi, \phi_2)|^2, \ldots$. That is, the mixture

$$U' = \sum_{n=1}^{\infty} |(\phi, \phi_n)|^2 P_{[\phi_n]}$$

obtains. More generally, the mixture U goes over into

$$U' = \sum_{n=1}^{\infty} (U\phi_n, \phi_n) P_{[\phi_n]}$$

(process **1.** in V.1.). Since the states go over into mixtures, the process is not causal.

The difference between these two processes $U \longrightarrow U'$ is a very fundamental one: aside from the different behaviors in regard to the principle of causality, they are also different in that the former is (thermodynamically) reversible, while the latter is not (cf. V.3.).

Let us now compare these circumstances with those which actually exist in nature or in its observation. First, it is inherently entirely correct that the measurement or the related process of the subjective perception is a new entity relative to the physical environment and is not reducible to the latter. Indeed, subjective perception leads us into the intellectual inner life of the individual, which is extra-observational by its very nature (since it must be taken for granted by any conceivable observation or experiment). (Cf. the discussion above.) Nevertheless, it is a fundamental requirement of the scientific viewpoint -- the so-called principle of the psycho-physical parallelism -- that it must be possible so to describe the extra-physical process of the subjective perception as if it were in reality in the physical world -- i.e., to assign to its parts equivalent physical processes in the objective environment, in ordinary space. (Of course, in this correlating procedure there arises the frequent necessity of localizing some of these processes at points which lie within the portion of space occupied by our own bodies. But this does not alter the fact of their belonging to the "world about us," the objective environment referred to above.) In a simple example, these concepts might be applied about as follows: We wish to measure a temperature. If we want, we can pursue this process numerically until we have the temperature of the environment of the mercury container of the thermometer, and then say: this temperature is measured by the thermometer. But we can carry the calculation further, and from the properties of the mercury, which can be explained in kinetic and molecular terms, we can calculate its heating, expansion, and the resultant length of the mercury column, and then say: this length is seen by the observer. Going still further, and taking the light source into consideration, we could find out the reflection of the light quanta on the opaque mercury column, and the path of the remaining light quanta into the eye of the observer, their refraction in the eye lens, and the formation of an image on the retina, and then we would say: this image is registered by the retina of the observer. And were our physiological knowledge more precise than it is today, we could go still further, tracing the chemical reactions which produce the impression of this image on the retina, in the optic nerve tract and in the brain, and then in the end say: these chemical changes of his brain cells are perceived by the observer. But in any case, no matter how far we calculate -- to the mercury vessel, to the scale of the thermometer, to the retina, or into the brain, at some time we must say: and this is perceived by the observer. That is, we must always divide the world into two parts, the one being the observed system, the other the observer. In the former, we can follow up all physical processes (in principle at least) arbitrarily precisely. In the latter, this is meaningless. The boundary between the two is arbitrary to a very large extent. In particular we saw in the four different possibilities in the example above, that the observer in this sense needs not to become identified with the body of the actual observer: In one instance in the above example, we included even the thermometer in it, while in another instance, even the eyes and optic nerve tract were not included. That this boundary can be pushed arbitrarily deeply into the interior of the body of the actual observer is the content of the principle of the psycho-physical parallelism -- but this does not change the fact that in each method of description the boundary must be put somewhere, if the method is not to proceed vacuously, i.e., if a comparison with experiment is to be possible. Indeed experience only makes statements of this type: an observer has made a certain (subjective) observation; and never any like this: a physical quantity has a certain value.

Now quantum mechanics describes the events which occur in the observed portions of the world, so long as they do not interact with the observing portion, with the aid of the process **2.** (V.1.), but as soon as such an interaction occurs, i.e., a measurement, it requires the application of process **1**. The dual form is therefore justified.[207] However, the danger lies in the fact that

---

[207]N. Bohr, Naturwiss. 17 (1929), was the first to point out that the dual description which is necessitated by the formalism of the quantum mechanical description of nature is fully justified by the physical nature of things that it may be connected with the principle of the psycho-physical parallelism.

the principle of the psycho-physical parallelism is violated, so long as it is not shown that the boundary between the observed system and the observer can be displaced arbitrarily in the sense given above.

In order to discuss this, let us divide the world into three parts: I, II, III. Let I be the system actually observed, II the measuring instrument, and III the actual observer.[208] It is to be shown that the boundary can just as well be drawn between I and II + III as between I + II and III . (In our example above, in the comparison of the first and second cases, I was the system to be observed, II the thermometer, and III the light plus the observer; in the comparison of the second and third cases, I was the system to be observed plus the thermometer, II the light plus the eye of the observer, III the observer, from the retina on; in the comparison of the third and fourth cases, I was everything up to the retina of the observer, II his retina, nerve tracts and brain, III his abstract "ego.") That is, in one case **2.** is to be applied to I , and **1.** to the interaction between I and II + III ; and in the other case, **2.** to I + II , and **1.** to the interaction between I + II and III . (In each case, III itself remains outside of the calculation.) The proof of this assertion, that both procedures give the same results regarding I (this and only this belongs to the observed part of the world in both cases), is then our problem.

But in order to be able to accomplish this successfully, we must first investigate more closely the process of forming the union of two physical systems (which leads from I and II to I + II) .

---

[208]The discussion which is carried out in the following, as well as that in VI.3., contains essential elements which the author owes to conversations with L. Szilard. Cf. also the similar considerations of Heisenberg, in the reference cited in Note 181.

2. COMPOSITE SYSTEMS

As was stated at the end of the preceding section, we consider two physical systems I, II (which do not necessarily have the meaning of the I, II above), and their combination I + II. In the classical mechanical method of description, I would have k degrees of freedom, and therefore the coordinates $q_1, \ldots, q_k$, in place of which we shall use the one symbol q; correspondingly, let II have l degrees of freedom, and the coordinates $r_1, \ldots, r_l$ which shall be denoted by r. Therefore, I + II has $k + l$ degrees of freedom and the coordinates $q_1, \ldots, q_k, r_1, \ldots, r_l$, or, more briefly, q, r. In quantum mechanics then, the wave functions of I have the form $\phi(q)$, those of II the form $\xi(r)$ and those of I + II the form $\Phi(q, r)$. In the corresponding Hilbert spaces $\mathfrak{R}^{\mathrm{I}}, \mathfrak{R}^{\mathrm{II}}, \mathfrak{R}^{\mathrm{I+II}}$, the inner product is defined by $\int \phi(q)\overline{\psi(q)}\, dq$, $\int \xi(r)\overline{\eta(r)}\, dr$ and $\iint \Phi(q, r)\Psi(q, r)dq\, dr$ respectively. The physical quantities of I, II, I + II are correspondingly the (hypermaximal) Hermitian operators A, A, and $\mathcal{A}$ in $\mathfrak{R}^{\mathrm{I}}, \mathfrak{R}^{\mathrm{II}}$, and $\mathfrak{R}^{\mathrm{I+II}}$ respectively.

Each physical quantity in I is naturally also one in I + II, and in fact its $\mathcal{A}$ is to be obtained from its A in this way: to obtain $\mathcal{A}\,\Phi(q, r)$ consider r as a constant and apply A to the q function $\Phi(q, r)$.[209] This rule of transformation is correct in any case for the coordinate and momentum operators $Q_1, \ldots, Q_k$ and $P_1, \ldots, P_k$, i.e.,

$$q_1, \ldots, q_k, \frac{h}{2\pi i}\frac{\partial}{\partial q_1}, \ldots, \frac{h}{2\pi i}\frac{\partial}{\partial q_k}$$

(cf. I.2.), and it conforms with the principles **I.**, **II.** in

---

[209]It can easily be shown that if A is Hermitian or hypermaximal, $\mathcal{A}$ is also.

IV.2.[210] We therefore postulate them generally. (This is the customary procedure in quantum mechanics.)

In the same way, each physical quantity in II is also one in I + II, and its A gives its $\mathcal{A}$ by the same rule: $\mathcal{A}\,\Phi(q, r)$ equals $A\Phi(q, r)$ if in the latter expression, q is taken as constant, and $\Phi(q, r)$ is considered as a function of r.

If $\phi_m(q)$ $(m = 1,2,\dots)$ is a complete orthonormal set in $\mathfrak{R}^{I}$ and $\xi_n(r)$ $(n = 1,2,\dots)$ one in $\mathfrak{R}^{II}$, then $\Phi_{m|n}(q, r) = \phi_m(q)\xi_n(r)$ $(m, n = 1,2,\dots)$ is clearly one in $\mathfrak{R}^{I+II}$. The operators A, A, $\mathcal{A}$ can therefore be represented by matrices $\{\mathsf{a}_{m|m'}\}$, $\{a_{n|n'}\}$, and $\{\alpha_{mn|m'n'}\}$ respectively $(m, n', n, n' = 1,2,\dots)$.[211] We shall make frequent use of this. The matrix representation means that

$$\mathsf{A}\phi_m(q) = \sum_{m'=1}^{\infty} \mathsf{a}_{m|m'}\phi_{m'}(q), \quad A\xi_n(r) = \sum_{n'=1}^{\infty} a_{n|n'}\xi_{n'}(r)$$

and

$$\mathcal{A}\Phi_{mn}(q, r) = \sum_{m',n'=1}^{\infty} \alpha_{mn|m'n'}\Phi_{m'n'}(q, r),$$

i.e.,

---

[210] For I. this is clear, and for II. also, so long as only polynomials are concerned. For general functions, it can be inferred from the fact that the correspondence of a resolution of the identity and a Hermitian operator is not disturbed in our transition $A \longrightarrow \mathcal{A}$.

[211] Because of the large number and variety of indices, we use this method of denoting the matrices, which differs somewhat from the notation used thus far.

$$A\,\phi_m(q)\xi_n(r) = \sum_{m'n'=1}^{\infty} \alpha_{mn|m'n'}\phi_{m'}(q)\xi_{n'}(r) \,.$$

In particular the correspondence $A \longrightarrow \mathcal{A}$ means that

$$\mathcal{A}\phi_m(q)\xi_n(r) = (A\,\phi_m(q))\xi_n(r) = \sum_{m'=1}^{\infty} a_{m|m'}\phi_{m'}(q)\xi_n(r) \,,$$

i.e.,

$$\alpha_{mn|m'n'} = a_{m|m'}\delta_{n|n'} \quad \left(\delta_{n|n'} \begin{cases} = 1 \,, & \text{for } n = n' \\ = 0 \,, & \text{for } n \neq n' \end{cases}\right) .$$

In an analogous fashion, the correspondence $A \longrightarrow \mathcal{A}$ implies that $\alpha_{mn|m'n'} = a_{n|n'}\delta_{m|m'}$.

A statistical ensemble in I + II is characterized by its statistical operator U or by its matrix $\{\upsilon_{mn|m'n'}\}$. This also determines the statistical properties of all quantities in I + II , and therefore the properties of the quantities in I also. Consequently there also corresponds to it a statistical ensemble in I alone. In fact, an observer who could perceive only I , and not II , would view the ensemble of systems I + II as one such of systems I . What is now the statistical operator $\mathcal{U}$ or its matrix $\{u_{m|m'}\}$, which belongs to this I ensemble? We determine it as follows: The I quantity with the matrix $\{a_{m|m'}\}$ has the matrix $\{a_{m|m'}\delta_{n|n'}\}$ as an I + II quantity, and therefore, by reason of a calculation in I , it has the expectation value

$$\sum_{m,m'=1}^{\infty} u_{m|m'}a_{m'|m} \,,$$

while the calculation in I + II gives

$$\sum_{m,n,m',n'=1}^{\infty} \upsilon_{mn|m'n'}\, a_{m'|m}\, \delta_{n'|n} = \sum_{m,m',n=1}^{\infty} \upsilon_{mn|m'n'}\, a_{m'|m}$$

$$= \sum_{m,m'=1}^{\infty} \left(\sum_{n=1}^{\infty} \upsilon_{mn|m'n} \right) a_{m'm} \;.$$

In order that both expressions be equal, we must have

$$u_{m|m'} = \sum_{n=1}^{\infty} \upsilon_{mn|m'n} \;.$$

In the same way, our I + II ensemble, if only II is considered and I is ignored, determines a II ensemble, with a statistical operator U and matrix $\{\mu_{n|n'}\}$. By analogy, we obtain

$$u_{n|n'} = \sum_{m=1}^{\infty} \upsilon_{mn|mn'} \;.$$

We have thus established the rules of correspondence for the statistical operators of I, II, I + II , i.e., U , U, $\mathcal{U}$. They proved to be essentially different from those which control the correspondence between the operators A , A, $\mathcal{A}$ of physical quantities.

It should be mentioned that our U , U, $\mathcal{U}$ correspondence depends only apparently on the choice of the complete orthonormal sets $\phi_m(q)$ and $\xi_n(q)$. Indeed it was derived from an invariant condition (which is satisfied by this arrangement alone): Namely, from the requirement of agreement between the expectation values of A and of $\mathcal{A}$, or of those of A and of $\mathcal{A}$.

$\mathcal{U}$ expresses the statistics in I + II , U and U those statistics restricted to I or II respectively. There now arises the question: do U, U determine $\mathcal{U}$ uniquely or not? In general one will expect a negative answer because all "probability dependencies" which may exist between the two systems disappear as the information is reduced to the sole knowledge of U and U, i.e., of the separated systems I and II. But if one knows the state of I precisely, as also that of II, "probability questions" do not arise, and then I + II, too, is precisely known. An exact mathematical discussion is, however, preferable to these qualitative considerations, and we shall proceed to this.

The problem is, then: For two given definite matrices $\{u_{m|m'}\}$ and $\{\mu_{n|n'}\}$, find a third definite matrix $\{\upsilon_{mn|m'n'}\}$, such that

$$\sum_{n=1}^{\infty} \upsilon_{mn|m'n} = u_{m|m'}, \quad \sum_{m=1}^{\infty} \upsilon_{mn|mn'} = u_{n|n'} .$$

(From

$$\sum_{m=1}^{\infty} u_{m|m} = 1, \quad \sum_{n=1}^{\infty} u_{n|n} = 1,$$

it then follows directly that

$$\sum_{m,n=1}^{\infty} \upsilon_{mn|mn} = 1,$$

i.e., the correct normalization is obtained.) This problem is always solvable, for example, $\upsilon_{mn|m'n'} = u_{m|m'} u_{m|n'}$ is always a solution (it can easily be seen that this matrix is definite), but the question arises as to whether this is the only solution.

We shall show that this is the case if and only if at least one of the two matrices $\{u_{m|m'}\}$, $\{u_{n|n'}\}$ is a state. First we prove the necessity of this condition, i.e., the existence of several solutions if both matrices correspond to mixtures. In such a case (cf. IV.2.)

$$u_{m|m'} = \alpha v_{m|m'} + \beta w_{m|m'}, \quad u_{n|n'} = \gamma v_{n|n'} + \delta w_{n|n'} .$$

($v_{m|m'}$, $w_{m|m'}$ definite and $v_{n|n'}$, $w_{n|n'}$ also, differing by more than a constant factor,

$$\sum_{m=1}^{\infty} v_{m|m} = \sum_{m=1}^{\infty} w_{m|m} = \sum_{n=1}^{\infty} v_{n|n} = \sum_{n=1}^{\infty} w_{n|n} = 1$$

$\alpha, \beta, \gamma, \delta > 0, \alpha + \beta = 1, \gamma + \delta = 1$).

We easily verify that each

$$\upsilon_{mn|m'n} = \pi v_{m|m'} v_{n|n'} + \rho w_{m|m'} v_{n|n'} + \sigma v_{m|m'} w_{n|n'} + \tau w_{m|m'} w_{n|n'}$$

with

$$\pi + \sigma = \alpha, \ \rho + \tau = \beta, \ \pi + \rho = \gamma, \ \sigma + \tau = \delta ,$$

$$\pi, \rho, \sigma, \tau > 0 ,$$

is a solution. Then π, ρ, σ, τ can be chosen in an infinite number of ways: Because of $\alpha + \beta = \gamma + \delta$ only three of the four equations are independent; therefore, $\rho = \gamma - \pi$, $\sigma = \alpha - \pi$, $\tau = (\delta - \alpha) + \pi$, and in order that all be > 0, we must require $\alpha - \delta = \gamma - \beta < \pi < \alpha, \gamma$, which is the case for infinitely many π. Now different π, ρ, σ, τ lead to different $\upsilon_{mn|m'n'}$, because the $v_{m|m'} \cdot v_{n|n'}, \ldots, w_{m|m'} \cdot w_{n|n'}$ are linearly independent, since the $v_{m|m'} \cdot w_{m|m'}$ are such, as well as the $v_{n|n'} \cdot w_{n|n'}$.

Next we prove the sufficiency, and here we may assume that $u_{m|m'}$ corresponds to a state (the other case is disposed of in the same way). Then $U = P_{[\phi]}$ and since the complete orthonormal set $\phi_1, \phi_2, \ldots$ was arbitrary, we can assume $\phi_1 = \phi$. $U = P_{[\phi_1]}$ has the matrix

$$u_{m|m'} \left\{ \begin{array}{l} = 1 \text{ , for } m = m' = 1 \\ \\ = 0 \text{ , otherwise} \end{array} \right| .$$

Therefore

$$\sum_{n=1}^{\infty} \upsilon_{mn|m'n} \begin{cases} = 1 \text{ , for } m = m' = 1 \\ = 0 \text{ , otherwise} \end{cases} .$$

In particular, for $m \neq 1$,

$$\sum_{n=1}^{\infty} \upsilon_{mn|mn} = 0 ,$$

but since all $\upsilon_{mn|mn} \geq 0$ because of the definiteness of $\upsilon_{mn|m'n'}$ $[\upsilon_{mn|mn} = (U \Phi_{mn}, \Phi_{mn})]$, therefore in this case $\upsilon_{mn|mn} = 0$. That is, $(U \Phi_{mn}, \Phi_{mn}) = 0$, and hence, because of the definiteness of U, $(U \Phi_{mn}, \Phi_{m'n'})$ also $= 0$ (cf. II.5., THEOREM 19.), where m', n' are arbitrary. That is, it follows from $m \neq 1$ that $\upsilon_{mn|m'n'} = 0$, and because of the Hermitian nature, this also follows from $m' \neq 1$. For $m = m' = 1$ however, this gives

$$\upsilon_{1n|1n'} = \sum_{m=1}^{\infty} \upsilon_{mn|mn'} = u_{n|n'} .$$

Consequently, as was asserted, the solution $\upsilon_{mn|m'n}$ is determined uniquely.

We can thus summarize our result as follows: A statistical ensemble in I + II with the operator $U = \{\upsilon_{mn|m'n'}\}$ is determined uniquely by the statistical ensembles determined by it in I and II individually, with the respective operators $U = \{u_{m|m'}\}$ and $U = \{u_{n|n'}\}$, if and only if the following two conditions are satisfied:

1. $\upsilon_{mn|m'n'} = v_{m|m'} v_{n|n'}$. (From

$$\operatorname{Tr} U = \sum_{m,n=1}^{\infty} \upsilon_{mn|mn} = \sum_{m=1}^{\infty} v_{m|m} \sum_{n=1}^{\infty} v_{n|n} = 1 ,$$

it follows that, by multiplication of $v_{m|m'}$ and $v_{n|n'}$ with two reciprocal constant factors, we can obtain

$$\sum_{m=1}^{\infty} v_{m|m} = 1 \,, \quad \sum_{n=1}^{\infty} v_{n|n} = 1$$

But then we see that $u_{m|m'} = v_{m|m'}$, $u_{n|n'} = v_{n|n'}$.)

2. Either $v_{m|m'} = \bar{x}_m x_{m'}$ or $v_{n|n'} = \bar{x}_n x_n$. (Indeed $U = P_{[\phi]}$ means that

$$\phi = \sum_{m=1}^{\infty} y_m \phi_m \,,$$

and therefore $u_{m|m'} = \bar{y}_m y_{m'}$ and correspondingly for $v_{m|m'}$; by analogy the same is true with $U = P_{[\xi]}$.)

We shall call U and U the projections of U in I and II respectively.[212]

We now apply ourselves to the states of I + II , $U = P_{[\Phi]}$. The corresponding wave functions $\Phi(q, r)$ can be expanded according to the complete orthonormal set $\Phi_{mn}(q, r) = \phi_m(q)\xi_n(r)$:

$$\Phi(q, r) = \sum_{m,n=1}^{\infty} f_{mn}\phi_m(q)\xi_n(r) \,.$$

We can therefore replace them by the coefficients f_{mn} ($m, n = 1,2,\ldots$) which are subject only to the condition that

$$\sum_{m,n=1}^{\infty} |f_{mn}|^2 = ||\Phi||^2$$

be finite.

---

[212]The projections of a state of I + II are in general mixtures in I or II ; cf. above. This circumstance was discovered by Landau, Z. Physik 45 (1927).

We can define two operators F, F^* by

$$F\phi(q) = \int \overline{\Phi(q, r)}\phi(q)dq$$

(**F.**)

$$F^*\xi(r) = \int \Phi(q, r)\xi(r)dr \quad .$$

These are linear, but have the peculiarity of being defined in $\mathfrak{R}^{I}$ and $\mathfrak{R}^{II}$ respectively, and of taking on values from $\mathfrak{R}^{II}$ and $\mathfrak{R}^{I}$ respectively. Their relation is that of adjoints, since obviously $(F\phi, \xi) = (\phi, F^*\xi)$ (the inner product on the left is to be formed in $\mathfrak{R}^{II}$ and that on the right is to be formed in $\mathfrak{R}^{I}$). Since the difference of $\mathfrak{R}^{I}$ and $\mathfrak{R}^{II}$ is mathematically unimportant, we can apply the results of II.11: then, since we are dealing with integral operators, $\Sigma(F)$ and $\Sigma(F^*)$ are equal to

$$\iint |\Phi(q, r)|^2 dqdr = \|\Phi\|^2 = 1 \quad (\|\Phi\| \text{ in } \mathfrak{R}^{I+II}!) ,$$

and are therefore finite. Consequently F, F^* are continuous, in fact are completely continuous operators, and F^*F as well as FF^* are definite operators, $\mathrm{Tr}\,(F^*F) = \Sigma(F) = 1$, $\mathrm{Tr}\,(FF^*) = \Sigma(F^*) = 1$.

If we again consider the difference between $\mathfrak{R}^{I}$ and $\mathfrak{R}^{II}$ then we see that F^*F is defined and assumes values in $\mathfrak{R}^{I}$, and FF^* similarly in $\mathfrak{R}^{II}$.

Since $F\phi_m(q)$ comes out equal to

$$\sum_{n=1}^{\infty} \overline{f}_{mn}\xi_n(r) \quad ,$$

F has the matrix $\{\overline{f}_{mn}\}$ [by use of the complete orthonormal sets $\phi_m(q)$ and $\overline{\xi}_n(r)$ respectively -- note that the latter is a complete orthonormal set along with $\xi_n(r)$], likewise F^* has the matrix $\{f_{mn}\}$ (with the same complete orthonormal systems). Therefore F^*F, FF^*

have the matrices

$$\left\{\sum_{n=1}^{\infty} \bar{f}_{mn} f_{m'n}\right\}$$

(using the complete orthonormal set $\phi_m(q)$ in $\mathfrak{R}^{\mathrm{I}}$) and

$$\left\{\sum_{n=1}^{\infty} \bar{f}_{mn} f_{mn'}\right\}$$

(using the complete orthonormal set $\overline{\xi_n(r)}$ in $\mathfrak{R}^{\mathrm{II}}$).

On the other hand, $U = P_{[\Phi]}$ has the matrix $\{\bar{f}_{mn} f_{m'n'}\}$ (using the complete orthonormal set $\Phi_{mn}(q, r) = \phi_m(q)\xi_n(r)$ in $\mathfrak{R}^{\mathrm{I+II}}$), so that its projections in I and II , U and U have the matrices

$$\left\{\sum_{n=1}^{\infty} \bar{f}_{mn} f_{m'n}\right\}$$

and

$$\left\{\sum_{m=1}^{\infty} \bar{f}_{mn} f_{mn'}\right\}$$

respectively (with the complete orthonormal sets given above).[213] Consequently

$$(\mathbf{U}.) \qquad \mathsf{U} = F^*F, \quad \mathrm{U} = FF^* .$$

Note that the definitions (**F**.) and the equations (**U**.) make no use of the ϕ_m, ξ_n -- hence they are valid independently of these.

The operators U , U are completely continuous, and by II.11. and IV.3., they can be written in the form

---

[213] The mathematical discussion is based on a paper by E. Schmidt, Math. Ann. 83 (1907).

$$U = \sum_{k=1}^{\infty} w_k' P_{[\psi_k]}, \quad U = \sum_{k=1}^{\infty} w_k'' P_{[\eta_k]},$$

in which the ψ_k form a complete orthonormal set in $\mathfrak{R}^{I}$, the η_k one in $\mathfrak{R}^{II}$ and all $w_k', w_k'' \geq 0$. We now neglect the terms in each of the two formulas with $w_k' = 0$ or $w_k'' = 0$ respectively, and number the remaining terms with $k = 1, 2, \ldots$. Then the ψ_k and η_k again form orthonormal, but not necessarily complete sets; the sums

$$\sum_{k=1}^{M'}, \quad \sum_{k=1}^{M''}$$

appear in place of the two

$$\sum_{k=1}^{\infty}$$

where M', M'' can be equal to ∞ or finite. Also, all w_k', w_k'' are now > 0.

Let us now consider a ψ_k. $U\psi_k = w_k'\psi_k$ and therefore $F^*F\psi_k = w_k'\psi_k$, $FF^*F\psi_k = w_k'F\psi_k$, $UF\psi_k = w_k'F\psi_k$. Furthermore

$$(F\psi_k, F\psi_l) = (F^*F\psi_k, \psi_l) = (U\psi_k, \psi_l)$$

$$= w_k'(\psi_k, \psi_l) \begin{cases} = w_k', & \text{for } k = l \\ = 0, & \text{for } k \neq l \end{cases},$$

therefore, in particular, $\|F\psi_k\|^2 = w'_k$. The $\frac{1}{\sqrt{w_k'}} F\psi_k$ then form an orthonormal set in R^{II} and they are eigenfunctions of U, with the same eigenvalues as the ψ_k for U (i.e., w_k'). That is, each eigenvalue of U is also one of U with at least the same multiplicity. Interchanging U, U shows that they have the same eigenvalues with the same multiplicities. The w_k' and w_k'' therefore coincide except for their order. Hence $M' = M'' = M$, and by re-enumeration of the w_k'' we can obtain $w_k' = w_k'' = w_k$. And if this occurs, then we can clearly choose

$$\eta_k = \frac{1}{\sqrt{w_k}} F\psi_k$$

in general. Then

$$\frac{1}{\sqrt{w_k}} F^*\eta_k = \frac{1}{w_k} F^*F\psi_k = \frac{1}{w_k} \mathsf{U}\psi_k = \psi_k .$$

Therefore

$$\eta_k = \frac{1}{\sqrt{w_k}} F\psi_k, \quad \psi_k = \frac{1}{\sqrt{w_k}} F^*\eta_k . \qquad \text{(V.)}$$

212

Let us now extend the orthonormal set $\psi_1, \psi_2, \cdots$ to a complete $\psi_1, \psi_2, \cdots, \psi_1', \psi_2', \cdots$ and likewise $\eta_1, \eta_2, \cdots$ to $\eta_1, \eta_2, \cdots, \eta_1', \eta_2', \cdots$ (each of the two sets $\psi_1', \psi_2', \cdots$ and $\eta_1', \eta_2', \cdots$ can be empty, finite or infinite, and in addition each set independently of the other set). We have observed before, that (**F.**), (**U.**) make no reference to the ϕ_m, ξ_n. We may therefore use (**V.**), as well as the above construction, and let them determine the choice of the complete orthonormal sets $\phi_1, \phi_2, \cdots$ and $\xi_1, \xi_2, \cdots$. Specifically we let these coincide with the $\psi_1, \psi_2, \cdots, \psi_1', \psi_2', \cdots$ and $\overline{\eta}_1, \overline{\eta}_2, \cdots, \overline{\eta}_1', \overline{\eta}_2', \cdots$ respectively. Now let ψ_k correspond to ϕ_{μ_k}, η_k to ξ_{ν_k} $(k = 1, \ldots, M)$ ($\mu_1, \mu_2, \cdots$ different from one another, $\nu_1, \nu_2, \cdots$ likewise). Then

$$F\phi_{\mu_k} = \sqrt{w_k}\, \xi_{\nu_k} ,$$

$$F\phi_m = 0 \quad \text{for} \quad m \neq \mu_1, \mu_2, \cdots .$$

Therefore

$$f_{mn} \begin{cases} = \sqrt{w_k}\ , \text{ for } m = \mu_k,\ n = \nu_k,\ k = 1, 2, \ldots \\ \\ = \ 0\ , \text{ otherwise} \end{cases} \Bigg| ,$$

or equivalently

$$\Phi(q, r) = \sum_{k=1}^{M} \sqrt{w_k}\ \phi_{\mu_k}(q)\xi_{\nu_k}(r) .$$

By suitable choice of the complete orthonormal sets $\phi_m(q)$ and $\xi_n(r)$ we have thus established that each column of the matrix $\{f_{mn}\}$ contains at most one element $\neq 0$ (that this is real and > 0 , namely $\sqrt{w_k}$, is unimportant for what follows). What is the physical meaning of this mathematical statement?

Let A be an operator with the eigenfunctions $\phi_1, \phi_2, \ldots$ and with only distinct eigenvalues, say $a_1, a_2, \ldots$; likewise B with $\xi_1, \xi_2, \ldots$ and $b_1, b_2, \ldots$. A corresponds to a physical quantity in I , B to one in II . They are therefore simultaneously measurable. It is easily seen that the statement "A has the value a_m and B has the value b_n" determines the state $\Phi_{mn}(q, r) = \phi_m(q)\xi_n(r)$, and that this has the probability $(P_{[\Phi_{mn}]}\Phi, \Phi) = |(\Phi, \Phi_{mn})|^2 = |f_{mn}|^2$ in the state $\Phi(q, r)$. Consequently, our statement means that A, B are simultaneously measurable, and that if one of them was measured in Φ , then the value of the other is determined by it uniquely. (An a_m with all $f_{mn} = 0$ cannot result, because its total probability

$$\sum_{n=1}^{\infty} |f_{mn}|^2$$

cannot be 0 , if a_m is ever observed -- therefore for

exactly one n, $f_{mn} \neq 0$; likewise for b_n .) That is, there are several possible A values in the state Φ (namely, those a_m for which

$$\sum_{n=1}^{\infty} |f_{mn}|^2 > 0 \quad ,$$

i.e., for which there exists an n with $f_{mn} \neq 0$ -- usually all a_m are such), and an equal number of possible B values (those b_n for which

$$\sum_{n=1}^{\infty} |f_{mn}|^2 > 0 \quad ,$$

i.e., for which there exists an m with $f_{mn} \neq 0$), but Φ establishes a one-to-one correspondence between the possible A values and the possible B values.

If we call the possible m values $\mu_1, \mu_2, \cdots$ and the corresponding possible n values $\nu_1, \nu_2, \cdots$, then

$$f_{mn} \begin{cases} = c_k \neq 0 , & \text{for } m = \mu_k,\ n = \nu_k,\ k = 1, 2, \cdots \\ = 0 , & \text{otherwise} \end{cases} \Bigg| ,$$

therefore (M finite or ∞)

$$\Phi(q, r) = \sum_{k=1}^{M} c_k \phi_{\mu_k}(q) \xi_{\nu_k}(r) \quad ,$$

hence

$$u_{mm'} = \sum_{n=1}^{\infty} \bar{f}_{mn} f_{m'n} \begin{cases} = |c_k|^2 , & \text{for } m = m' = \mu_k,\ k = 1, 2, \cdots \\ = \quad 0 , & \text{otherwise} \end{cases} \Bigg| ,$$

$$u_{nn'} = \sum_{m=1}^{\infty} \bar{f}_{mn} f_{mn'} \begin{cases} = |c_k|^2 \text{ , for } n = n' = \nu_k, \ k = 1, 2, \ldots \\ = 0 \text{ , otherwise} \end{cases}$$

and therefore

$$\mathrm{U} = \sum_{k=1}^{M} |c_k|^2 P_{[\phi_{\mu_k}]} , \quad U = \sum_{k=1}^{M} |c_k|^2 P_{[\xi_{\nu_k}]} .$$

Hence, when Φ is projected in I or II , it in general becomes a mixture, while it is a state in I + II only. Indeed, it involves certain information regarding I + II which cannot be made use of in I alone or in II alone, namely the one-to-one correspondence of the A and B values with each other.

For each Φ we can therefore so choose A, B , i.e., the ϕ_m and the ξ_n , that our condition is satisfied; for arbitrary A, B , it may of course be violated. Each state Φ then establishes a particular relation between I and II , while the related quantities A, B depend on Φ . How far Φ determines them, i.e., the ϕ_m and the ξ_n , is not difficult to answer. If all $|c_k|$ are different and $\neq 0$, then U, U (which are determined by Φ) determine the respective ϕ_m, ξ_n uniquely (cf. IV.3.). The general discussion is left to the reader.

Finally, let us mention the fact that for $M \neq 1$ neither U nor U is a state (because all $|c_k|^2 > 0$). For $M = 1$ they both are: $\mathrm{U} = P_{[\phi_{\mu_1}]}$, $U = P_{[\xi_{\nu_1}]}$. Then $\Phi(q, r) = c_1 \phi_{\mu_1}(q) \xi_{\nu_1}(r)$. We can absorb c_1 in $\phi_{\mu_1}(q)$. Therefore U, U are states if and only if $\Phi(q, r)$ has the form $\phi(q)\xi(r)$, and in that case they are equal to $P_{[\phi]}$ and $P_{[\xi]}$ respectively.

On the basis of the above results, we note: If

I is in the state $\phi(q)$ and II in the state $\xi(r)$, then I + II is in the state $\Phi(q, r) = \phi(q)\xi(r)$. If on the other hand I + II is in a state $\Phi(q, r)$ which is not a product $\phi(q)\xi(r)$, then I and II are mixtures and not states, but Φ establishes a one-to-one correspondence between the possible values of certain quantities in I and in II.

3. DISCUSSION OF THE MEASURING PROCESS

Before we complete the discussion of the measuring process in the sense of the ideas developed in VI.1. (with the aid of the formal tools developed in VI.2.), we shall make use of the results of VI.2. to exclude a possible explanation often proposed for the statistical character of the process **1.** (V.1.). This rests on the following idea: Let I be the observed system, II the observer. If I is in a state $U = P_{[\phi]}$ before the measurement, while II on the other hand is in a mixture

$$U = \sum_{n=1}^{\infty} w_n P_{[\xi_n]} ,$$

then I + II is a uniquely determined mixture $\mathcal{U}$, and in fact, as we can easily calculate from VI.2.,

$$\mathcal{U} = \sum_{n=1}^{\infty} w_n P_{[\Phi_n]}, \quad \Phi_n(q, r) = \phi(q)\xi_n(r)$$

If now a measurement of a quantity A takes place in I, then this is to be regarded as an interaction of I and II. This is a process **2.** (V.1.), with an energy operator $\mathbf{H}$. If it has the time duration t, then we obtain

$$\mathcal{U}' = e^{-\frac{2\pi i}{h} t\mathbf{H}} \mathcal{U} e^{\frac{2\pi i}{h} t\mathbf{H}}$$

from U , and in fact,

$$U' = \sum_{n=1}^{\infty} w_n P_{\left[e^{-\frac{2\pi i}{h} tH} \Phi_n \right]}$$

If now each

$$e^{-\frac{2\pi i}{h} tH} \Phi_n(q, r)$$

were of the form $\psi_n(q)\eta_n(r)$, where the ψ_n are the eigenfunctions of A , and the η_n any fixed complete orthonormal set, then this intervention would have the character of a measurement. For it transforms each state ϕ of I into a mixture of the eigenfunctions ψ_n of A . The statistical character therefore arises in this way: Before the measurement I was in a (unique) state, but II was a mixture -- and the mixture character of II has, in the course of the interaction, associated itself with I + II , and in particular, it has made a mixture of the projection in I . That is, the result of the measurement is indeterminate, because the state of the observer before the measurement is not known exactly. It is conceivable that such a mechanism might function, because the state of information of the observer regarding his own state could have absolute limitations, by the laws of nature. These limitations would be expressed in the values of the w_n , which are characteristic of the observer alone (and therefore independent of ϕ) .

At this point, the attempted explanation breaks down. For quantum mechanics requires that $w_n = (P_{\psi_n}\phi, \phi) = |(\phi, \psi_n)|^2$, i.e., w_n dependent on ϕ ! There might exist another decomposition

$$U' = \sum_{n=1}^{\infty} w_n' P_{[\Phi_n']} ,$$

(the $\Phi'_n(q, r) = \psi_n(q)\eta_n(r)$ are orthonormal) but this is of no use either; because the w'_n are (except for order) determined uniquely by U' (IV.3.), and are therefore equal to the w_n.[214]

Therefore, the non-causal nature of the process **1.** is not produced by any incomplete knowledge of the state of the observer, and we shall therefore assume in all that follows that this state is completely known.

Let us now apply ourselves again to the problem formulated at the end of VI.1. I, II, III shall have the meanings given there, and, for the quantum mechanical investigation of I, II, we shall use the notation of VI.2., while III remains outside of the calculations (cf. the discussion of this in VI.1.). Let A be the quantity (in I) actually to be measured, $\phi_1(q), \phi_2(q), \ldots$ its eigenfunctions. Let I be in the state $\phi(q)$.

If I is the observed system, II + III the observer, then we must apply the process **1.**, and we find that the measurement transforms I from the state ϕ into one of the states ϕ_n $(n = 1, 2, \ldots)$, the probabilities for which are respectively $|(\phi, \phi_n)|^2$ $(n = 1, 2, \ldots)$. Now, what is the method of description if I + II is the observed system, and only III the observer?

In this case we must say that II is a measuring instrument which shows on a scale the value of A (in I): the position of the pointer on this scale is a physical quantity B (in II) which is actually observed by III (if II is already within the body of the observer, we have the corresponding physiological concepts in place of the scale and pointer, e.g., retina and image on the retina, etc.) Let A have the values $a_1, a_2, \ldots$, B the values $b_1, b_2, \ldots$, and let the numbering be such that a_n is associated with b_n.

---

[214]This approach is capable of still more variants, which must be rejected for similar reasons.

Initially, I is in the (unknown) state $\phi(q)$ and II in the (known) state $\xi(r)$, therefore I + II is in the state $\Phi(q, r) = \phi(q)\xi(r)$. The measurement (so far as it is performed by II on I) is, as in the earlier example, carried out by an energy operator H (in I + II) in the time t: This is the process **2.**, which transforms the Φ into

$$\Phi' = e^{-\frac{2\pi i}{h} tH} \Phi .$$

Viewed by the observer III, one has a measurement only if the following is the case: If III were to measure (by process **1.**) the simultaneously measurable quantities A, B (in I or II respectively, or both in I + II), then the pair of values a_n, b_n would have the probability 0 for $m \neq n$, and the probability w_n for $m = n$. That is, it suffices "to look at" II, and A is measured in I. Quantum mechanics then requires in addition $w_n = |(\phi, \phi_n)|^2$.

If this is established, then the measuring process so far as it occurs in II, is "explained" theoretically, i.e., the division of I | II + III discussed in VI.1. is shifted to I + II | III.

The mathematical problem is then the following. A complete orthonormal set $\phi_1, \phi_2, \cdots$ is given in I. Such a set $\xi_1, \xi_2, \cdots$ in $\mathfrak{R}^{II}$ as well as a state ξ in R^I, also an (energy) operator H in $\mathfrak{R}^{I+II}$, and a t, are to be found so that the following holds. If ϕ is an arbitrary state in R^I and

$$\Phi(q, r) = \phi(q)\xi(r), \quad \Phi'(q, r) = e^{-\frac{2\pi i}{h} tH} \Phi(q, r) ,$$

then $\Phi'(q, r)$ must have the form

$$\sum_{n=1}^{\infty} c_n \phi_n(q) \xi_n(r)$$

(the c_n are naturally dependent on ϕ). Therefore $|c_n|^2 = |(\phi, \phi_n)|^2$. (That the latter is equivalent to the physical requirement formulated above was discussed in VI.2.)

In the following we shall use a fixed set $\xi_1, \xi_2, \ldots$ and a fixed ξ along with the fixed $\phi_1, \phi_2, \ldots$, and shall investigate the unitary operator

$$\Delta = e^{-\frac{2\pi i}{h} tH}$$

instead of H.

The mathematical problem leads us back to the problem solved in VI.2.: there the quantity corresponding to our present Φ was given, and we showed the existence of c_n, ϕ_n, ξ_n. Now ϕ_n, ξ_n are fixed and Φ, c_n are given dependent on ϕ, and it remains so to determine a fixed Δ that for $\Phi' = \Delta\Phi$ these c_n, ϕ_n, ξ_n result.

We shall show that such a determination of Δ is indeed possible. In this case only the principle is of importance to us, i.e., the existence of any such Δ. The further question, whether the

$$\Delta = e^{-\frac{2\pi i}{h} tH}$$

corresponding to simple and plausible measuring arrangements also have this property, shall not concern us. Indeed, we saw that our requirements coincide with a plausible intuitive criterion of the measurement character in an intervention. Furthermore the arrangements in question are to possess the characteristics of the measurement. Hence quantum mechanics, as applied to observation would be in blatant contradiction with experience, if these Δ did not satisfy the requirements in question (at least approximately).[215] Therefore, in the following, only an abstract

[215] The corresponding calculation for the case of the posi-

Δ which satisfies our conditions exactly, shall be given.

Therefore, let the ϕ_m $(m = 0, \pm 1, \pm 2, \ldots)$ and the ξ_n $(n = 0, \pm 1, \pm 2, \ldots)$ respectively be two given complete orthonormal sets in $\mathfrak{R}^{I}$ and $\mathfrak{R}^{II}$ respectively. (We do not let m, n run over $1, 2, \ldots$, but over $0, \pm 1, \pm 2, \ldots$. This is purely for technical convenience, and is in principle equivalent to the former). Let the state ξ be, for simplicity, ξ_0. We define the operator Δ by

$$\Delta \sum_{m,n=-\infty}^{\infty} x_{mn}\phi_m(q)\xi_n(r) = \sum_{m,n=-\infty}^{\infty} x_{mn}\phi_m(q)\xi_{m+n}(r) ,$$

since the $\phi_m(q)\xi_n(r)$ as well as the $\phi_m(q)\xi_{m+n}(r)$ form a complete orthonormal set in $\mathfrak{R}^{I+II}$, this Δ is unitary. Now

$$\phi(q) = \sum_{m=-\infty}^{\infty} (\phi, \phi_m)\cdot\phi_m(q), \quad \xi(r) = \xi_0(r) ,$$

therefore

$$\Phi(q, r) = \phi(q)\xi(r) = \sum_{m=-\infty}^{\infty} (\phi, \phi_m)\cdot\phi_m(q)\xi_0(r) ,$$

$$\Phi'(q, r) = \Delta\Phi(q, r) = \sum_{m=-\infty}^{\infty} (\phi, \phi_m)\cdot\phi_m(q)\xi_m(r) .$$

Hence our purpose is accomplished. We have in addition $c_n = (\phi, \phi_n)$.

A better overall view of the mechanism of this process can be obtained if we exemplify it by concrete Schrödinger wave functions, and give H in place of Δ.

The observed object, as well as the observer

---

tion measurement discussed in III.4. is contained in a paper by Weizsäcker, Z. Physik 70 (1931).

(i.e., I and II respectively) may be characterized by a single variable q and r respectively, running continuously from $-\infty$ to $+\infty$. That is, let both be thought of as points which can move along a line. Their wave functions then have always the form $\psi(q)$ and $\eta(r)$ respectively. We assume that their masses m_1 and m_2 are so large that the kinetic energy portion of the energy operator (i.e., $\frac{1}{2m_1}\left(\frac{h}{2\pi i}\frac{\partial}{\partial q}\right)^2 + \frac{1}{2m_2}\left(\frac{h}{2\pi i}\frac{\partial}{\partial r}\right)^2$) can be neglected. Then there remains of H only the interaction energy part which is decisive for the measurement. For this we choose the particular form $\frac{h}{2\pi i} q \frac{\partial}{\partial r}$.

The Schrödinger time dependent differential equation then is (for the I + II wave functions $\psi_t = \psi_t(q, r)$):

$$\frac{h}{2\pi i}\frac{\partial}{\partial t}\psi_t(q, r) = -\frac{h}{2\pi i} q \frac{\partial}{\partial r}\psi_t(q, r) ,$$

$$\left(\frac{\partial}{\partial t} + q\frac{\partial}{\partial r}\right)\psi_t(q, r) = 0 ,$$

i.e.,

$$\psi_t(q, r) = f(q, r - tq) .$$

If, for $t = 0$, $\psi_0(q, r) = \Phi(q, r)$, then we have $f(q, r) = \Phi(q, r)$, and therefore

$$\psi_t(q, r) = \Phi(q, r - tq)$$

In particular, if the initial states of I, II are represented by $\phi(q)$ and $\xi(r)$ respectively, then, in the sense of our calculation scheme (if the time t appearing therein is chosen to be 1)

$$\Phi(q, r) = \phi(q)\xi(r) ,$$

$$\Phi'(q, r) = \psi_1(q, r) = \phi(q)\xi(r - q) .$$

We now wish to show that this can be used by II for a position measurement of I, i.e., that the coordinates are tied to each other. (Since q, r have continuous spectra, they are therefore measurable with only arbitrary precision, but not with absolute precision. Hence this can be accomplished only approximately.)

For this purpose, we wish to assume that $\xi(r)$ is different from 0 only in a very small interval $-\epsilon < r < \epsilon$ (i.e., the coordinate r of the observer before the measurement is very accurately known), in addition ξ should of course be normalized:

$$||\xi|| = 1 \text{ , i.e., } \int |\xi(r)|^2 dr = 1 \text{ .}$$

The probability therefore that q lies in the interval $q_o - \delta < q < q_o + \delta$, and r in the interval $r_o - \delta' < r < r_o + \delta'$ is

$$\int_{q_o-\delta}^{q_o+\delta}\int_{r_o-\delta}^{r_o+\delta} |\Phi'(q, r)|^2 dq dr = \int_{q_o-\delta}^{q_o+\delta}\int_{r_o-\delta}^{r_o+\delta} |\phi(q)|^2 |\xi(r - q)|^2 dq dr \text{ .}$$

If q_o, r_o are to differ by more than $\delta + \delta' + \epsilon$, then this is 0, i.e., q, r are so very closely tied to each other that the difference can never be greater than $\delta + \delta' + \epsilon$. And for $r_o = q_o$ this is, equal to

$$\int_{q_o-\delta}^{q_o+\delta} |\phi(q)|^2 dq \text{ ,}$$

if we choose $\delta' \geq \delta + \epsilon$, because of the assumptions on ξ. But since we can choose $\delta, \delta', \epsilon$ arbitrarily small (they must be different from zero, however), this means that q, r are tied to each other with arbitrary closeness, and the probability density has the value furnished by quantum mechanics, $|\phi(q)|^2$.

That is, the relations of the measurement, as we had discussed them in IV.1., and in this section, are realized.

The discussion of more complicated examples, say of an analog to our four-term example of IV.1., or the control determination of the validity of a measurement which II carried out on I , effected by a second observer III , can also be carried out in this fashion. It is left to the reader.